Current Topics in Flavours and Fragrances

Towards a New Millennium of Discovery

Edited by

Karl A.D. Swift

Quest International,
Ashford, Kent,
U.K.

SPRINGER SCIENCE+BUSINESS MEDIA, B.V.

CURRENT TOPICS IN FLAVOURS AND FRAGRANCES

A C.I.P. Catalogue record for this book is available from the Library of Congress.

ISBN 978-94-010-5775-2 ISBN 978-94-011-4022-5 (eBook)
DOI 10.1007/978-94-011-4022-5

Printed on acid-free paper

The cover illustration displays the solid-state conformation of α-cycloaltrin (α-CA) in different forms: the molecule adopts a conformation of 3-fold rotational symmetry with a unique alternating sequence of $^4C_1$ and $^1C_4$ altropyranose chair geometries (upper left model, space group $P6_3$), resulting in banana-shaped disaccharide units, of which one is represented as a space-filling CPK-type model (lower left entry). In the crystal lattice the compact molecules are stacked in transposed layers (upper right, water molecules left off for clarity). Each α-CA molecule is embedded into a matrix of 21 water molecules (lower right, view along the hexagonal *c*-axis). The space filling models reveal α-CA to be devoid of a central 'through-going' cavity. The graphics were generated using Brickmann's MOLCAD molecular modeling program; for further information see the web-pages at 'http://caramel.oc.chemie.tudarmstadt.de/immel/molcad/Gallery.html'or 'S. Immel, G.E. Schmitt, and F.W. Lichtenthaler: α-Cycloaltrin: Conformation and Properties in the Solid-State and Aqueous Solution.', pp. 41–48 of this volume.

CONTENTS

PREFACE

This book is designed to give the reader up to date information on some of the more exciting developments that have taken place at the leading edge of fragrance and flavour research.

Chapter one gives the reader a rapid excursion through the chronological landmarks of fragrance and flavour materials and sets the scene for the remaining nine chapters which cover topics that are at the forefront of modern research. Chapter two looks at the total synthesis of synthetically interesting perfumery natural materials. This chapter aims to highlight the creative and elegant chemistry that has been performed by some of the worlds greatest chemists in their quest to synthesise one of the five natural products reviewed in the chapter. The chapter fits in with the forward looking theme of the book as it will hopefully inspire other chemists that are interested in synthesising natural products to produce elegant new, or industrially applicable routes to these and other perfumery materials. Chapter three looks at the growing area of interest in asymmetric fragrance materials. The chapter focuses on the use of the metal-BINAP catalytic system for the preparation of fragrance and flavour ingredients. Environmental considerations are now an integral and vital part of planning any new industrial chemical process. Chapter four aims to give the reader an insight into the wide-ranging and often readily applicable chemistry that is currently available for the installation of environmentally friendly chemical processes. Our final chapter (chapter five) in the fragrance area concerns the area of chemistry relating to the manufacture and synthesis of macrocyclic molecules. The chapter aims to look at and review the progress of the most recent advances in the industrial synthesis of medium to large ring molecules.

The analysis of flavour and fragrance molecules has always been a hard task due to their often-complex composition in high dilution in a no less complex non-volatile matrix. Chapter six, the first of our flavour orientated chapters looks at the newer methods and at the recent advances in known methods for the analysis of volatiles in the flavour area. Chapter seven looks at the area of flavour matrix interactions, whilst chapter eight concerns itself with biotransformations in relation to the flavour industry. Biotransformations are becoming more frequently a commercially viable option for the preparation of some materials of interest to the flavour and fragrance industry. Chapter nine looks at lipid derived flavours. The chapter brings all of the current literature together and looks not only at the processes involved in producing lipid-derived flavours, but also at the applications of these flavours, whilst our final chapter (ten) looks at the current safety and legislation issues surrounding the flavour area.

Finally, I must thank Lucy Swift and Kim Yarwood who both helped with the proof reading of some of the chapters. I must also thank all of the authors, firstly, for their excellent manuscripts and secondly, for their patience considering the turmoil that the transition from Thomson Science to Kluwer Academic publishers caused. Finally I would like to thank the publisher, Kluwer Academic for publishing the book.

INTRODUCTION

Karl Swift
Quest International
Ashford, Kent, TN24 0LT
United Kingdom

Before the advent of the modern day flavour and fragrance industry, perfumes and flavour ingredients had been part of the everyday life of mankind since the dawn of civilisation.

Ancient Assyrian tablets tell the tale of incense being offered to the God of the Sun in Ninive. The Egyptians used incense and a whole range of other oils, and the art of extracting plants and flowers was persued with vigour during the Greek and Roman civilisations.

Around the 9th Century AD distillation techniques were improved markedly by the Arabs, and this period is generally acclaimed to be the beginning of the production era for modern odorants. In the 10th Century the physician *Avicenna* introduced rose oil and water. The scale of production of these odorants soon escalated due to export demand from around the world. This extraction technology soon lead to many essential oils being produced this way.

The advancement of fragrance and flavour chemistry then plateaued for a considerable period of time until the field of organic chemistry started its rapid development. This development continues unabated to the present day. Some of the highlights of achievements from this period of time are described in the following paragraphs.

It was in the early 1800's that analysis and synthetic techniques started to break new ground and in 1833 Dumas reported the empirical formula of a number of terpene compounds including, camphor, anethole, and borneol. He followed this in 1834 by isolating the important flavouring ingredient cinnamaldehyde from cinnamon oil along with Péligot. Liebig in 1837 showed that by hydrolysing amygadalin (from bitter almond seed) that one could isolate benzaldehyde. Cahous, who in 1844 noticed that it was a major constituent of wintergreen oil, made the use of methyl salicylate as a flavouring agent possible.

A major milestone for perfumery came about from Bertagnini's discovery of aliphatic aldehydes as potential perfumery ingredients in 1853. Also during this year, Bertagnini discovered the now classical method for purification of aldehydes utilising bisulphite. This was a busy period of time, because it was also around this time that Cannizzaro

K.A.D. Swift (ed.), Current Topics in Flavours and Fragrances, 1–4.

discovered the reaction, which is now named after him. This classical reaction involves treating benzalehyde with an alkali to yield benzyl alcohol and sodium benzoate.

Cinnamaldehyde may have been first isolated in 1834, but it was not until 1856 that Chiozza demonstrated the first potentially commercial synthesis of this important material/intermediate. 1856 also saw Perkin's synthesis of cinnamic acid from benzaldehyde and acetic anhydride make its debut. It was another twenty-one years before he applied this reaction to salicylic aldehyde to yield coumarin. Perkin's synthesis of coumarin in 1877 was followed by von Pechman's coumarin synthesis that utilised the condensation of a phenol with an β -ketoester. This opened up the doorway for the preparation of various substituted coumarin's.

Kolbe made a major discovery in 1860. He reacted sodium phenolate with carbon dioxide to give salicylic acid. This process was then utilised for the large-scale commercial manufacture of salicylates, which are still some of the biggest volume ingredients used within our industry today.

Over the next few years several more key ingredients succumbed to the synthetic efforts of chemists. Notable examples are the preparation of benzaldehyde from benzyl chloride, water, and alkali by Cahours, and the discovery of the famous Reimer-Tiemann synthesis of aldehydes in the mid 1870's. Tiemann and Reimer had prepared synthetic vanillin from guaicol, chloroform, and alkali. Vanillin mixed with coumarin is responsible for the 'sweet oriental' note in the famous Guerlain perfume 'Jicky'. Vanillin is still widely used in perfumes and is used globally as a major flavour ingredient.

1871 saw the exact structure of heliotropin elucidated by Barth. Until the 1880's the cost of this molecule was prohibitively expensive. This was because it was only obtainable from the oxidation of piperic acid a substance that was prepared from pipenine (isolated from pepper). It was not until Eykman discovered the relationship between heliotropin and safrole that the price of heliotropin fell to make the molecule widely accessible to the fragrance industry.

Also around this period of time there was a great deal of knowledge gained in the area of terpenes by the now famous names such as, Tilden, Berthelot, Barbier, Kekule, and Bouchardat. The late 1880's saw the analysis of terpenes being tackled by Wallach. In particular, in 1888 he established the identity of dl-limonene. He went on to make significant contributions to both terpene chemistry and organic chemistry in general.

In 1888 whilst working on explosives, Baur made a major discovery which was to have a major impact on the perfumery industry for many years to come. He discovered the group of musks, which we call the nitromusks. These have been very widely used throughout the perfumery industry for at least one hundred years.

Ionones were discovered by accident by Tiemann in 1893 whilst he was working on the constitution of irones. This work also had a profound influence on later work directed towards the structure of vitamin A.

Just before the turn of the century synthetic research saw a surge in activity involving the chemistry of terpenes. In 1895 Tiemann and Semmler elucidated the structure of terpineol. It was also around this time that terpenes such as geraniol, and linalool were beginning to be produced commercially from natural essential oils. Infact, by the turn of the century the industry was commercially producing many materials from natural sources. Examples included menthol, santalol, borneol, citronellol, and cinnamic alcohol. There were also several synthetics being commercially produced at this time. Examples are cinnamic aldehyde, methyl/amyl salicylate, benzyl alcohol, and benzyl acetate.

The dawn of the twentieth century witnessed a sequence of important discoveries. Moureu and Delange discovered methyl heptine carbonate and its octine analogue. Both molecules have intense odours reminiscent of violet leaves. Darzens discovered the 'glycidic method' of aldehyde synthesis in 1904, and utilised this to prepare methylnonylacetaldehyde. Also in 1904 another present day important perfumery material succumbed to a commercial process. The material in question is phenylethyl alcohol, and Bouveault Blanc discovered the process.

Another chemist who has made a notable contribution to both perfumery and organic chemistry is Ruzicka. In the early 1920's he synthesised nerolidol and farnesol, elucidated the structures of the natural macrocyclic musks, and the components of violet flower and leaf oil.

The prospect of a commercially viable synthesis for the macrocyclic lactone musks was brought a step closer by the work of Carothers in the 1930's. His work on polyesters relating to the synthesis of nylon brought these desirable molecules a step closer.

The modern day industry.

The vast majority of aroma chemicals, that are available to the modern day fragrance industry, are the result of an intensive research effort that was made from around the period of the 1950's and continues to this day. This effort although mostly driven by the perfume industry has, in the earlier years, also been contributed to significantly by other major speciality chemical manufacturers especially those involved with utilising terpene feedstocks, and those that produce vitamins.

An example of a fragrance molecule that was discovered during this productive period is methyl dihydrojasmonate. Discovered in 1962 by the Swiss company Firmenich, it has since been incorporated into many well known fine fragrances such as 'Eau Sauvage', 'Chanel No 19', and 'Anaïs Anaïs'.

The research effort to find new molecules goes on unabated still. The research department of a flavour and fragrance company of the late 1990's relies on the interaction of a large number of scientific disciplines such as, analytical chemistry, computational chemistry, organic chemistry, process chemistry, physical chemistry, chemical engineering, biochemistry, microbiology, molecular biology, food science, materials science, sensory science, and toxicology, etc.

Requirements for new flavour and fragrance materials have diverged over the last decade or two. The flavour and food requirements have moved towards natural and nature identical ingredients whereas fragrance requirements have moved towards added functionality, better environmental acceptance, and improved performance in products.

As for the future? Who knows what fragrance and flavour molecules are still awaiting discovery? One thing is for sure, these molecules when discovered will have to meet an ever growing list of attributes which they must possess in order to make it as a 'must have' fragrance or flavour material.

Literature Referenced:

i. P.Z. Bedoukian, Perfumery and Flavouring Synthetics, Allured Publishers, 1986.
ii. K Bauer, D. Garbe, H. Surburg, Common Fragrance and Flavour Materials, VCH Publishers, 1990.
iii. G. Ohloff, Scent and Fragrance, Springer-Verlag Publishers, 1990.
iv. E-J. Brunke, 'Synthetic Aroma Chemicals Trends and New Findings', Dragoco Report No.3, 1993.
v. P. J. Teisseire , Chemistry of Fragrant Substances, VCH Publishers, 1994.

THE TOTAL SYNTHESIS OF SYNTHETICALLY INTERESTING PERFUMERY NATURAL PRODUCTS

Karl Swift
Quest International
Ashford, Kent, TN24 0LT
United Kingdom

"The organic chemist is more than a logician and strategist; he is an explorer strongly influenced to speculate, to imagine, and to create. These added elements provide the touch of artistry which can be included in a cataloging of the basic principles of synthesis but they are very real and extremely important." *E. J. Corey, 1967 [1].*

1. Introduction

Synthetic organic chemists are always looking for new challenges. The synthesis of natural products from readily available starting materials has always attracted, and continues to attract, the best chemical minds. In the classical era, total synthesis was used to confirm a natural product's chemical structure. This is rarely necessary now as chemists have a whole range of sophisticated analytical techniques that they can call upon to aid in structure identification. Total synthesis is still vital for a number of different reasons:

i. The discovery of new synthetic chemistry.
ii. To make available quantities of material which might not be available by other means such as extraction.
iii. The identification of potential large-scale processes to a given material.
iv. The creativity that the work brings out in chemists.

The total synthesis of perfumery interesting natural products began in the early 1900's with the synthesis of simple terpenes such as camphor (Komppa, 1903; Perkin, 1904) [2] and α-terpineol (Perkin 1904) [3]. It was in the 1950's and the 1960's that the total synthesis of natural products took off in a big way. During this period a vast number of perfumery related natural products were synthesised for the first time.

There are many natural products routinely used within the fragrance industry. Some of the simpler ones, for example, carvone, menthol, and geraniol are readily extracted from nature as single compounds or can be made synthetically.

K.A.D. Swift (ed.), Current Topics in Flavours and Fragrances, 5–32.

There are many essential oils that are still used within our industry. This is because the natural products which are contained in these oils continue to elude an economical industrial synthesis, or that the natural product mixture in question is just too complex to reproduce synthetically.

In this chapter we will be looking at the total or formal synthesis attempts directed towards some of the natural products that can be found in the following essential oils; vetiver (khusimone and β-vetivone), orris (the irones), patchouli (patchouli alcohol) and osmanthus (the theaspirones). It should also be noted that the natural products chosen are not necessarily the most important odour contributors in their particular oil. They are also not necessarily even a major component of the oil. They have been chosen because they all represent challenging synthetic targets.

The chapter is not only intended to be a review of attempts to synthesise a given natural product, it is designed to also give an insight into some of the more creative and innovative work that has gone into the synthesis of perfumery natural product targets. Apologies must also be made for the lack of detailed text in some areas of this chapter, this has been done deliberately in order to accommodate as much chemistry as possible in the limited space available.

2. Vetiver Oil

2.1 INTRODUCTION

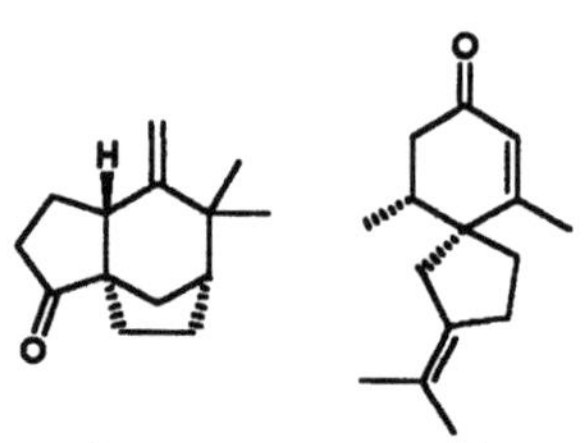

1 2

Let us begin our investigation with a look at two synthetically interesting terpenes that are found in vetiver oil which is obtained from the grass *vetiveria zizanoides.* The two terpenes in question are khusimone (1), and β-vetivone (2). These two molecules are not necessarily the main odour imparting molecules in vetiver oil, but each has unique characteristics and challenges in terms of its synthesis.

2.1.1 Khusimone (1)

Figure 1 shows an overview of some of the pathways that chemists have taken towards the synthesis of khusimone.

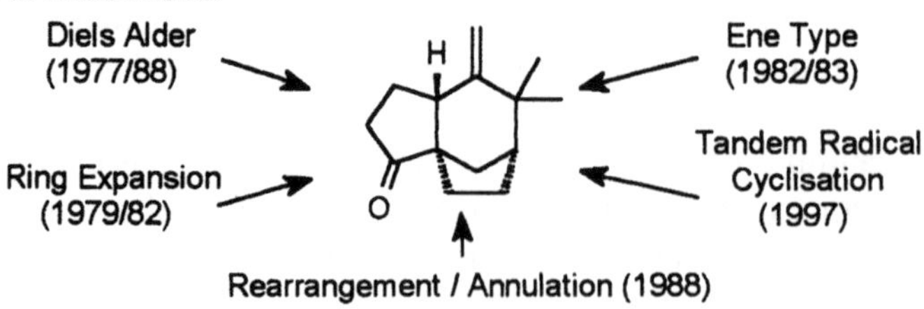

Figure 1. Overview of total synthesis routes to Khusimone

The Diels-Alder approach was used in the first synthesis of khusimone by G. Büchi and co-workers (Figure 2)[4]. His retrosynthetic analysis route was designed to mimic the proposed biogenetic pathway for the formation of 1 from γ-curcumene. There are two Diels-Alder reactions in Büchi's synthesis. The first is between the ketene equivalent, alpha-chloro acrylonitrile and isoprene. This, after elimination of HCl and addition of the necessary alkyl chain, sets up the diene system for the crucial step which is the second Diels-Alder. This intramolecular Diels-Alder yields a 3:1 mix of compounds 2a and 2b respectively. They then separated these isomers using chromatography and treated each of the isomers with acid. The acid initiates two Wagner-Meerwein rearrangements which set up the skeleton of khusimone (3) (figure 3) in the case of 2b, but as well as producing 3, it also produces a significant quantity of nor-cedrenone from 2a. Three further synthetic steps transform 3 into khusimone (70%) 1, and *epi*-khusimone (30%).

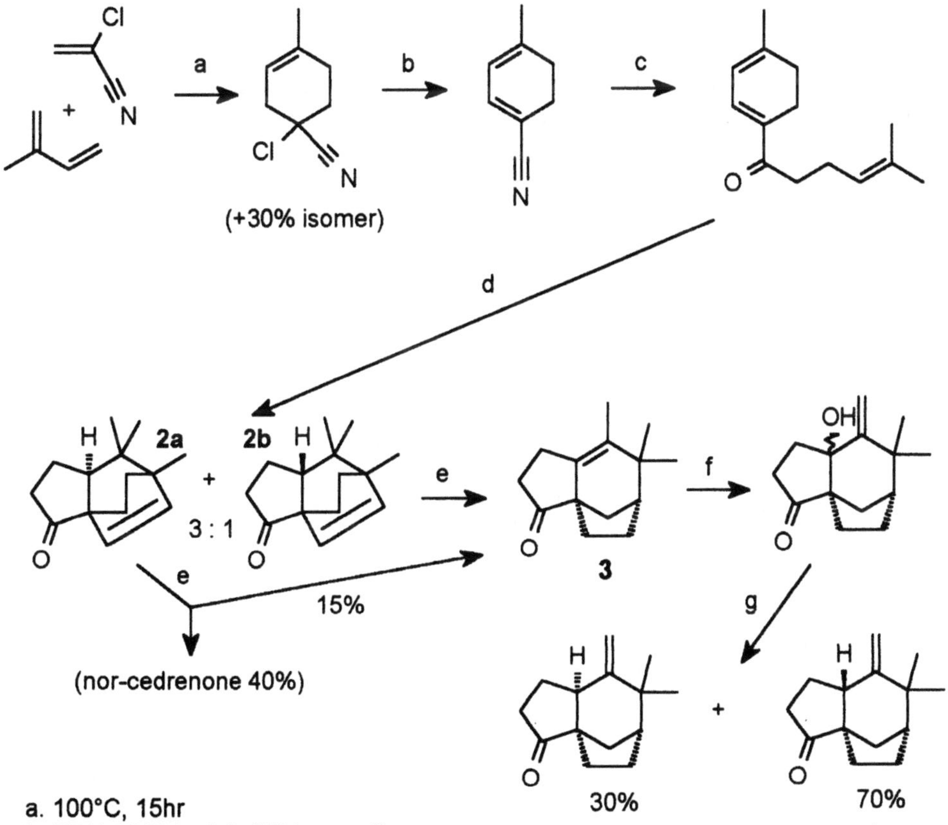

a. 100°C, 15hr
b. DBN, THF (a+b is 55% overall)
c. i. $Me_2CCH(CH_2)_2Li$, ii. HCl,H_2O, Et_2O (75%)
d. i. $CH_2(OH)CH(OH)Me$, H_3PO_4 (91%), ii. 250°C, iii. $HClO_4$, H_2O, THF (55%)
e. Heat, pTsOH
f. i. O_2, Rose bengal, EtOH, H_2O, ii.$PO(OEt)_3$ (77%)
g. Zn, HCl (g)

Figure 2. G. Büchi's synthesis of Khusimone (1977)

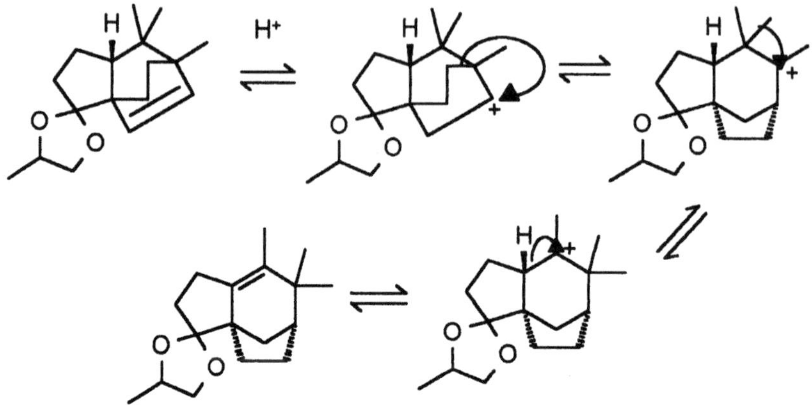

Figure 3. Detail of the Wagner-Meerwein rearrangement in Büchi's synthesis of Khusimone

(-)-Khusimone has also been prepared via a fifteen step reaction strategy which included a crucial Diels-Alder reaction as the opening reaction [5].

Figure 4 shows Oppolzer's approach to the synthesis of 1 [6]. His route involves an intramolecular Type II magnesium-Ene reaction which sets up the main bicyclic skeleton. Khusimone is prepared in an overall yield of 11% over the 13 steps. He later refined this synthesis route to prepare enantioselective (-)-khusimone [7] using the same reaction scheme as shown in figure 4, but this time he focused on the dienolate addition / alkylation step (step 'a' in Figure 4).

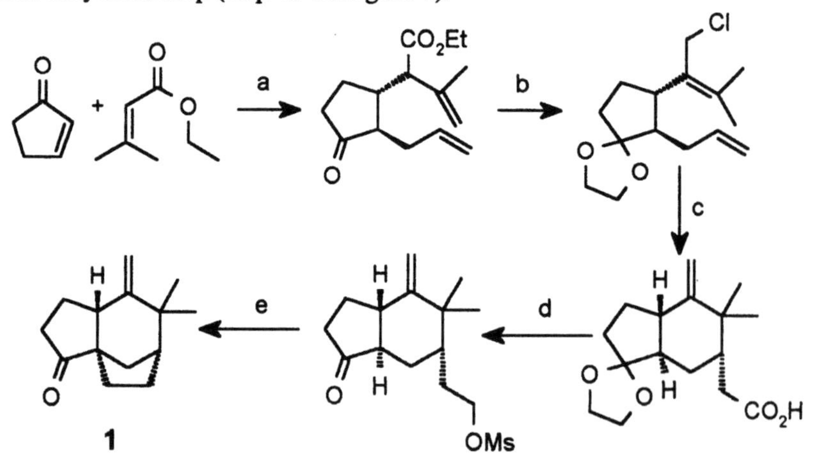

a i. LDA ii. Allyl bromide, HMPA, THF (50%)

b i. $HOCH_2CH_2OH$, pTSA (97%) ii. NaOEt (74%) iii. LAH (92%) iv. MsCl, Pyr then LiCl (10% aq.) (51%)

c i. Mg, THF ii. 60°C, 17hr iii. CO_2 (82%)

d i. LAH (92%) ii. MsCl, NEt_3 iii. HCl, H_2O, Et_2O (ii + iii 93%)

e tBuOK, tBuOH, PhH (98%)

Figure 4. Oppolzer's approach to Khusimone (1982/3)

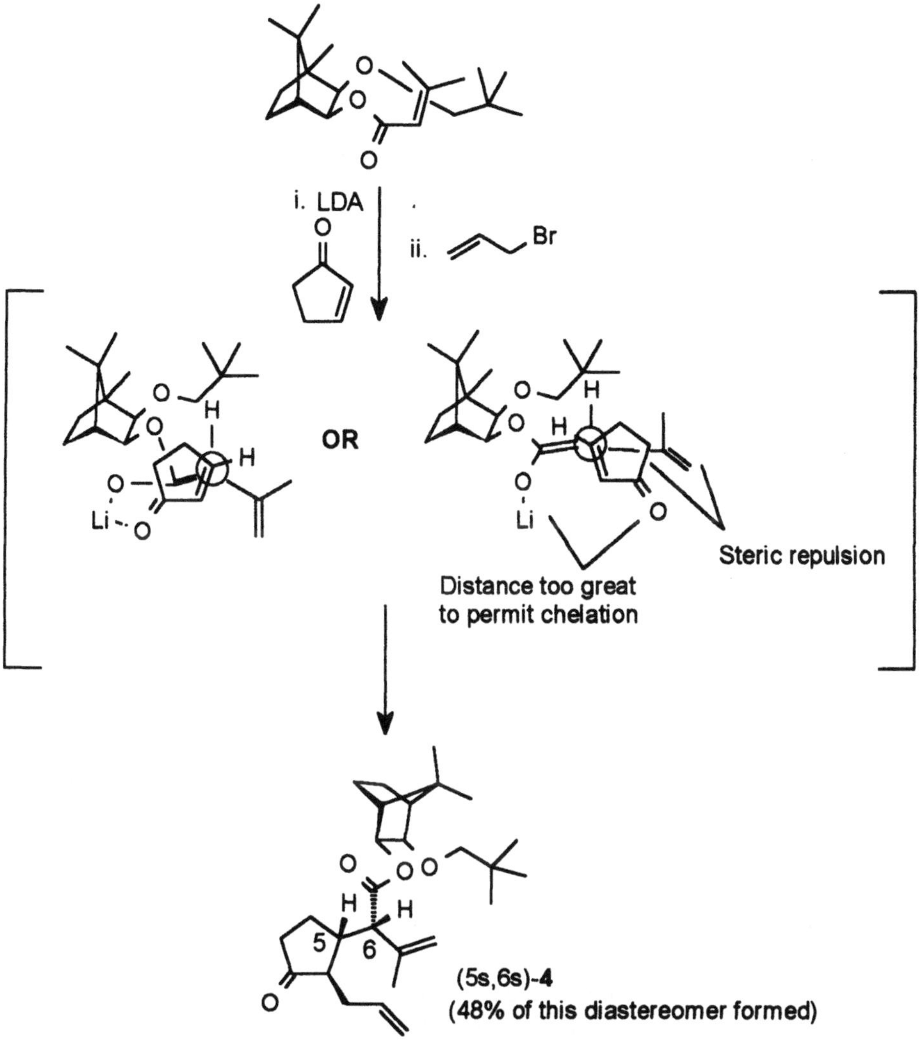

Figure 5 The rationale behind Oppolzer's synthesis of (-)-1

Figure 5 shows the rationale behind his thoughts. For further detail the reader is referred to the original literature [7]. (-)-Khusimone was prepared from 4 in an overall yield of 14%.

Liu and Chan prepared (-)-khusimone utilizing the ring expansion of the cyclobutyl-ketone 5 (Figure 6)[8]. Their starting material, (-)-α-campholenic acid was obtained from the ammonium salt of *l*-10-camphorsulphonic acid in a yield of 50% by fusion with solid potassium hydroxide according to the published literature procedure of Sauers *et al.* [9]. After ozonolysis and cyclisation, the photocycloaddition proceeded with complete regioselectivity to give an inseparable mixture of isomers. Hydrolysis of the ketal yielded two diastereomers (6,7) as the only two products. Although only diastereomer 7 was directly usable in the synthesis of (-)-khusimone, the other

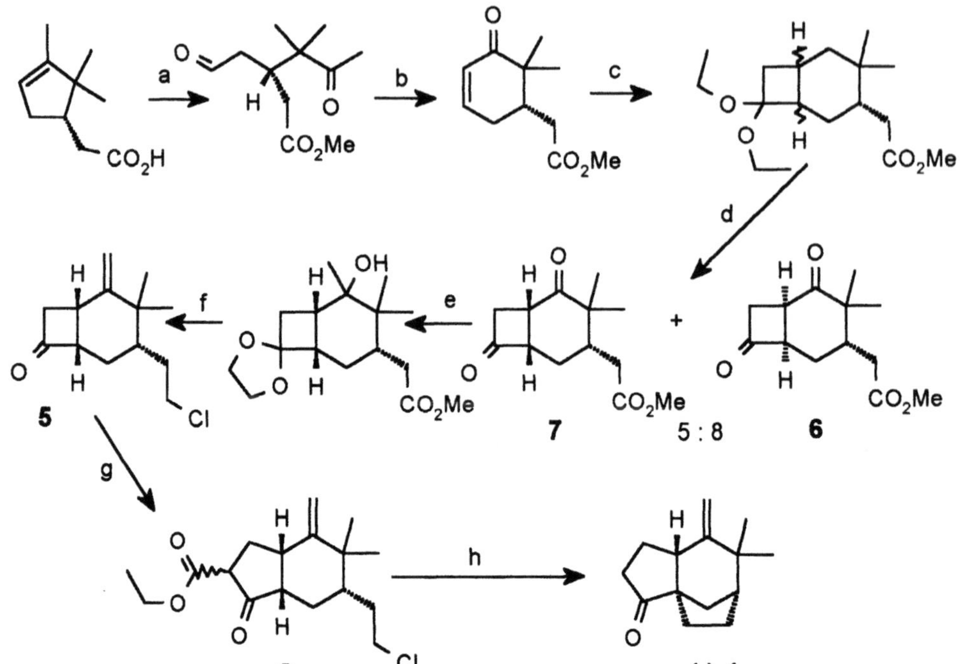

a i. MeI, K_2CO_3 ii. O_3, PPh_3 b. pTsOH, PhH (70% from a) c. $CH_2C(OEt)_2$, hυ,rt. d HCl, Me_2CO (80% from c) e. i. 2-Ethyl-2-methyl-1,3-dioxolane, H^+ ii. NaOH (2N), MeOH iii. NaH, MeMgBr (5eq) iv. CH_2N_2 (71%) f. i. $SOCl_2$, Pyr (80%) ii LAH iii. $POCl_3$, Pyr (70% for ii and iii) iv. HCl (aq) (100%) g. BF_3OEt_2, $EtO_2CCH_2N_2$ (59%) h. NaOH (1N) (61%)

Figure 6. Liu and Chan's synthesis of (-)-1 (1979)

diastereomer 6 was not wasted. Treatment of 6 with pyridinium bromide perbromide in acetic acid followed by reduction using zinc in acetic gave a 2:3 mixture of 6 and 7. The compound 7 produced could then be utilised directly into the main synthetic scheme, whilst the lesser quantity of 6 could be further recycled. Khusimone's five membered ring was set up by the highly regioselective ring expansion of the cyclobutane ring system. The final ring closure was readily achieved by reaction of 8 with dilute base. This also removed the ethyl ester group.

This approach was also utilized in the synthesis of other zizaane sesquiterpenes by the same authors [10].

A relatively simple (in terms of the chemistry involved) formal synthesis of khusimone has been published by Wu and Burnell [11]. They claim an overall yield of 35% for iso-khusimone starting from nor-camphor (figure 7). Their key steps involve the formation of the 1,3-cyclopentadione moiety (step f) using Kuwajima's published method [12], and the use of a McMurry coupling (step g) to form the tricyclic system.

Although their reaction scheme is very demanding in terms of the equivalents of reagents used it is

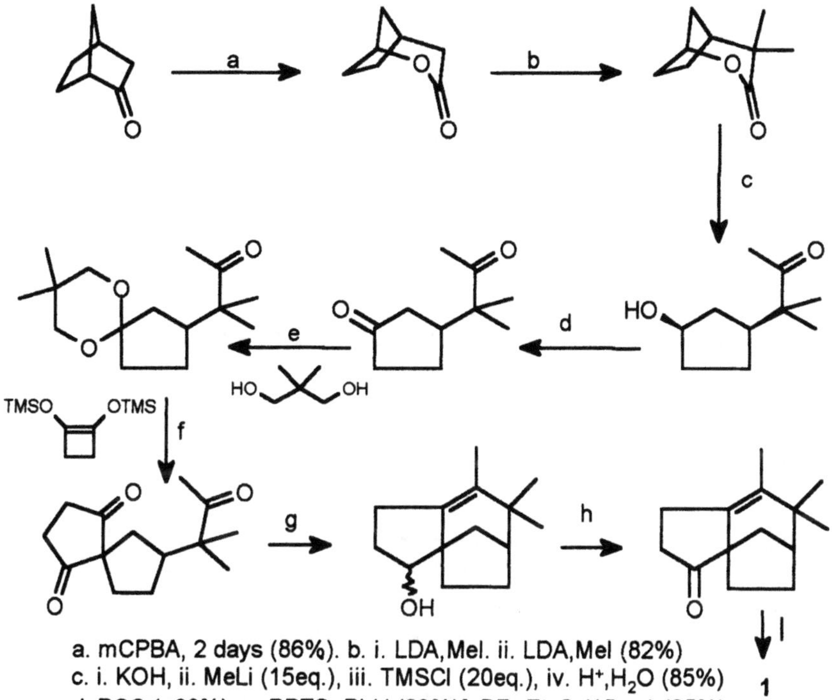

a. mCPBA, 2 days (86%). b. i. LDA,MeI. ii. LDA,MeI (82%)
c. i. KOH, ii. MeLi (15eq.), iii. TMSCl (20eq.), iv. H^+,H_2O (85%)
d. PCC (>99%). e. PPTS, PhH (89%)f. $BF_3.Et_2O$ (15eq.) (85%)
g. $TiCl_3$ (4eq.), Zn-Cu (21eq.) h. PCC (>99%) I. See figure
2 step f&g.

Figure 7. Wu and Burnell's formal synthesis of khusimone (1988)

interesting for its novel approach towards the formal synthesis of 1. This constitutes a formal synthesis of khusimone since step 'I' in figure 7 is part of Büchi's khusimone synthesis [4].

An even more recent formal synthesis of (±)-khusimone by Kim and Cheong [13] utilises a consecutive carbon-carbon bond formation approach which is based on the radical cyclisation of *N*-aziridinylimines. Figure 8 shows their approach.

Their starting material was readily prepared in 78% yield by the tert-butyldimethylsilyl triflate (TBSOTf) promoted conjugate addition of the piperidine enamine of isobutyl aldehyde to 2-cyclopenten-1-one in methylene chloride at -78°C. Addition of the selenoalkyl lithium followed by oxidation, hydrolysis and selective *N*-aziridinylimine formation all proceeded in high yield to give 9. Reaction of 9 with tributyltin hydride under high dilution conditions promotes the formation of the khusimone skeleton. Four further steps yield compound 10. Again this constitutes a formal synthesis of khusimone since step 'f' in figure 8 is part of Büchi's khusimone synthesis [4].

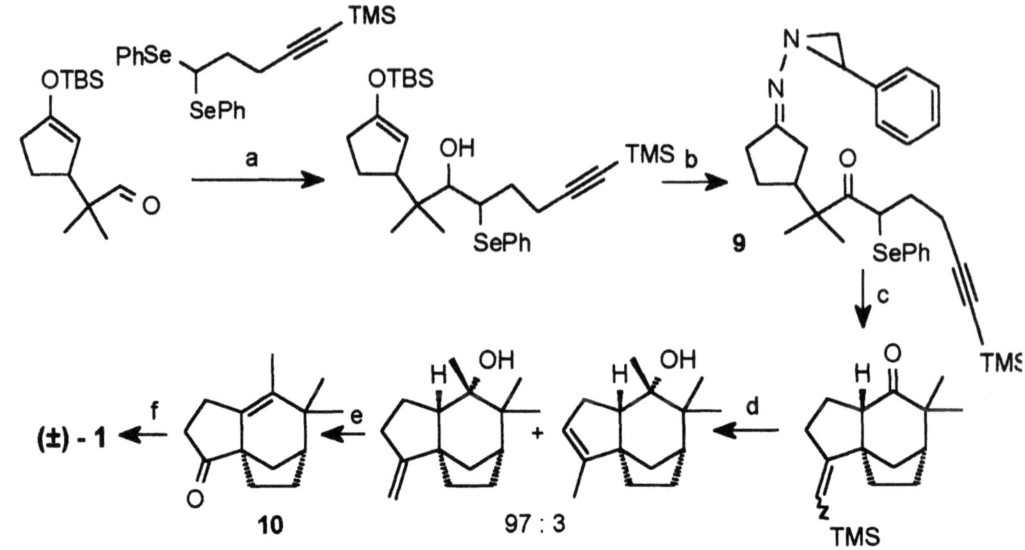

a. nBuLi (79%) b. i. Swern Oxidation ii. 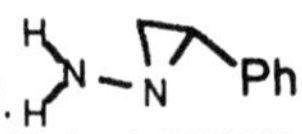 iii. HCl (aq) (87%)
c. AIBN, Bu_3SnH (78%) d. i. MeLi ii. HCl (aq) (97%) e. i. RuO_2, $NaIO_4$ (73%)
ii. $SOCl_2$, Pyr. (88%), f. See figure 2 reaction step f&g.

Figure 8. Kim and Cheong's approach to khusimone (1997)

Although their synthesis is relatively complex and undesirable in terms of the toxicity of some of the reagents, it is still elegant with respect to the tandem cyclisation that occurs. This cyclisation is shown in more detail in figure 9.

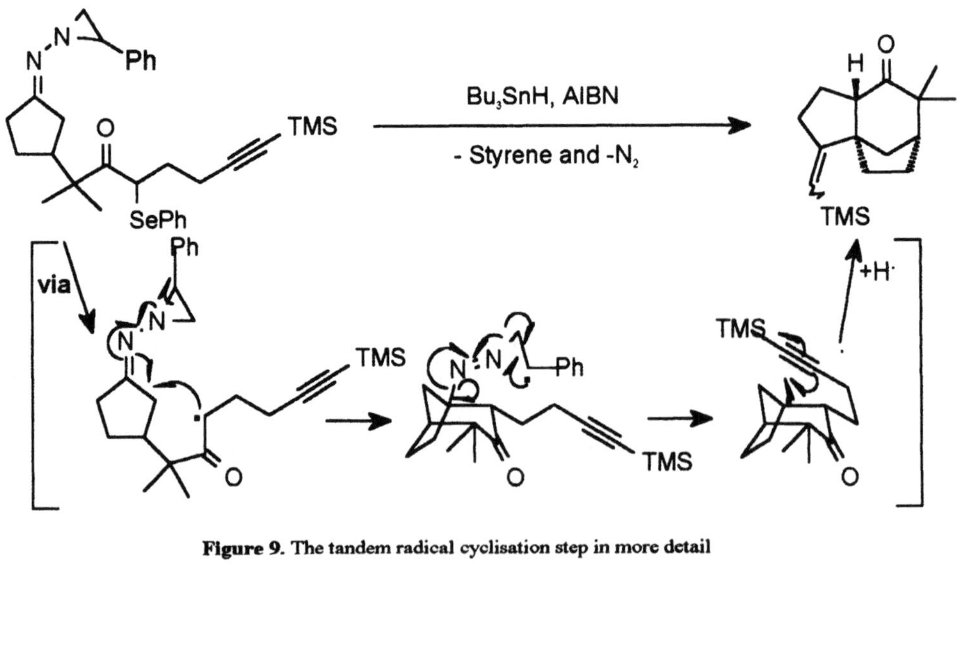

Figure 9. The tandem radical cyclisation step in more detail

2.1.2. β-Vetivone (2)

β-Vetivone is a terpene which has commanded a considerable amount of interest from chemists because its spiro[4,5]decane structure. There are over twenty total or formal syntheses of 2 reported in the chemical literature. Figure 10 shows an overview to some of the routes taken to synthesise 2.

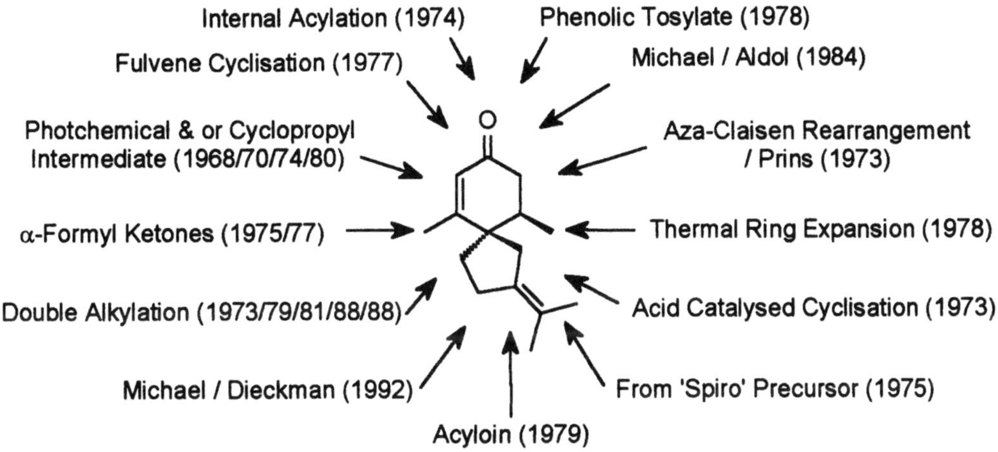

Figure 10. Overview of total synthesis routes to β-vetivone

The earliest total synthesis of (±)-β-vetivone was by Marshall and Johnson in 1968 [14], who had also earlier proposed the now accepted spiro[4,5]decane structure of 2 [15].

The double alkylation approach is the most often used approach to the synthesis of the 'spiro' centre in 2. Of the many syntheses in this area [16], we shall only look at a couple.

In 1973 Stork *et al.* published a short and elegant synthesis of (±)-2 [16i] (figure 11). Compound 11 was alkylated via its kinetic enolate, to yield initially the mono alkylated product 12. They postulated that the stereochemistry of the first alkylation is as shown in 12 because the first alkylation had to involve the allylic halide. They then postulated that the subsequent enolate ion geometry would force the ring methyl into an axial conformation, thus one would expect to get ring closure trans to the methyl group. This was indeed the case.

A second example in this area is from the work of Asaoka *et al.* [16v] who prepared (-)-2 by the route shown in figure 12 which involves a stereoselective intramolecular alkylation.

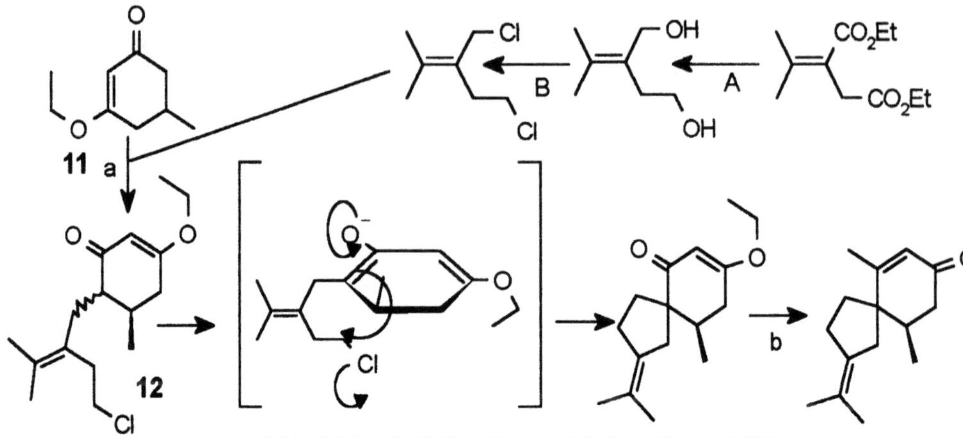

a. i. HMPA (1eq), THF, LDA (1eq) ii. LDA (2eq) (45%)
b. i. MeLi ii. HCl (60%)

A. Lithium aluminium ethoxy hydride (90%)
B. MeLi, HMPA, MsCl, LiCl (35%)

Figure 11. Stork's approach to β-vetivone (1973)

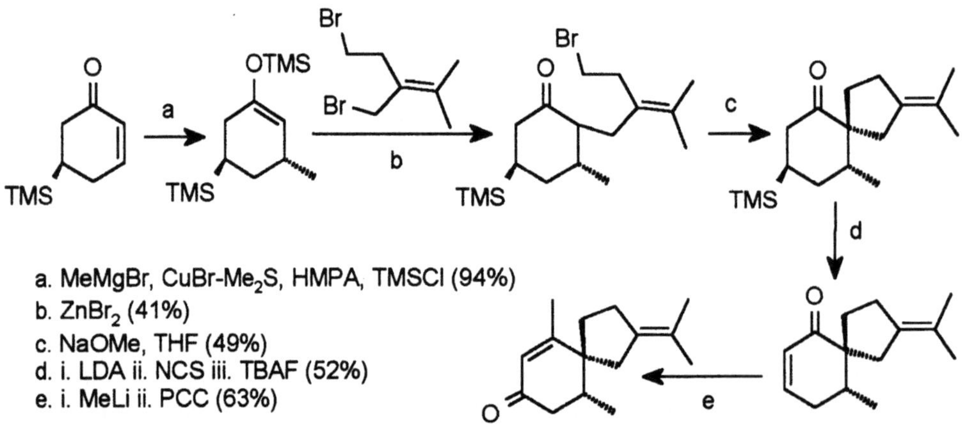

a. MeMgBr, CuBr-Me_2S, HMPA, TMSCl (94%)
b. $ZnBr_2$ (41%)
c. NaOMe, THF (49%)
d. i. LDA ii. NCS iii. TBAF (52%)
e. i. MeLi ii. PCC (63%)

Figure 12. Asaoka's approach to (-)-β-vetivone (1988)

(-)-2 has been prepared by the group of Posner *et al.* Utilising an enantiomerically pure vinyl sulphoxide as the key starting material[16iv].

We shall now take a chronological look at some of the other reported routes to 2.

The well known Aza-Claisen rearrangement was utilised by McCurry and Singh in their synthesis of (±)-β-vetivone [17]. The pyrrolidine enamine 13 in the presence of trans-crotyl bromide undergoes the aza-Claisen rearrangement as shown in figure 13 to yield a 55:45 mix of the diastereomers 14.

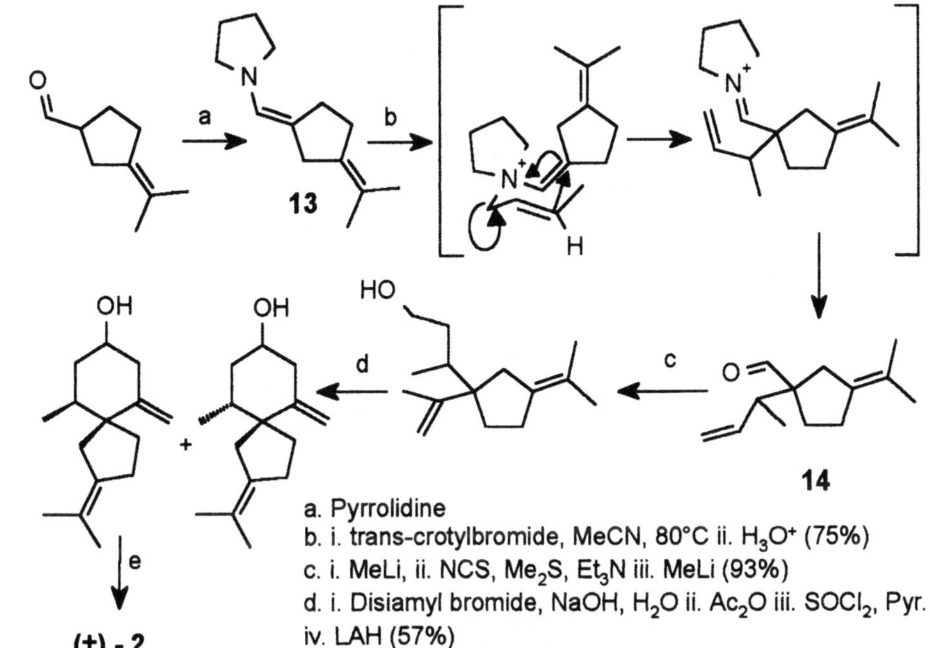

a. Pyrrolidine
b. i. trans-crotylbromide, MeCN, 80°C ii. H_3O^+ (75%)
c. i. MeLi, ii. NCS, Me_2S, Et_3N iii. MeLi (93%)
d. i. Disiamyl bromide, NaOH, H_2O ii. Ac_2O iii. $SOCl_2$, Pyr.
iv. LAH (57%)
e. i. Oxidation ii. $SnCl_2$,

Figure 13. McCurry and Singh's approach to (±)-2

The final ring closure is accomplished using a Prins reaction which gives the cyclised products as a mixture of only two diastereomers, even though four diastereomers are theoretically possible. Oxidation yields a mixture of (±)-β-vetivone and (±)-10-epi-β-vetivone.

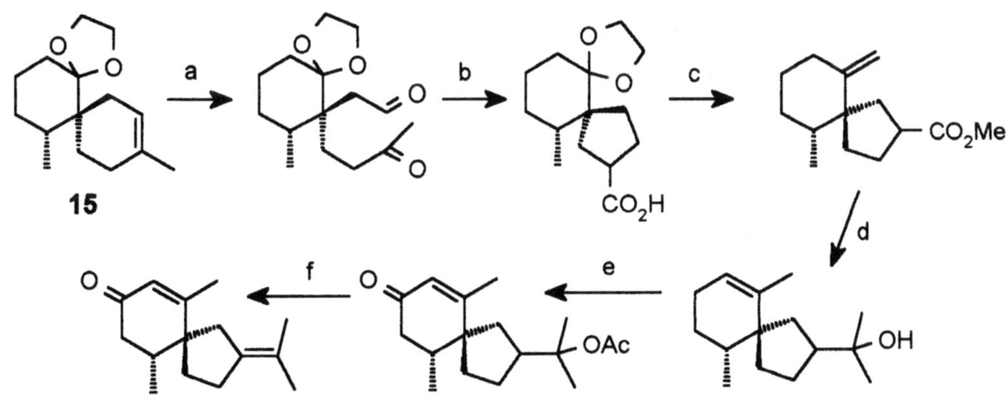

a. O_3, Me_2S b. i. KOH ii. H_2/Pd iii. Br_2, NaOH c. i. CH_2N_2 ii. HCl (3N) iii. Wittig
d. i. pTSA, PhH, reflux ii. MeLi e. Ac_2O, NaOAc ii. Na_2CrO_3, HOAc, Ac_2O
f. i. BF_3OEt_2 ii. Column over $AgNO_3$

Figure 14. Synthesis of (-)-2 by Ramage and co-workers

The earliest synthesis of (-)-2 was by Ramage and co-workers in 1975 [18]. His reaction sequence is shown in figure 14 and starts from the chiral 'spiro' compound 15.

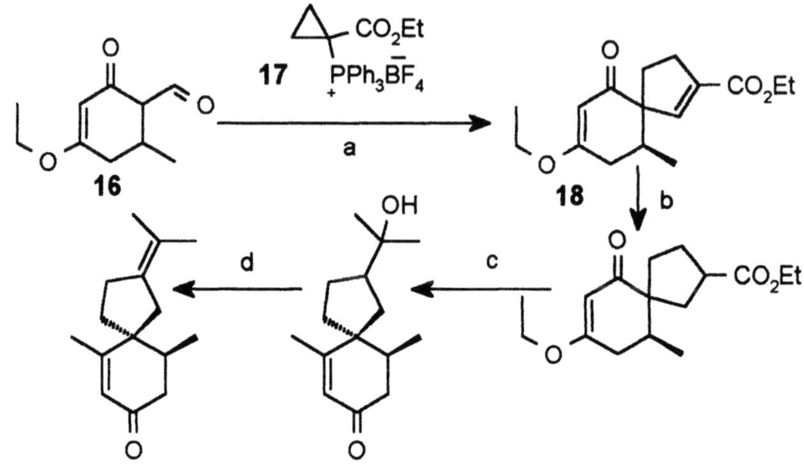

a. HMPT, NaH, 13 (38%)
b. H_2, Pd/C, EtOH
c. i. MeLi (excess) ii. HCl (12N) (b+c = 64%)
d. i. Ac_2O ii. BF_3OEt_2 (50%)

Figure 15. Dauben and Hart's approach to (±)-2

An interesting approach towards the synthesis of 2 was published by Dauben and Hart [19]. The spiro centre in their synthesis is set up by reaction of an α-formyl ketone (16) with carboethoxycyclopropyltriphenylphosphonium tetrafluoroborate (17) (figure 15). Compound 18 is very versatile and has been utilized in the synthesis of many other spirovetivanes [20].

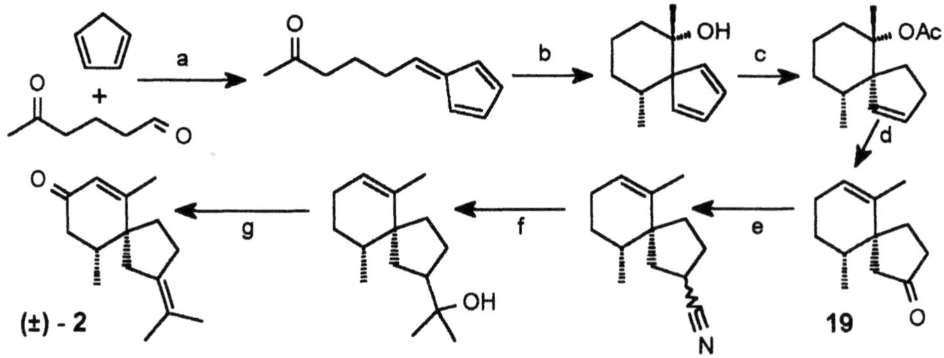

a. Et_2NH (70%) b. Me_2CuLi, -20°C (90%) c. Diimide (NH_2NH_2, H_2O, EtOH) (57%)
d. i. NaOH, EtOH (95%), ii. AcO_2H (95%), iii. BuLi, Et_2O (72%), iv. pTSA, PhH (90%)
e. pToluenesulphonylisocyanide (75%) f. i. MeMgBr (80%) ii. MeLi (90%)
g. i. CrO_3, Pyr. (70%) ii. pTSA, PhH (80%)

Figure 16. Bŭchi's approach to (±)-2

1976 also saw an elegant synthetic route to (±)-2 published by Büchi *et al.* [21]. The actual publication is directed towards the synthesis of 19 (figure 16) which is also a component of vetiver essential oil. Figure 16 shows his synthesis of 19, and how it was readily transformed into (±)-2.

The ingenious parts of Büchi's synthesis are the initial stages. He condenses 5-oxohexanal with cyclopentadiene to yield the corresponding fulvene. This is then reacted with lithium dimethylcuprate to yield the lithium cyclopentadienide which then cyclises by reacting with the ketone carbonyl to form the corresponding spiro cyclohexanol product. Ten further steps yield (±)-2.

β-Vetivone has also been prepared from aromatic precursors. One such synthesis was that of Torii and co-workers [22] who prepared (±)-2 and epi-(±)-2 via a seven step synthesis starting from the aromatic precursor 20 as shown in figure 17. The spiro centre is constructed from the base-catalysed cyclisation of the appropriate phenolic tosylate (step c, figure 17).

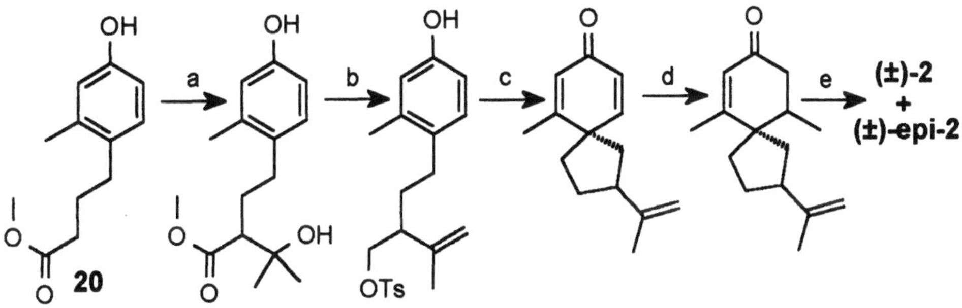

a. LDA, Me_2CO (86%) b. i. $SOCl_2$ ii. LAH iii. TsCl, Pyr. (68%)
c. tBuOK, tBuOH (74%) d. $LiCuMe_2$ (75%) e. $RhCl_3$, EtOH (80%)

Figure 17. Torii and co-workers approach to (±)-2

The 1970's was clearly the most prolific time in terms of the number of publications concerning the synthesis of 2. Several other publications from this decade are not included here [23].

In contrast the 1980's saw only a few publications directed towards the synthesis of 2 [24], the most notable of which will now be described.

Murai, Masamune, and Sato's approach to 2 utilized an initial Diels-Alder reaction, and a subsequent π-cyclisation as the key synthetic step [25]. Although the total reaction scheme was very lengthy it is worth showing the key cyclisation step (figure 18).

To conclude the section on β-vetivone we will look at the synthesis of (±)-2 by Posner *et al.* [26]. He utilized a one-flask, 3-component Michael-Michael-Dieckman protocol as the key step to construct the three new carbon bonds A, B, and C in 21. Figure 19 shows his entire synthesis of (±)-2.

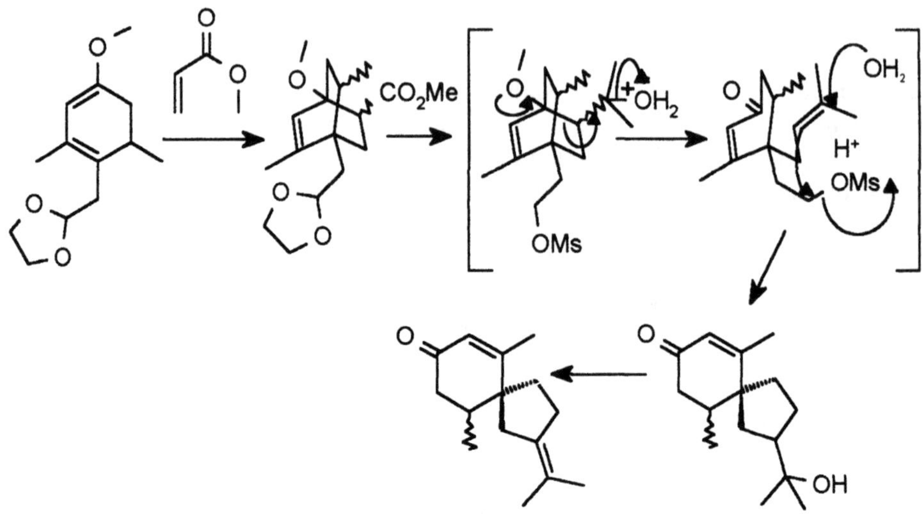

Figure 18. Murai, Masamune, and Sato's key cyclisation step in their synthesis of (±)-2

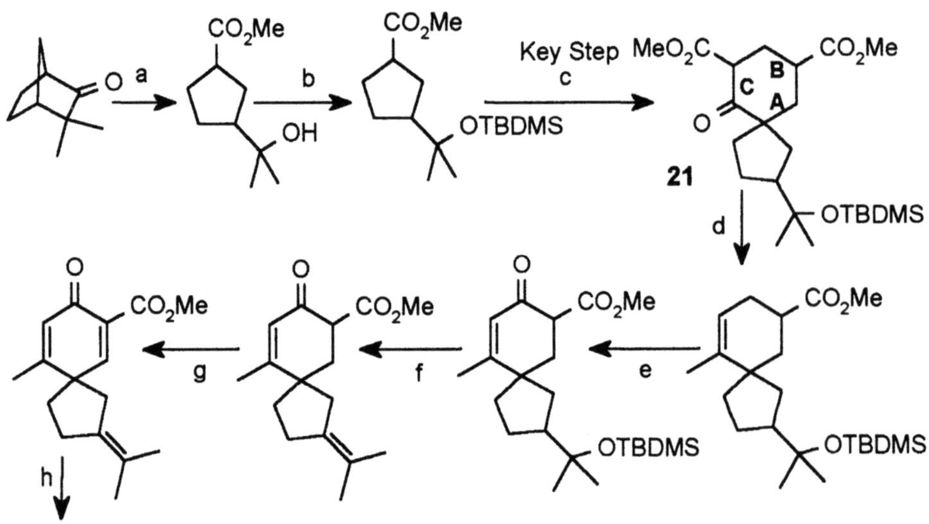

(±)-2

a. i. mCPBA, CF_3CO_2H (79%) ii. LiOMe (85%) b. $TfOSiMe_2tBu$ (96%)
c. i. LDA ii. H_2CCHCO_2Me (2eq) (51%) d. i. NaCl, DMSO (81%) ii. MeMgBr Et_2O (75%), iii. $SOCl_2$, DMAP, Pyr. (72%) e. i. PDC, tBuOOH, PhH (47%)
f. BF_3OEt_2, CH_2CH_2 (65%) g. Me_2CuLi, BF_3OEt_2 (66%)
h. i. Me_2CuLi, BF_3OEt_2 (81%) ii. NaCl, DMSO (77%)

Figure 19. Posner's approach towards (±)-2

3. Orris Oil

3.1 IRONES

The constituents of Orris (Iris) oil that we shall be concerning ourselves with in this chapter are the irones (22-24). Orris oil comes from the steam distillation of the rhizomes of *Iris pallida* (which contains the major irone components; (-)-trans-α-irone, (+)-cis-γ-irone, (+)-cis-α-irone, (+)-β-irone, and (+)-trans-γ-irone) or *Iris germanica*. Both species of iris contain the same irones except that the irones in *Iris germanica* are the optical antipodes of those found in *Iris pallida* [27]. The three basic irone structures are shown in figure 20.

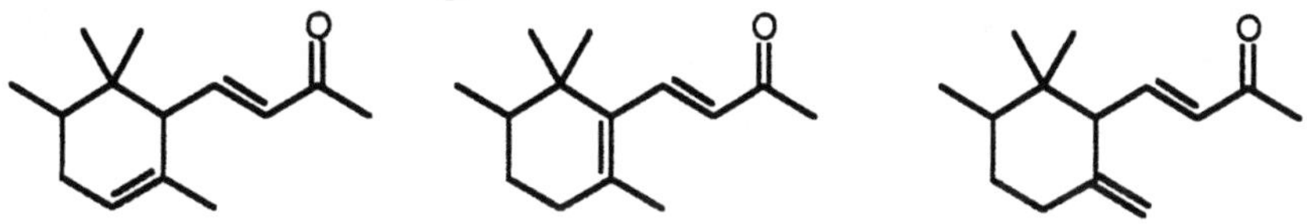

22 **23** **24**

Figure 20. α-Irone (22), β-Irone (23), and γ-Irone (24)

The irones have been challenging synthetic targets for chemists since the 1940's. They are challenging not only for the crucial stereochemistry necessary throughout the molecules' chiral centres and carbon-carbon double bonds, but also for the financial reward that a simple, low cost, and chiral synthesis could deliver. Orris oil is one of the most expensive essential oils that is available on the market.

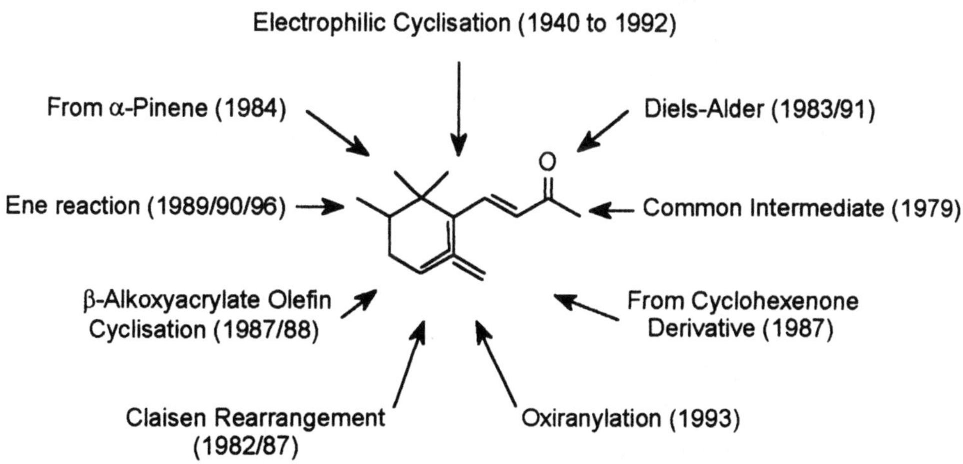

Figure 21. Overview of synthesis routes to the irones

The majority of early syntheses in this area produced irones from the electrophilic cyclisation of 9-methyl-pseudoionone or other intermediates [28]. Figure 21 shows the approaches that have been taken towards the irones.

Two notable electrophilic cyclisation approaches came from Eshinazi [29], and Barton [30]. These elegant syntheses are shown in figures 22 and 23 respectively. A chiral variant on Eshinazi's synthesis was published in 1984 [31].

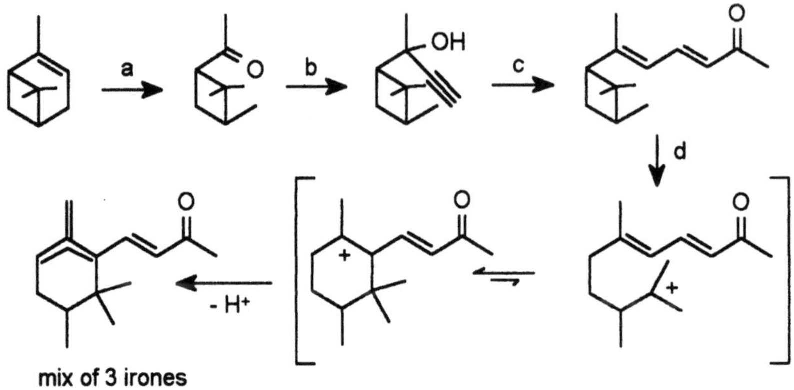

a. i. O_3 (70%) ii. Pd 200°C (80%) b. HCCH, NaCCH, NH_3 (99-100%)
c. $CH_3C(O)CH_2CO_2Et$ (65%) d. H_3PO_4 (50%)

Figure 22. The approach taken by Eshinazi *et al.* towards irones

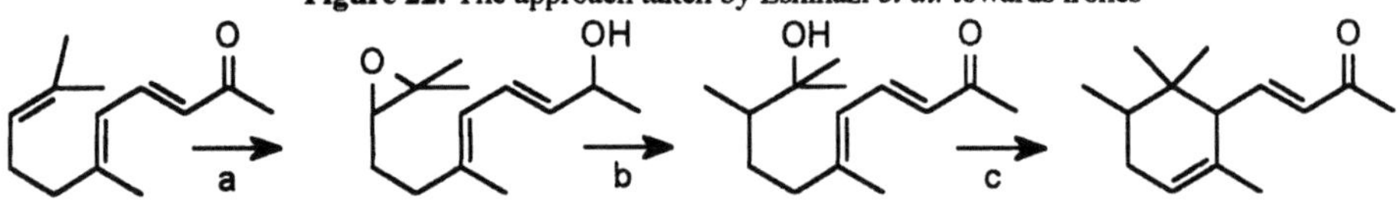

a. i. monoperphthalic acid, Et_2O ii. $NaBH_4$, MeOH
b. i. MeMgBr ii. MnO_2 c. H_3PO_4

Figure 23. The approach taken by Barton *et al.* towards irones

The 1970's saw a much reduced level of publications related to the synthesis of irones, when compared to the vibrant period of the 1940's to the 1960's. One synthesis which was simplistic in its approach was that of Garnero and Joulain [32]. This route which yields (±)-trans-2,6-γ-irone (25) is shown in figure 24.

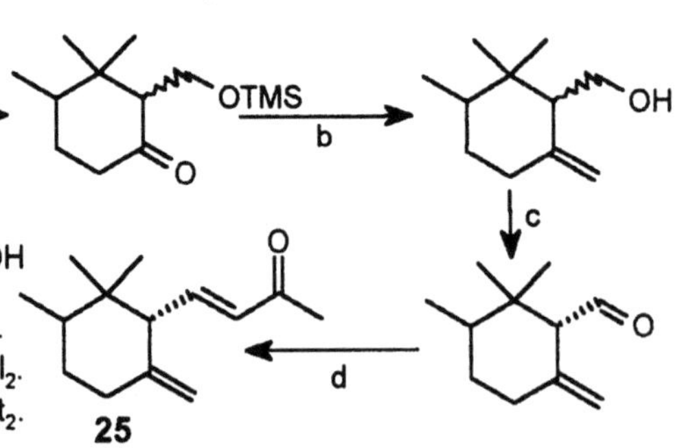

a. i. MeMgI,CuI ii. HCOH
iii. TMSCl, Pyr.
b. i. CH_2I_2, Mg ii. EtOH.
c. PCC, AcONa, CH_2Cl_2.
d. $MeC(O)CH_2P(O)OEt_2$.

Figure 24. Garnero and Joulains synthesis of (±)-trans-2,6-γ-irone

1982 saw the first enantioselective synthesis of any of the irones. This work by Yoshikoshi and co-workers [33] utilised (+)-citronellal as the key chiral building block. Their 19-20 step route yielding (+)-trans-γ-irone as the major product is shown in figure 25. This approach also yielded the irones; (-)-trans-α-irone and (+)-β-irone.

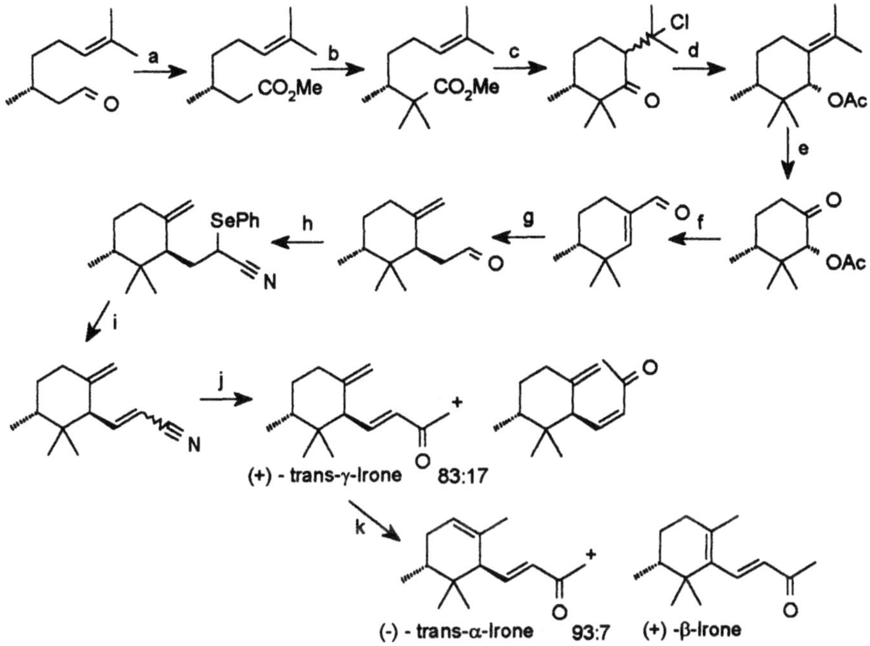

a. i. Jones Oxidation ii. Esterification. b. LDA, MeI, THF. c. i. KOH, EtOH. ii. $(COCl)_2$, PhH. iii. $SnCl_4$, CH_2Cl_2. d. i. Li_2CO_3, DMA. ii. LAH, Et_2O. iii. Ac_2O, Pyr. e. O_3. f. i. $Ph_3P{=}CHOMe$, PhMe. ii. $HClO_4$, Al_2O_3. g. i. LAH, Et_2O. ii. $EtOCHCH_2$, $Hg(OAc)_2$. iii. Heat. h. i. KCN, HOAc, EtOH. ii. MsCl, Pyr. iii. PhSeNa, EtOH. i. H_2O_2, Pyr. j. MeLi, Et_2O. k. H_3PO_4.

Figure 25. Yoshikoshi and co-workers approach to (+)-trans-γ-irone, (-)-trans-α-irone, and (+)-β-irone

In the late 1980's Georg Fráter and Cornelius Nussbaumer of Givaudan published a relatively short and elegant synthesis of (±)-cis-α-irone [34].

Their key step involves a β-alkoxyacrylate olefin cyclisation which sets up the cis configuration of the molecule (figure 26). In a later publication they also used intermediate 26 to prepare (±)-cis-γ-irone (figure 26)[35].

A Japanese research group adapted their own chemistry which had been directed towards the synthesis of the taxane terpene skeleton, to prepare a mixture of 64.4% (+)-cis-α-irone and 15.6% (+)-trans-α-irone in nine steps from (-)-27 (figure 27) [36]. Compound (-)-27 is itself prepared in seven steps from the diene (shown in figure 27) and maleic anhydride, and is obtained chirally pure after chromatography.

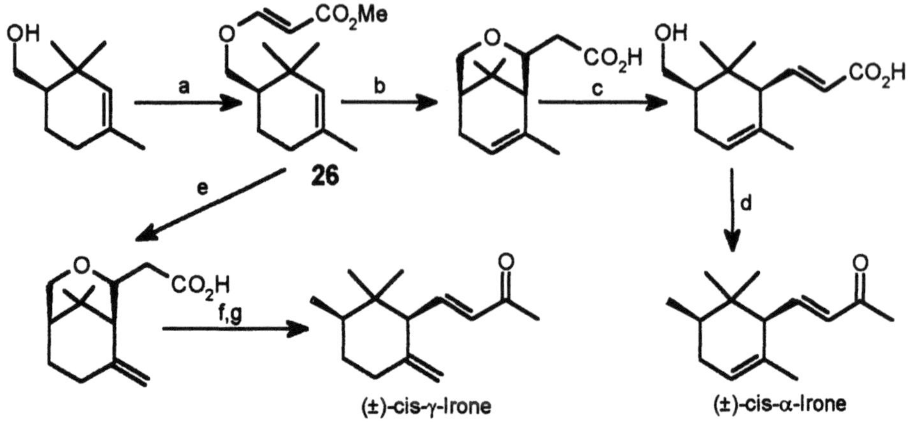

a. Methylpropiolate, *N*-methylmorpholine (88%).
b. i. $MeSO_3H$, CH_2Cl_2 (94%). ii. NaOH, MeOH, H_2O (78%).
c. i. LDA (2equiv.). ii. H_3O^+ (88%). iii. $MeSO_2Cl$, Pyr (94%).
d. i. Zn, NaI, DME (76%). ii. MeLi (89%).
e. i. PhMe, 300°C. ii. NaOH, MeOH, H_2O.
f. i. LDA (2equiv.). ii. H_3O^+. iii. $MeSO_2Cl$, Pyr.
g. i. Zn, NaI, DME. ii. MeLi.

Figure 26. Nussbaumer, and Fráter's synthesis of (±)-cis-α-irone and (±)-cis-γ-irone

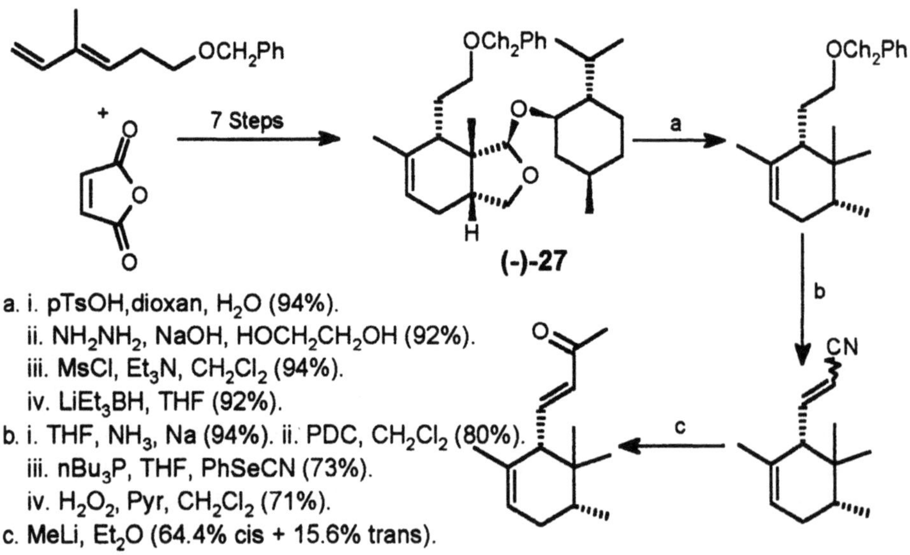

a. i. pTsOH,dioxan, H_2O (94%).
ii. NH_2NH_2, NaOH, $HOCH_2CH_2OH$ (92%).
iii. MsCl, Et_3N, CH_2Cl_2 (94%).
iv. $LiEt_3BH$, THF (92%).
b. i. THF, NH_3, Na (94%). ii. PDC, CH_2Cl_2 (80%).
iii. nBu_3P, THF, PhSeCN (73%).
iv. H_2O_2, Pyr, CH_2Cl_2 (71%).
c. MeLi, Et_2O (64.4% cis + 15.6% trans).

Figure 27. Preparation of (+)-cis-α-irone (major product) by Ohtsuka, Itoh, and Oishi

The final synthesis in the area of the irones that we shall look at is that by Monti *et al.*[37]. They prepared (+)-(2*R*,6*R*)-trans-γ-irone (28) by a relatively simple route starting from (+)-(*R*)-3,4-dimethylcyclohex-2-en-1-one (29). This in turn was prepared from the readily available (+)-(2*R*,5*R*)-trans-dihydrocarvone in four steps. This synthesis is different from all of its predecessors because it yields a completely enantioselective single product. The key part of their synthesis is the H-ene reaction which is promoted cleanly using the mild Lewis acid, dimethylaluminium chloride. They also take advantage of a well known side reaction that readily occurs with allylsilanes. The side reaction in question is the protodesilylation which readily occurs with Lewis acids such as titanium tetrachloride. It was applied to compound 30 and the desired product (28) was obtained as a single diastereomer with no double bond isomerisation being observed. Their synthesis is shown in figure 28.

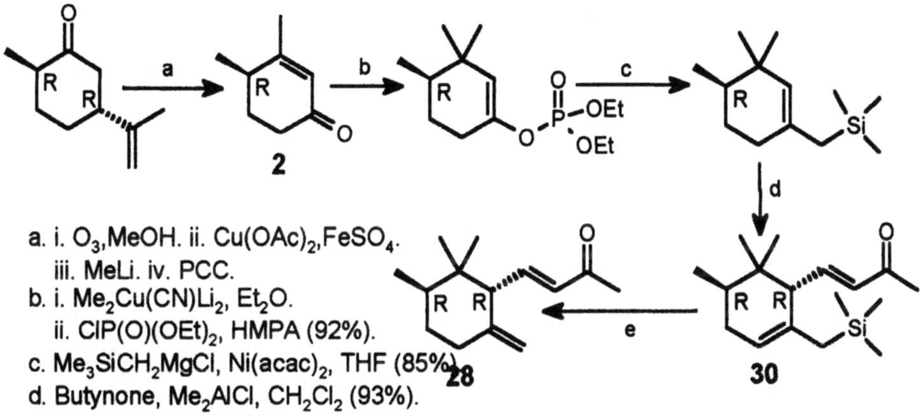

Figure 28. Preparation of (+)-(2*R*,6*R*)-trans-γ-irone by Monti *et al.*

There are several other irone syntheses which have not been included here. If the reader is interested they can be found listed under reference 38.

4. Patchouli Oil

4.1 PATCHOULI ALCOHOL (31)

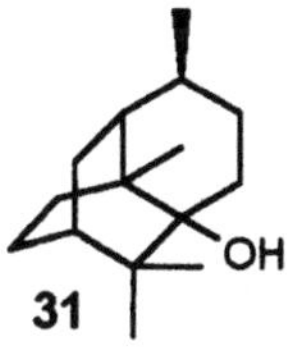

31

Patchouli oil obtained from *Pogostemon cablin* is widely used within the fragrance industry. The major component in patchouli oil is (-)-patchouli alcohol, 31. This sesquiterpene has attained much interest from synthetic organic chemists over the years because of its challenging chemical structure, namely the tricyclo[5.3.1.0]undecane carbon skeleton with its five asymmetric carbon centres of which four are contiguous. Figure 29 shows the approaches that have been taken towards patchouli alcohol.

The first synthesis of patchouli alcohol was by Büchi *et al.* in 1964 [39]. Since that time the most popular approach towards the synthesis of 31 has involved the Diels-Alder reaction [40a-f]. Out of these, we will look at the syntheses of (±)-31 by Bertrand, Teisseire, and Pelerin [40e](figure 30), the synthesis of (±), (+), and (-)-31 by Näf, Ohloff and co-workers [40c/d](figure 31), and the synthesis of (±)-31 by Stork *et al.* [40f] (figure 32).

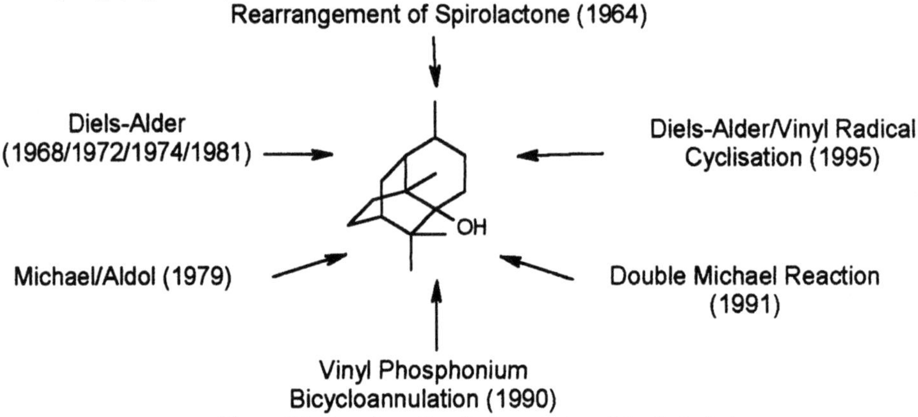

Figure 29. Overview of synthesis routes to patchouli alcohol

Bertrand, Teisseire, and Pelerin's short approach to patchouli alcohol starts with the Diels-Alder reaction between the diene 32 and methyl vinyl ketone (figure 30). The endo bicyclic diketone produced is readily rearranged to the corresponding exo bicyclic which is then readily transformed into (±)-31 in four further steps. The key step in their synthesis is the final cyclisation involving the methoxymethylether derived from the corresponding δ-ethylenic γ-hydroxyketone.

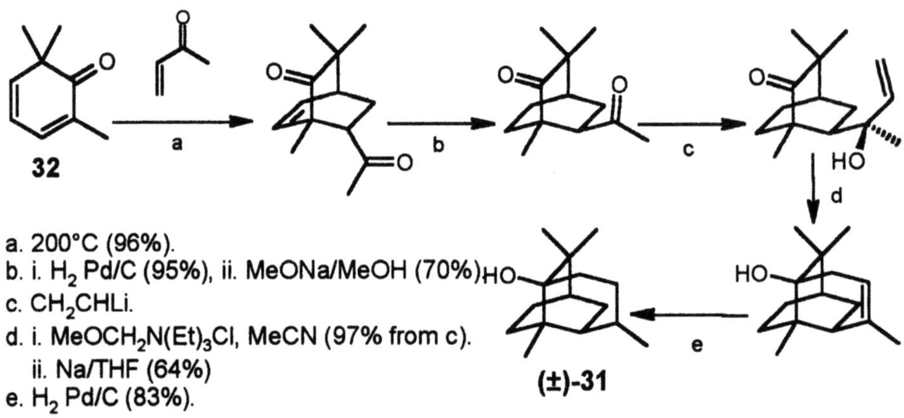

Figure 30. Bertrand, Teisseire, and Pelerin's synthesis route to (±)- patchouli alcohol

The elegant approach to (-), and (+)-31 taken by Näf, Ohloff and co-workers is shown in figure 31. They also produced (±)-31 by an almost identical route, but this is not shown here, the reader is referred to the original literature [40c/d]. Their route to (-)-31 starts with (+)-33 which was obtained from a commercial source (it is synthesised from (+) α-pinene). The chiral bromide (+)-34 was synthesised in five steps respectively from (+)-33 in an overall yield of about 20% and with an enantiomeric purity of about 70%. The bromide was then converted to the corresponding organo-lithium species and reacted with 2,6,6-trimethylcyclohexa-2,4-dien-1-one to yield the respective tertiary alcohol as a mix of two diastereoisomers 35a/b. The key step in this synthesis is the intramolecular Diels-Alder reaction. They found that a trace of potassium tert-butoxide in the reaction mixture was essential. Without this only an aromatic compound (2,3,6-trimethyl-1-(3-methyl-4-penten-1-yl)benzene) was obtained from the reaction, this being formed from the aromatisation of the diene. Under the Diels-Alder reaction conditions they found that only one diastereomer (35a) from the crude mix of diastereomers 35a/b reacted. This observation allowed then to use the crude diastereomer mix directly thus cutting out the need for any further purification. They attribute this selectivity to the presence of a highly disfavoured 1,3-diaxial interaction in the transition state of 35b. The intramolecular Diels-Alder reaction was also found to be completely regioselective. These observations were also identical for the diastereomer mix 36a/b.

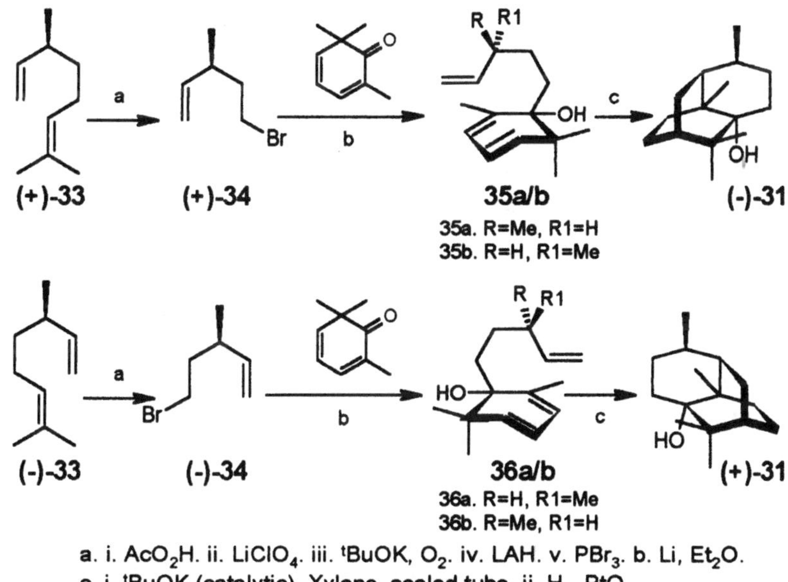

Figure 31. Näf, Ohloff and co-workers approach to (-) and (+) patchouli alcohols

Hydrogenation of the patchoulenol produced from 35a yielded (-)-31, and hydrogenation of the patchoulenol produced from 36a yielded (+)-31 respectively.

The final synthesis of (±)-31 utilising a Diels-Alder step that we shall look at is that of Stork *et al.* Starting, also from 2,6,6-trimethylcyclohexa-2,4-dien-1-one (figure 32), they react this with phenyl vinyl sulphoxide (an acetylene equivalent) to yield the symmetrical bicyclic compound 37. Addition of lithiated 2-butyne gave the unstable compound 38 which was then selectively reduced on the carbon-carbon double bond which is *syn* to the hydroxyl group. An explanation of this selective reduction can be found in the original paper [40f]. The alkyne group was then migrated to the terminal position to give 39 which then underwent a 6-*exo-trig* vinyl radical cyclisation to yield

(±)-31 after protodestannylation over silica and hydrogenation (via rearrangement to the endocyclic alkene).

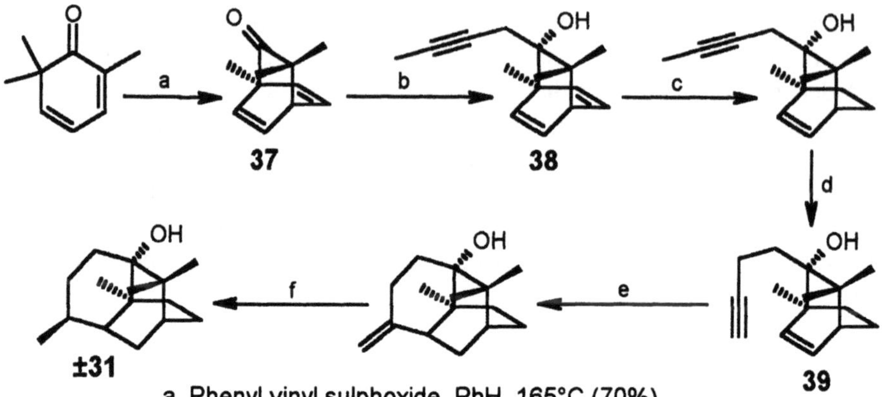

a. Phenyl vinyl sulphoxide, PhH, 165°C (70%).
b. 1-lithio-2-butyne, $MgBr_2$-OEt_2, -78°C to r.t.(82%).
c. Ca (3equiv.), 0.05M in NH_3:Et_2O (3:1), $Mg(ClO_4)_2$, -78°C (72%).
d. ${}^tBuLi$ (15equiv.), Et_2O, 10-30°C (77%).
e. i. n-Bu_3SnH, AIBN, 0.01M in PhH, reflux. ii. SiO_2 (80%).
f. H_2 10%Pd/C, EtOAc (85%).

Figure 32. Stork *et al.* approach to (±)- patchouli alcohol

A different and multi-step approach to (±)-31 was taken by Yamada and co-workers [41]. They used the base catalysed cyclisation of the conjugated cyclohexenone 40 to give the intermediate 41 which was then transformed to (±)-31 and also into (±)-seychellene (a sesquiterpene also isolated from patchouli oil). Figure 33 shows the key part of their synthesis.

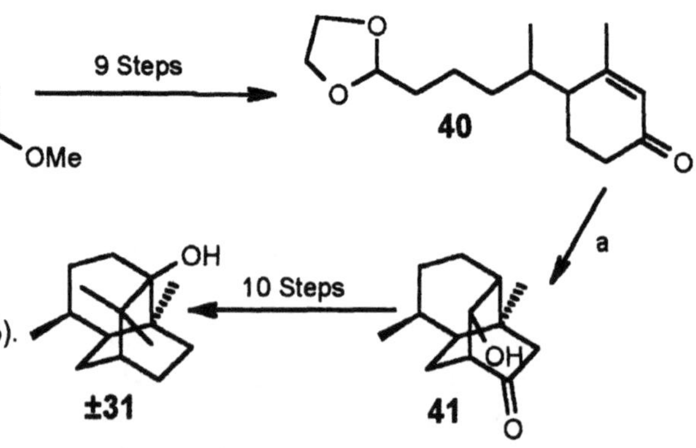

a. i. AcOH, H_2O (98%).
ii. ${}^tBuOK$, ${}^tBuOH$ (26%).

Figure 33. Yamada and co-workers approach to (±)- patchouli alcohol

To conclude our look at the literature syntheses of patchouli alcohol (31) we will take a look at one of the two formal syntheses of (-)-31 which start from (-)-carvone. The first one was by Cory, Bailey, and Tse in 1990 [42]. The second one was by Zhao and Wu in 1991 [43]. Of the two formal syntheses, Cory and co-workers is the most elegant and can be seen in figure 34. From (-)-carvone to the common intermediate 42, he obtained a reasonable overall yield of about 50%. Intermediate 42 has been taken through to 31 previously [40a,e]. The key step in their synthesis is the vinyl phosphonium bicycloannulation of methylcarvone to yield 43 (step 'b' figure 34), which was used to synthesise not only (-)-31, but also the sesquiterpenes (-)-seychellene, and (-)-cycloseychellene.

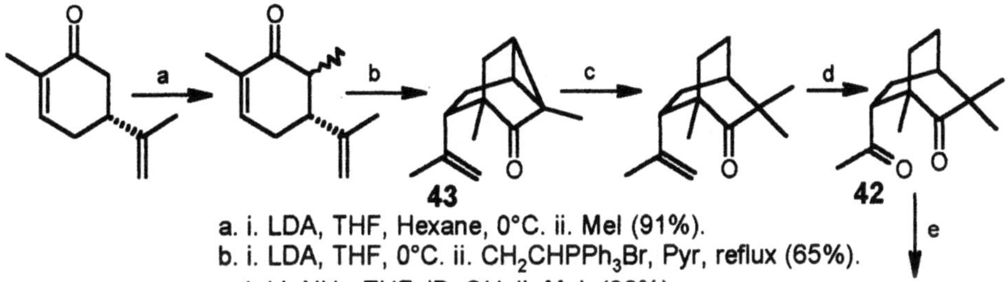

a. i. LDA, THF, Hexane, 0°C. ii. MeI (91%).
b. i. LDA, THF, 0°C. ii. CH_2CHPPh_3Br, Pyr, reflux (65%).
c. i. Li, NH_3, THF, ${}^tBuOH$. ii. MeI. (90%).
d. $RuCl_3$, $NaIO_4$, MeCN, CCl_4, H_2O (94%).
e. See references [40a and 40e].

Figure 34. Cory, Bailey, and Tse's formal synthesis of (-)-patchouli alcohol

Zhao and Wu's synthesis is similar in length and also ultimately yields the intermediate 42. Their key step was the reaction of methylcarvone with methyl acrylate in a double Michael reaction to give the basic bicyclic framework needed to prepare 42. For details of their synthetic route the reader is referred to the original literature [43].

5. Osmanthus oil

5.1 THEASPIRONES (44)

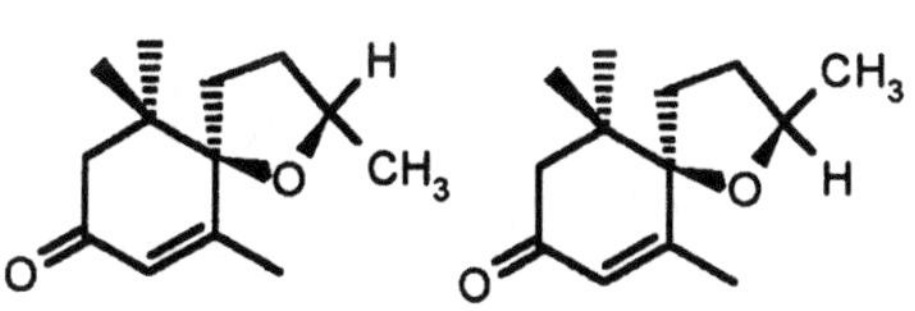

***trans*-44** ***cis*-44**

Osmanthus oil is a complex mixture of many different natural products obtained from the oil of the tree *Osmanthus fragrans*. The theaspirones (44) which are degradation products derived from the ionones make up only a small proportion of the oil but are interesting because they impart unique odour characteristics to the oil. The two theaspirones commonly found in osmanthus oil are (-)-theaspirone A ((-)-cis-44), and (-)-theaspirone B ((-)-trans-44). It is also worth noting that the theaspirones were originally isolated from black tea [44]. They are of interest to organic chemists because of the challenging spiro-ether moiety which has an asymmetric carbon centre alpha to the ether oxygen on either side, thus increasing the synthetic task. Figure 35 shows the synthetic approaches that have been taken towards the synthesis of theaspirones.

From looking at figure 35 it can be seen from the extra broad arrow that the majority of theaspirone syntheses have been derived from α or β-ionone. Of the many references in this area [45], we shall take a look at three examples.

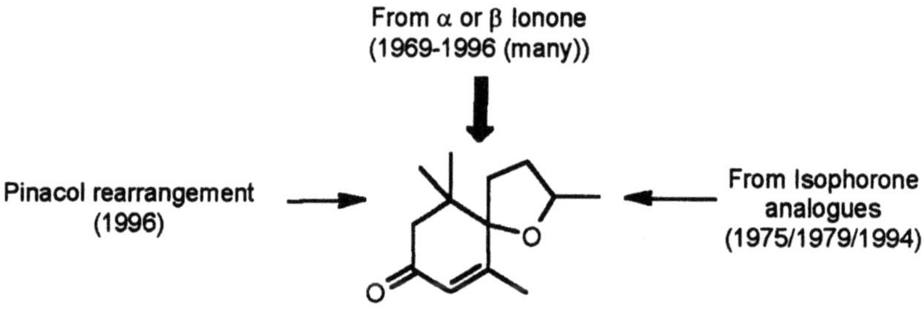

Figure 35. Synthetic approaches towards the theaspirones

The first synthesis that we shall look at is that of Nakanishi *et al.*[45d]. Their reaction scheme is shown in figure 36 and starts from α-ionone. Their publication is important as it describes the elucidation of the absolute configuration of natural (-)-cis-44 (as found in black tea).

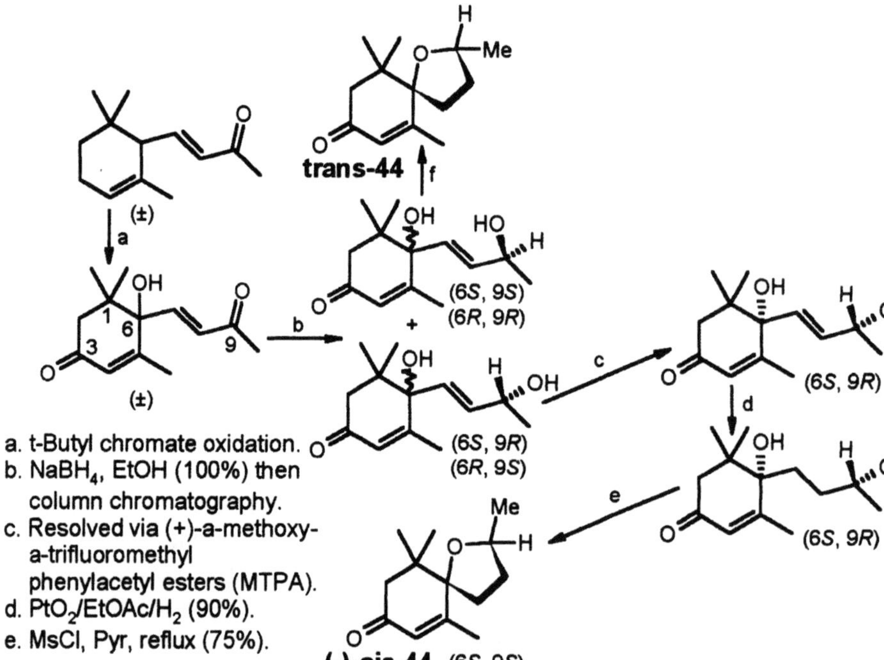

Figure 36.. Preparation of (-)-cis-theaspirone, and trans-theaspirone by Nakanishi *et al*

The second synthesis that we will look at is that of Könst, Apeldoorn, and Boelens [45g]. Their route was an attempt to avoid the drawbacks associated with some of the other ionone derived routes, namely the poor yields often obtained when adding the carbonyl group in the later stages of a synthesis, and not using a diol for the spiro ether forming step. The diol ring closing procedure often uses dimethylsulphoxide which makes it virtually impossible to obtain good odour quality products, because of the trace sulphur compounds that would be present after the reaction. Their route is shown in figure 37.

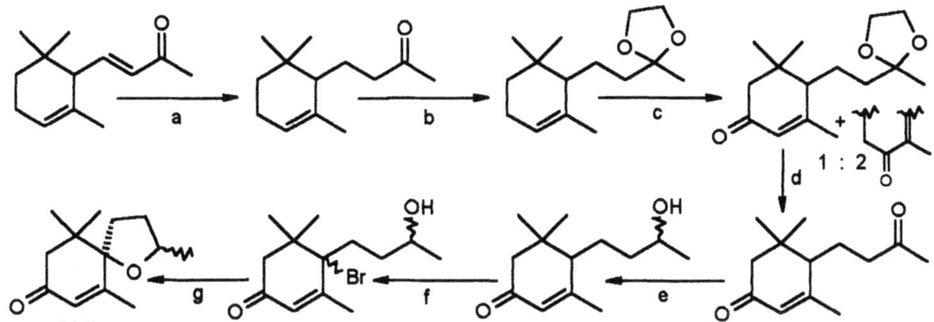

(44)

a. Raney Nickel, H_2 (90%). b. Ethylene glycol, PhH, H^+, reflux (80%).
c. t-Butyl Chromate oxidation (53% overall, but minor product is desired product).
d. Acetone/H_2O/H^+ (100%). e. $NaBH_4$ (1 equiv.), EtOH, 0°C (75%).
f. NBS, CCl_4 (100%). g. KOAc, Ac_2O, 110°C (56% of 1:1 mix cis:trans).

Figure 37.. Preparation of racemic-theaspirone by Könst, Apeldoorn, and Boelens

The final route that we will be look at is that by Bäckvall *et al.* [45i,j,k]. Starting from β-ionone they produced racemic 44 in six steps utilising a palladium (II) catalysed oxaspirocyclisation as the key step. Under these conditions this reaction was found not to be diastereoselective but their earlier work using stoichiometric amounts of palladium to prepare theaspirones was found to have a good diastereoselectivity [45j,k]. Their palladium catalysed route is shown in figure 38.

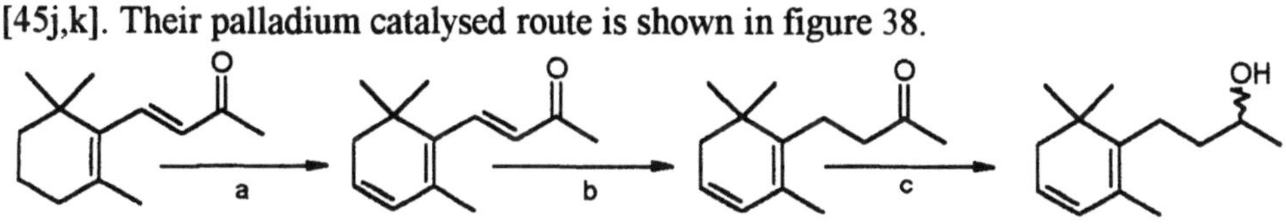

a. i. NBS, *hv*, CCl_4. ii. Na_2CO_3 DMF(77%).
b. Bu_3SnH, $Pd(PPh_3)_4$ (3-5 mol%), THF/H_2O/NH_4Cl (90%).
c. $NaBH_4$, 0°C (95%).
d. cat. $Pd(OAc)_2$, Benzoquinone, H_2O/HOAc (4:1)

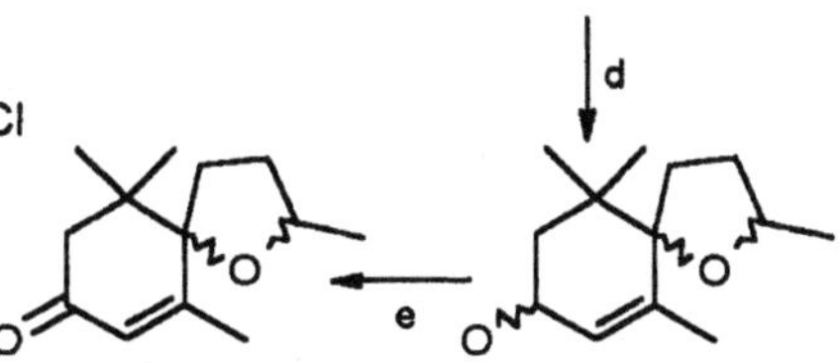

(44)

Figure 38.. Preparation of racemic-theaspirone by Bäckvall *et al.*

Theaspirones can also be synthesised from derivatised isophorone type molecules. The current literature shows at least three publications relating to this synthetic approach [46a,b,c]. We will look at the synthesis of (-)-cis-44 by Mayer in 1979. Mayer's route is shown in figure 39 and in its later stages shows some similarities with the earlier work of Nakanishi *et al.* [45d].

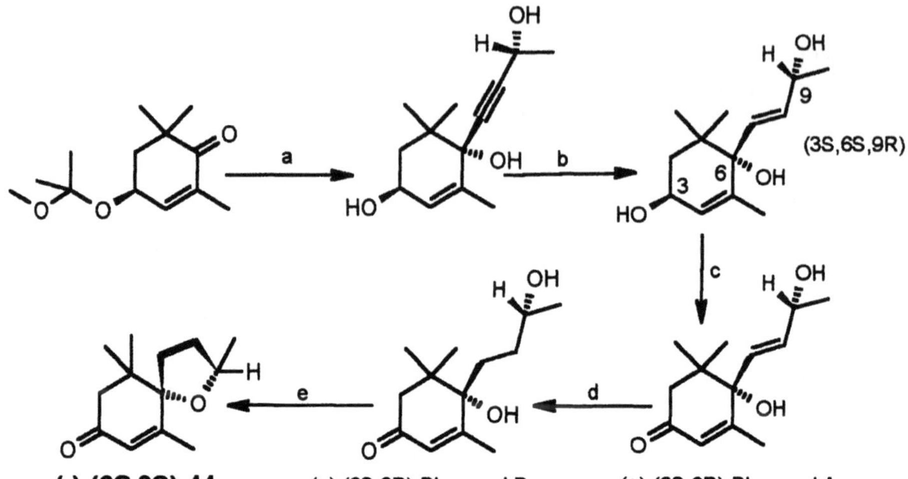

(-)-(6S,9S)-44 (+)-(6S,9R)-Blumenol B (+)-(6S,9R)-Blumenol A

a. Grignard. b. Li/NH_3. c. DDQ. d. H_2 Pd/C. e. MsCl/Pyr.

Figure 38.. Preparation of (-)-(6*S*,9*S*)-theaspirone by Mayer

The final synthesis of theaspirones that we shall look at is that of Paquette and co-workers [47]. Their elegant approach to the theaspirones is completely novel and is shown in figure 39. Starting from the readily available γ-valerolactone their synthetic route produces both cis-44 and trans-44 in ten steps. The key step in their approach is the acid catalysed oxonium ion-induced pinacol rearrangement of 45 which yields a mixture of the diastereomers 46 and 47 which are readily separated by chromatography.

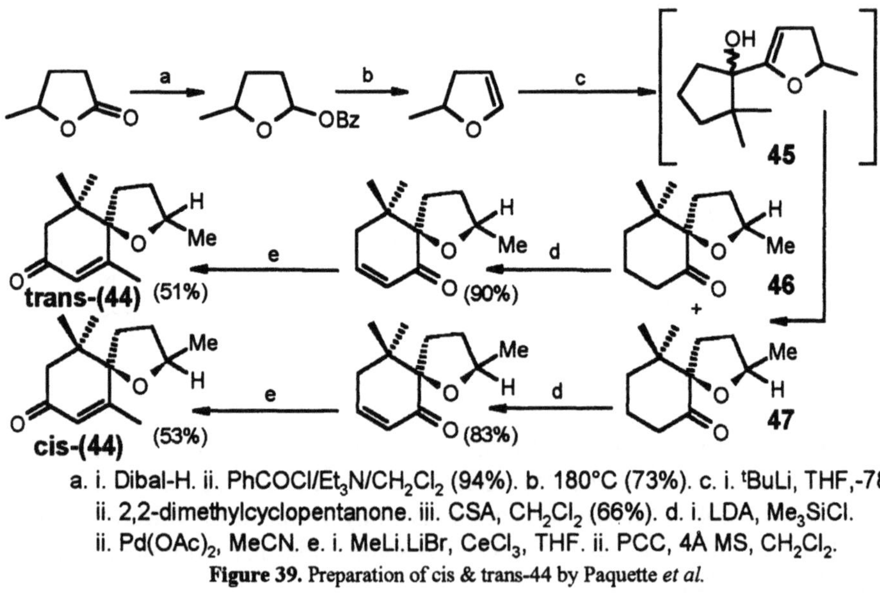

a. i. Dibal-H. ii. PhCOCl/Et_3N/CH_2Cl_2 (94%). b. 180°C (73%). c. i. ^tBuLi, THF,-78°C. ii. 2,2-dimethylcyclopentanone. iii. CSA, CH_2Cl_2 (66%). d. i. LDA, Me_3SiCl. ii. $Pd(OAc)_2$, MeCN. e. i. MeLi.LiBr, $CeCl_3$, THF. ii. PCC, 4Å MS, CH_2Cl_2.

Figure 39. Preparation of cis & trans-44 by Paquette *et al.*

6. Abbreviations

acac	acetylacetonyl
AIBN	2,2'-azobisisobutyronitrile
CSA	10-camphorsulphonic acid
DBN	1,5-diazabicyclo[4.3.0]non-5-ene
DDQ	2,3-dichloro-5,6-dicyano-1,4-benzoquinone
DMA	*N,N*-dimethylacetamide
DMAP	4-dimethylaminopyridine
DME	ethylene glycol dimethyl ether
DMF	*N,N*-dimethylformamide
DMSO	dimethylsulphoxide
HMPA	hexamethylphosphoramide
HMPT	hexamethylphosphorus triamide
LAH	lithium aluminium hydride
LDA	lithium diisopropylamide
mCPBA	*m*-chloroperoxybenzoic acid
Ms	mesyl (methanesulphonyl)
MS	molecular sieve
NBS	*N*-bromosuccinimide
NCS	*N*-chlorosuccinimide
PCC	pyridinium chlorochromate
PDC	pyridinium dichromate
PhH	benzene
PPTS	pyridinium p-toluenesulphonate
pTSA	*p*-toluenesulphonic acid
Pyr	pyridine
TBAF	tetra-*n*-butylammonium fluoride
Tf	trifluoromethanesulphonyl
THF	tetrahydrofuran
TMSCl	trimethylsilyl chloride
Ts	tosyl (*p*-toluenesulphonyl)

7. References

1. See, ref. 4, *Classics in Total Synthesis* (Ed. K.C. Nicolaou, E. J. Sorensen), VCH, (1996), **1**, 18.
2. See, A. F. Thomas, *The total synthesis of natural products*, (Ed. J. ApSimon), John Wiley and Sons, NY,(1973), **2**, p149-154.
3. W.H. Perkin, *J. Chem. Soc.*, (1904), p654.
4. G. Büchi, A. Hauser, J. Limacher, *J. Org. Chem.*, (1977), **42**(20), 3323.
5. K. Sakurai, T. Kitahara, K. Mori., *Tetrhedron*, (1988), **44**, 6581.
6. W. Oppolzer, R. Pitteloud, *J. Am. Chem. Soc.*, (1982), **104**, 6478.
7. W. Oppolzer, R. Pitteloud, G. Bernardinelli, K. Baettig, *Tetrahedron Lett.*, (1983), **24**, 4975.
8. H. J. Liu, W. H. Chan, *Can. J. Chem.*, (1979), **57**, 708.
9. R. R. Sauers, *J. Am. Chem. Soc.*, (1959), **81**, 925.
10. H. J. Liu, W. H. Chan, *Can. J. Chem.*, (1982), **60**, 1081.
11. Y-J. Wu, D.J. Burnell, *Tet. Lett.*, (1988), **29**(35), 4369.
12. J. Shimada, K. Hashimoto, B.H. Kim, E. Nakamura, I. Kitagawa, *J. Am. Chem. Soc.*, (1984), **106**, 1759. E. Nakamura, I. Kitagawa, *Org., Synth.*, (1987),**65**, 17.
13. S. Kim, J. H. Cheong, *Synlett*, (1977), 947.
14. J. A. Marshall, P. C. Johnson, *Chemical Communications*, (1968), 391.
15. J. A. Marshall, N. H. Anderson, P. C. Johnson, *J. Am. Chem. Soc.*, (1967), **89**, 2748., J. A. Marshall, P. C. Johnson, *J. Am. Chem. Soc.*, (1967), **89**, 2750.
16. i. G. Stork, R. L. Danheiser, B. Ganem, *J. Am. Chem. Soc.*, (1973), **95**, 3414., ii. P. Johnson, V. Vais, *J. Chem. Soc. Chem. Comm.*, (1979), 817., iii. R. G. Eileman, B. J. Willis, *J. Chem. Soc. Chem. Comm.*,

(1981), 30., iv. G. H. Posner, T. G. Hamil, *J. Org. Chem.*, (1988), **53**, 6031., v. M. Asaoka, K. Takenouchi, H. Takei, *Chem. Lett.*, (1988), 1225.
17. P. M. McCurry, Jr., R. K. Singh, *Tetrahedron Lett.*, (1973), 3325.
18. M. Deighton, C. R. Hughes, R. Ramage, *J. Chem. Soc. Chem. Commun.*, (1975), 662.
19. W. G. Dauben, D. J. Hart, *J. Am. Chem. Soc.*, (1975), **97**(6), 1622.
20. W. G. Dauben, D. J. Hart, *J. Am. Chem. Soc.*, (1977), **99**(22), 7307.
21. G. Büchi, D. Berthet, R. Decorzant, A. Grieder, A. Hauser, *J. Org. Chem.*, (1976), **41**(19), 3208.
22. S. Torii, K. Uneyama, K. Okamoto, *Bull. Chem. Soc. Japan*, (1978), **51**(12), 3590.
23. J. A. Marshall, P. C. Johnson, *J. Org. Chem.*, (1970), **35**(1), 192., K. Yamada, H. Nagase, Y. Hayakawa, K. Aoki, Y. Hirata, *Tetrahedron Lett.*, (1973), 4963., G. Bozzato, J-P Bachmann, M. Pesaro, *J. Chem. Soc. Chem. Comm.*, (1974), 1005, E. Wenkert, B. L. Buckwalter, A. A. Craveiro, E. L. Sanchez, S. S. Sathe, *J. Am. Chem. Soc.*,(1978), 1267., T. Ibuka, K. Hayashi, H. Minakata, Y. Ito, Y. Inubushi, *Can. J. Chem.*, (1979), **57**, 1579.
24. D. Caine, C-Y Chu, S. L. Graham, *J. Org. Chem.*, (1980), **45**, 3790., R. O. Hutchins, N. R. Natale, I. M. Taffer, R. Zipkin, *Synthetic Communications*, (1984), 445., see also ref. 16iii, 16iv, 16v, and 25.
25. A. Murai, S. Sato, T. Masamune, *Tetrahedron Lett.*, (1981), **22**, 1033., A. Murai, S. Sato, T. Masamune, *Bull. Chem. Soc. Japan*, (1984), **57**, 2276.
26. G. H. Posner, E. M. Schulman-Roskes, *Tetrahedron*, (1992), **48**(23), 4677.
27. W. Krick, F.-J. Marner, L. Jaenicke, *Helv. Chim. Acta*, 1984, **67**, 318.
28. See reference 6 in K. H. Schulte-Elle, H. Passingle, A. P. Uijttewaal, R. L. Snowden, Helvetica Chimica Acta., 1992, **75**, 759.
29. H. E. Eschinazi, J. Am. Chem. Soc., 1959, **81**, 2905.
30. D. H. R. Barton, M. Mousseron-Canet, J. Chem. Soc., 1960, 271.
31. V. Rautenstrauch, B. Willhalm, W. Thommen, G. Ohloff, Helvetica Chimica Acta., 1984, **67**, 325.
32. J. Garnero, D. Joulain, Bull. Soc. Chim. France, pt II, 1979, 15.
33. M. Miyashita, N. Makino, M. Singh, A. Yoshikoshi, *J. Chem. Soc. Perkin Tans. I*, 1982, 1303.
34. C. Nussbaumer, G. Fráter, *J. Org. Chem.*, 1987, **52**, 2096.
35. C. Nussbaumer, G. Fráter, *Helv. Chim. Acta.*, 1988, **71**, 619.
36. Y. Ohtsuka, F. Itoh, I. Oishi., *Chem. Pharm. Bull.*, 1991, **39**(10), 2540.
37. H. Monti, G. Audran, J-P. Monti, G. Leandri., *J. Org. Chem.*, 1996, **61**, 6021.
38. S. Totii, K. Uneyama, S. Matsunami, *J. Org. Chem.*, 1980, **45**, 16., T. Kitahara, K. Tanida, K. Mori, *Agric. Biol. Chem.*, 1983, **47**(3), 581., F. Leyendecker, M-T. Comte, *Tetrahedron*, 1987, **43**, 85., T. Kawanobe, M. Iwamoto, K.Kogami, M. Matsui, *Agric. Biol. Chem.*, 1987, **51**(3), 791., D. Hemlinger, G. Fráter, *Helv. Chim. Acta.*, 1989, **72**, 1515., A. M. Moiseenkov, V. V. Veselouskii, V. A. Dragan, G. A. Stashina, V. M. Zhulin, *Akademia Nauk SSSR Bulletin Div. Chem. (Eng. Trans.)*, 1990, **38**(11), 2394., C. Chapuis, R. Brauchli, *Helv. Chim. Acta.*, 1993, **76**, 2070.
39. G. Büchi, W. D. Macleod, Jr., J. Padilla O., *J. Am. Chem. Soc.* 1964, **86**, 4438.
40. a. S. Danishefsky, D. Dumas, *J. Chem. Soc. Chem. Commun.*, 1968, 1287., b. R. N. Mirrington, K. J. Schmalzl, *J. Org. Chem.*, 1972, **37**(18), 2871., c. F. Näf, G. Ohloff, *Helv. Chim. Acta.*, 1974, **57**(6), 1868., d. F. Näf, R. Decorzant, W. Giersch, G. Ohloff, *Helv. Chim. Acta.*, 1981, **64**(5), 1387., e. M. Bertrand, P. Teisseire, G. Pelerin, *Tet. Lett.*, 1980, 2055., f. T. V. Magee, G. Stork, P. Fludzinski., *Tet. Lett.*, 1995, **36**, 7607.
41. K. Yamada, Y. Kyotani, S. Manabe, M. Suzuki, *Tetrahedron*, 1979, **35**, 293.
42. R. M. Cory, M. D. Bailey, D. W. C. Tse, *Tet. Lett.*, 1990, **31**, 6839.
43. R-B. Zhao, Y-L. Wu, *Chinese Journal of Chemistry*, 1991, **9**(4), 377.
44. K. Ina, Y. Sakato, H. Fukami, *Tet. Lett.*, 1968, 2777.
45. a. Y. Nakatani, T. Yamanishi, *Tet. Lett.*, 1969, 1995., b. R. A. Heckman, D. L. Roberts, *Tet. Lett.*, 1969, 2701., c. K. Ina, T. Takano, Y. Imai, Y. Sakato, *Agric. Biol. Chem.*, 1972, **36**, 1033., d. G. Weiss, M. Koreeda, K. Nakanishi, *J. Chem. Soc. Chem. Commun.*, 1973, 565., e. R. Kaiser, A. Kappeier, D. Lamparsky, *Helv. Chim. Acta.*, 1978, **61**, 387., f. H. Mayer, *Pure Appl. Chem.*, 1979, **51**, 535., g. W. M. B. Könst, W. Apeldoorn, H. Boelens, *Synth. Commun.*, 1980, **10**, 899., h. H. Etoh, K. Ina, M. Iguchi, *Agr. Biol. Chem.*, 1980, **44**, 2871., i. J.-E. Bäckvall, P. G. Andersson, *J. Org. Chem.*, 1991, **56**, 2274. j. P. G. Andersson, Y. I. M. Nilsson, J.-E. Bäckvall, *Tetrahedron*, 1994, **50**, 559. k. Y. I. M. Nilsson, A. Aranyos, P. G. Andersson, J.-E. Bäckvall, J.-L. Parrain, C. Ploteau, J.-P. Quintard, *J. Org. Chem.*, 1996, **61**, 1825.
46. a. J. N. Marx, *Tetrahedron*, 1975, **31**, 1251. b. H. Mayer, *Pure and Appl. Chem.*, 1979, **51**, 535., c. P. Weyerstahl, T. Meisel. *Liebigs Ann. Chem.*, 1994, 425.
47. L. A. Paquette, J. C. Lanter, H.-L. Wang, *J. Org. Chem.* 1996, **61**, 1119.

ADVANCES IN THE TRANSITION METAL (Rh, Ru)-BINAP-CATALYZED ASYMMETRIC SYNTHESIS OF CHIRAL COMPOUNDS FOR FLAVOURS AND FRAGRANCES AND THEIR ASSOCIATED SENSORY PROPERTIES

Takeshi Yamamoto
Central Research Laboratory
Takasago International Corporation
4-11, 1-Chome, Nishiyahata, Hiratsuka City
Kanagawa Prefecture 254-0073
Japan

1. Introduction

The variable biophysical responses to enantiomers are quite well documented in the field of medicine and human taste perception. For example, the *l*-form of ephedrine is effective against asthma but the *d*-form is not, and the *d*-form of aspartame tastes sweet, while the *l*-form bitter. Since the olfactory epithelium, like all biological tissues, is composed of bio-molecules consisting of L-amino acids and has a specific chiral environment, it should be able to distinguish one chiral molecule from its antipode. However, discussion on the olfactory differences of enantiomers has hitherto not been enthusiastic because it is difficult to obtain an enantiomerically pure material for olfactory evaluation. Furthermore, such discussions are problematical because the sense of smell is something abstract and more difficult to assess quantitatively than the other senses such as sight and sound.

Nevertheless, the recent remarkable progress in the preparative separation and synthesis methodologies of chiral molecules has opened a new path, benefiting not only the fine chemical, pharmaceutical and agro-chemical industries but also the flavour and fragrance industry [1,2].

During the last decade an increasing interest in the separation and determination of olfactory characteristics of natural chiral molecules has resulted in a large number of publications. These are mainly due to the development of direct enantiomer resolution by means of chiral stationary phase chromatographic technology [3].

Summarizing these results, chiral compounds, which sometimes exist in enantiomerically pure forms in nature, often show great differences between the respective enantiomers and contribute significantly to the tastes and smells of plants, flowers, fruits, etc. Consequently, provided their costs are reasonable, chiral compounds may well become irresistible strategic tools for flavourists and perfumers, who have traditionally been working with racemic forms in order to reproduce natural tastes and

K.A.D. Swift (ed.), Current Topics in Flavours and Fragrances, 33–58.

smells. This growing awareness of the importance of enantiodifferentiation is also deeply indebted to the rapid progress of the methodologies and technologies that have arisen to produce such chiral compounds for flavours and fragrances during the last two decades. Among them is transition metal-catalysed asymmetric synthesis in which remarkable progress has been made [4].

This paper will describe the asymmetric synthesis of some chiral compounds for flavours and fragrances catalyzed by metal (Ru and Rh)-BINAP complexes and their sensory as well as other properties.

2. Asymmetric synthesis of chiral compounds for flavours and fragrances catalyzed by (Rh, Ru)-BINAP complexes, and their associated sensory properties

2.1. Rh-BINAP-CATALYZED ASYMMETRIC ISOMERISATION OF ALLYLAMINE (SYNTHESIS OF *l*-MENTHOL AND CITRONELLYL DERIVATIVES)

2.1.1. l-Menthol

l-Menthol (1) is the most important of the optically active flavour and fragrance substances which are produced synthetically on an industrial scale. Its unique organoleptic properties, refreshing taste and strong cooling effect, have been excellently demonstrated by comparing the flavour profiles of the eight diastereomers as shown in table 1 [5]. Among them, only one diastereomer (1) having the 1R, 3R and 4S configuration shows excellent properties.

	l-isomers	*d*-isomers		Flavour threshold (ppm)	Cooling threshold (ppm)	Bitter threshold (ppm)
n-	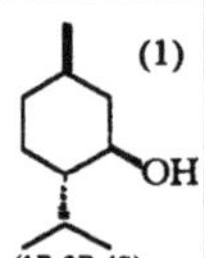 (1) (1R,3R,4S)	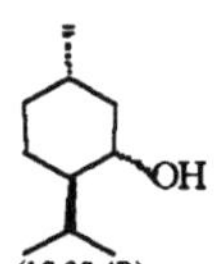 (1S,3S,4R)	*l*-	0.4	0.8	10-20
			d-	0.3	3	20-30
neo-	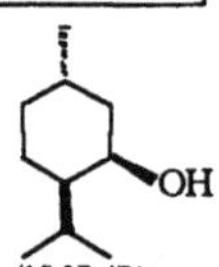 (1S,3R,4R)	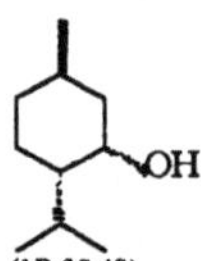 (1R,3S,4S)	*l*-	0.7	20-30	30
			d-	0.6	3	20-30
iso-	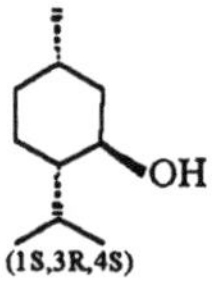 (1S,3R,4S)	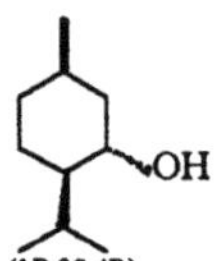 (1R,3S,4R)	*l*-	0.4	30	50
			d-	0.6	7	30
neoiso-	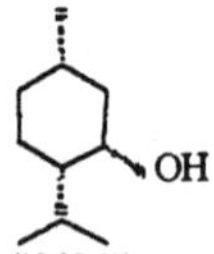 (1S,3S,4S)	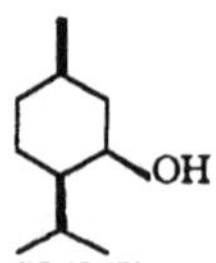 (1R,3R,4R)	*l*-	1.0	0.2	> 25
			d-	0.2	> 25	> 25

Table 1. Flavour, cooling, and bitter threshold values of diastereomers of menthol

With an estimated worldwide consumption of 11,800 metric tons in 1998 [6], *l*-menthol is widely used in many consumer products such as cigarettes, toothpaste, chewing gum, pharmaceutical and personal-care products. At present, about 8,000 metric tons of natural *l*-menthol, obtained mostly from the oil of *Mentha arvensis*, is produced annually, mainly in India and China. Annual production of synthetic *l*-menthol through optical resolution methods and asymmetric synthesis totals around 3,000 metric tons. It is manufactured in Germany, the United States and Japan.

In April 1984, Takasago successfully started industrial production of *l*-menthol by a new process (Scheme 1). This has made Takasago the world leader in the production of chiral compounds by transition metal-catalyzed asymmetric synthesis. Universities in the cities of Osaka and Nagoya collaborated with Takasago's R&D department in achieving this breakthrough. The background of the invention and its technological details are described elsewhere [8,9].

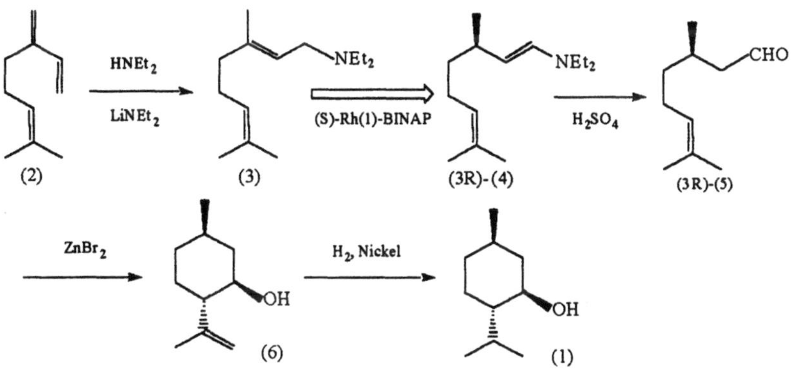

Scheme 1. Synthesis process for *l*-menthol

The key reaction is an asymmetric isomerisation of geranyldiethylamine (3), which is synthesized with high yield and high selectivity from myrcene using lithium diethylamide as catalyst, to provide (+)-citronellal enamine.

Essential in this new process is the use of a chiral rhodium-phosphine complex as the catalyst to isomerize an allylic amine to a corresponding optically active enamine. This key process is also capable of using neryldiethylamine (8) as substrate, which can be synthesized by the telomerisation of isoprene (7) with diethylamine using lithium diethylamide as the catalyst.

The relationship between the geometry of geranyldiethylamine (3) and neryldiethylamine (8) on one hand, and the chirality of the Rh-BINAP catalyst and that of the product enamine (4) on the other, is shown in Scheme 2. The co-ordination of a chiral ligand of BINAP {2,2'-bis (diphenylphosphino)-1,1'-binaphthyl} to Rh(I) species produced an extremely useful catalyst for this isomerisation.

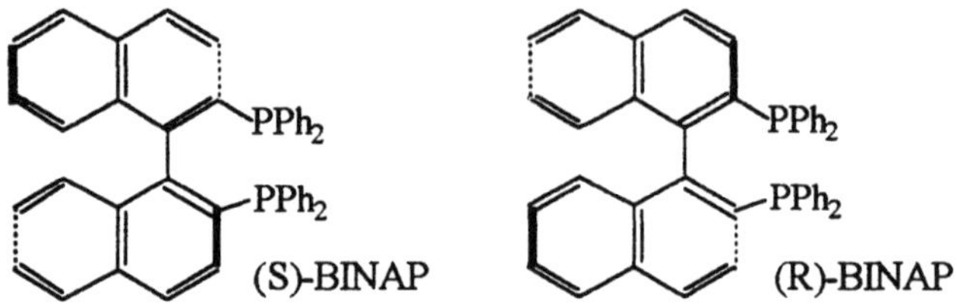

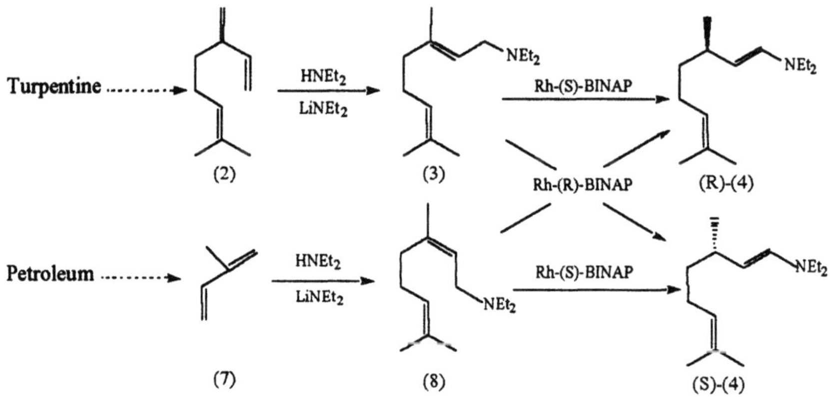

Scheme 2

Both chemical and enantioselectivities exceed 98% in the reaction. In addition, the catalyst is easily recoverable from the reaction mixture and can be reused. The result is that 1 mole of catalyst produces 400,000 mole of the enamine, thus making the process very cost-effective.

The hydrolysis of chiral enamine (4) gives chiral citronellal (5) with a very high yield (above 98 %e.e.). Zinc bromide catalyzed cyclization of *d*-citronellal (3R)-(+)-(5) to *l*-isopulegol (6), namely, an intramolecular ene reaction, yields selectively all equatorial isopulegol in a yield of 92-94 % (ordinary Lewis acids; 65 %). Enantiomerically and chemically pure *l*-isopulegol (1R,3R,4S)-(6) is obtained by deep cooling crystallization at -50° C, and this is converted to pure *l*-menthol (1) by hydrogenation. Today, Takasago produces more than 1,000 metric tons per year of *l*-menthol by this method as well as a combined total of more than 500 metric tons per year of many other flavour and fragrance raw materials processed from the *l*-menthol synthesis intermediates.

2.1.2 l-Citronellol

Enantiomerically pure *l*-citronellol (3S)-(-)-(9) exists in rose and geranium as the key and main odorous component. In contrast, most of the commercially available citronellols are racemates synthesized from the hydrogenation of geraniol.

Although *l*-citronellol (3S)-(-)-(9) has an odour much closer to natural rose oil, and is much better than the *d*-form or the racemate in creating or reconstituting rose-type fragrances, it has not been widely used because of the absence of an economically viable synthesis. The successful asymmetric synthesis of *l*-citronellol (9) has made it possible to supply this important fragrance raw material at a reasonable price (Scheme 3).

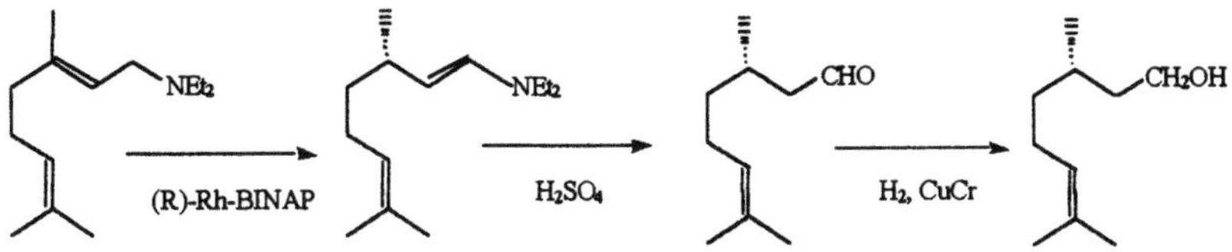

(3) (3S)-(4) (3S)-(5) (3S)-(9)

Scheme 3

The olfactory properties of the citronellol (9) enantiomers and their derivatives (10) – (14) obtained from the enantiomers of citronellal (5) (98 %e.e.) are shown in Table 2.

Compound	(3S)-form		(3R)-form	
	Odour	Threshold (ppb)	Odour	Threshold (ppb)
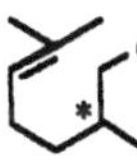 CH_2OH (9)	Very fresh, light and clean rosy-leafy petal-like	50	Slightly oily light rosy-leafy, peatal-like odour with irritating top note	50
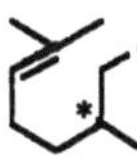 CH_2OAc (10)	Fresh lime-citrus odour with camphoraceous note	250	Fresh citrus-lime odour with dirty aldehydic note	250
 CH_2OEt (11)	Clean sweet rosy odour with powerful fresh top note	500	Floral rosy odour	250
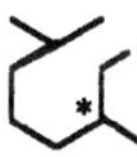 CH_2OH (12)	Floral odour like clean rose	250	Chemical, waxy and dry rosy	100
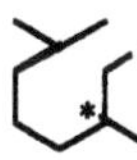 CH_2OAc (13)	Clean citrus and muguet floral	250	Sweet citrus and muguet floral	250
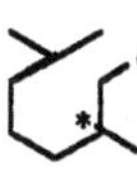 CH_2OEt (14)	Clean rosy	1000	Fresh rosy with clean fatty top note	500

Table 2. Odour profiles of the enantiomeric citronellyl derivatives (9)-(14)

In studies of these enantiomers, all the (3S)-forms show milder, cleaner top notes than their (3R) counterparts, and such top notes are much preferable to those of the latter, although differences in the threshold values and olfactory qualities are small [10].

Alternatively, *l*-citronellol (9) is also obtainable from the asymmetric hydrogenation of geraniol (15) or nerol (16), catalyzed by a chiral Ru-BINAP complex [11] (Scheme 4).

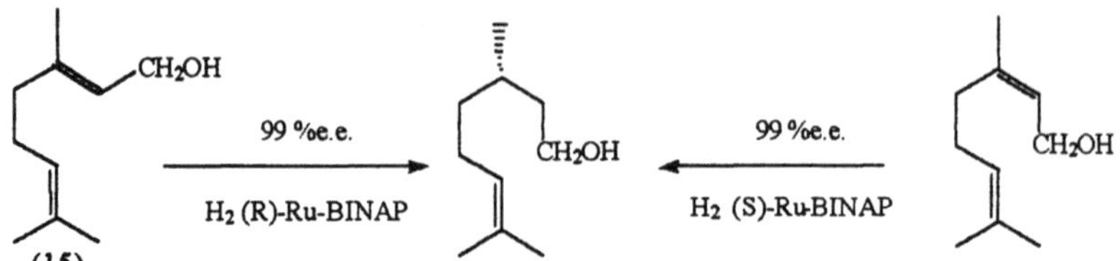

(15) *l*-(9) (16)

Scheme 4

Looking at the differences in the properties other than the odours of the enantiomeric pair (the dextrorotatory and laevorotatory) of citronellol, interaction of (3S)- and (3R)-citronellol (98 %e.e.) with the surface of a chiral substrate has been compared by a triangle method similar to that reported by Acree *et al.* [12] The result suggested that the *l*-form has a stronger affinity for the surface of cellulose than the *d*-form [10].

2.1.3. l-Laurinal™ (l-7-Hydroxydihydrocitronellal) and l-Aurantiol

Racemic Laurinal™ has a lily of the valley-type odour and is widely used in many fragranced products. However, the IFRA Guidelines recommend that this racemate should not be used in consumer products in levels exceeding 1% because of its allergenic potential. This is equivalent to 5 % in a fragrance compound used at 20% in consumer products.

The enantiomers of Laurinal™ (17) were obtained by hydration and hydrolysis of the corresponding two chiral citronellal diethylenamines (98 %e.e.).

As for the olfactory properties of the enantiomeric pure Laurinal™ (17), the *l*-Laurinal™ (3S)-(-)-(17) (clean, green, lily of the valley) shows a cleaner tone than the *d*-isomer (3R)-(+)-(17) (powerful, irritating, rosy, lily of the valley) and is superior in fine fragrance use to its antipode (3R)-(+)-(17). As the olfactory cell can recognize the chirality of fragrance molecules, various immune cells would also be expected to recognize the chirality of allergens. The allergenic potential of the enantiomeric pair of Laurinal™ by the contact allergenic dermatitis test in humans showed that the *l*-form is a much weaker allergen than the *d*-isomer. Also, the *l*-form causes almost no cross-reaction of the allergy that is associated with the *d*-form [13].

A similar observation was made on chiral aurantiol (18) (7-hydroxydihydrocitronellal-methyl anthranilate schiff's base), namely, *l*-aurantiol (3S)-(-)-(18) which shows a cleaner and more fresh odour, is a weaker allergen than the *d*-form [14].

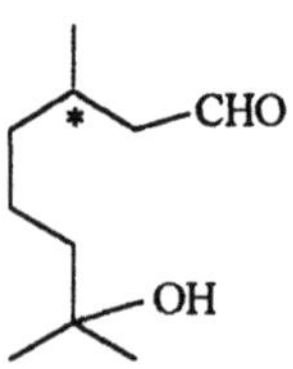

(17)

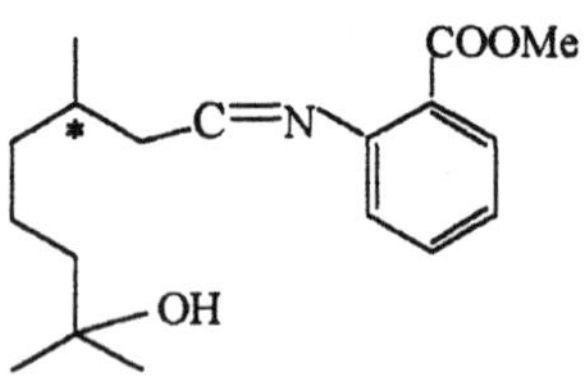

(18)

2.1.4. Damascone-related compounds

α-Damascone is a key odoriferous component of Bulgarian rose. Pickenhagen has reported the sensory properties of optically active α-damascone stating that the (R)-(-)-enantiomer has a cork and green apple animal note, whilst the (S)-(+)-antipode has a more floral odour reminiscent of rose petals and also has a winy character. The threshold values in water were found to be 1.5 ppb for the (S)-(-)- and 100 ppb for the (R)-form [15].

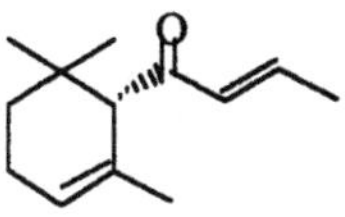

Cork and green apple animal note

Threshold ; 1.5 ppb

(S)-(-)-α-damascone

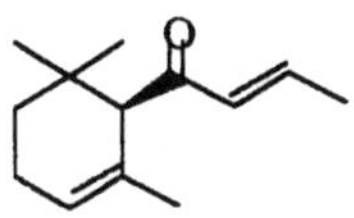

More floral reminiscent of rose petal having winy character

Threshold ; 100 ppb

(R)-(+)-α-damascone

As damascone-related compounds, the diastereomers of dihydrodamascone {trans-(E)-1-(2,2,6-trimethylcyclohexyl)-2-buten-1-one}(E)-(19) [16] and trans- ethyl 2,2,6-trimethylcyclohexylcarboxylate (20) [17] were synthesized from enantiomeric citronellal (98 %e.e.), and their odour properties (Table 3) were evaluated. As for the odour properties of racemic (E)-dihydrodamascone (19), its trans-(E)-form (iris, mimosa, tobacco and fancy bouquet) is allegedly reported to be olfactorily superior to the cis-(E)-form (very weak earthy, mint) [18].

As for the olfactory properties of the chiral trans-(E)-dihydrodamascone, the (1S,6R)-(E)-form shows a slightly camphoraceous fruity odour like β-damascone whereas the (1R,6S)-(E)-form exhibits a ripe fruity odour with a fresh rosy note. The difference in the threshold values between the enantiomeric pair is not so large compared with that of α-damascone. The (1R,6S)-(E)-(19) shows a threshold value seven times lower than the (1S,6R)-(E)-(19), while (S)-α-damascone is evaluated as having a threshold value sixty-seven times lower than the (R)-antipode. In the case of ethyl 2,2,6-trimethylcyclohexylcarboxylate (20), the (1R,6S)-form gives a stronger and better odour than the (1S, 6R)-form.

( ) ; Odour threshold

(1R,6S)- form	Odour	(1S,6R)-form	Odour
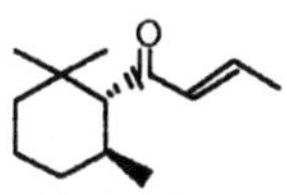 (1R,6S)-(19)	Ripe fruity odour with fresh rosy note (5 ppb)	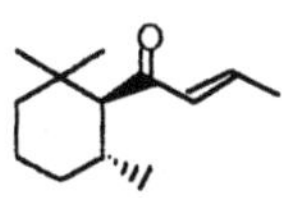 (1S,6R)-(19)	Slightly camphoric fruity odour like β-damascone (0.7 ppb)
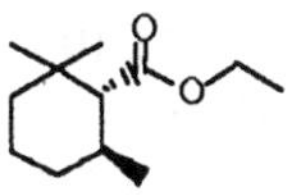 (1R,6S)-(20)	Clean and sweet floral odour (20 ppb)	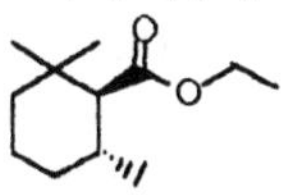 (1S,6R)-(20)	Dirty and heavy fatty floral odour (100 ppb)

Table 3. Odour profiles of the diastereomeric damascone related compounds (19) and (20)

Under the IFRA Guidelines, it is recommended that damascones should not be used in consumer products at levels that exceed 0.02 % on account of their sensitising potentials. This is equivalent to 0.1% in a fragrance compound used at 20% in consumer products.

The chiral dihydrodamascones, (1R,6S)-(E)-and (1S,6R)-(E)-(19), show a weaker sensitising potential than damascones, owing to the lack of a double bond in the cyclohexane ring. The enantiodependency of the sensitising potential of dihydrodamascones on guinea pigs was examined. This resulted in a finding that showed the (1R,6S)-(E)-(19) is safer than the corresponding antipode [10,19].

2.1.5. l-cis-Rose oxide, Dihydroroseoxide

Four diastereomers of rose oxide (21) have been identified in white wine, cognac, grape, green tea, rose, citronella oil, geranium and tropical fruit, and are assumed to be responsible for the floral green top note of rose. The multidimensional GC showed that the *l*-isomers {*l*-cis-(21) and *l*-trans-(21)} occur more profusely in nature. The composition ratio of *l*-cis-(21) vs. *l*-trans-(21) in Bulgarian rose oil is about 75 / 25. Ohloff [20] has already studied the odour properties of rose oxide (21) stating that the *d*-cis-(21) has a sweet scent, whereas the *l*-cis-(21) has a powerful fruity odour. Further odour evaluation of the four diastereomers synthesized from the enantiomeric citronellol (9) (98 %e.e.) has been conducted and the results are shown in Table 4 [21].

l-form	Odour	Threshold (ppb)	*d*-form	Odour	Threshold (ppb)
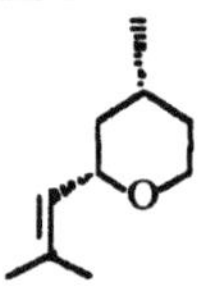 *l*-cis-(21) (2S,4R)	**Floral-Green** clean, sharp, metallic light, rose green diffusive, strong	0.5	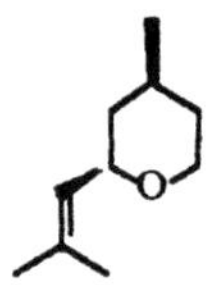 *d*-cis-(21) (2R,4S)	**Herbal-Green-Floral** hay green, earthy heavy	50
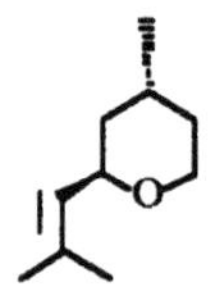 *l*-trans-(21) (2R,4R)	**Floral-Green** green herbal (minty) fruity	160	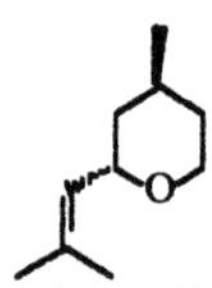 *d*-trans-(21) (2S,4S)	**Herbal-Green-Floral** fruity, herbal rose citrus (bitter peel)	80

Table 4. Odour profiles of the diastereomeric rose oxides (21)

Among them, the *l*-cis-isomer shows the lowest odour threshold value (1/100 to 1/300 compared with the other isomers), and is greatly superior in olfactory quality to the other isomers, all of which have herbal nuances in common. The stereoselective synthesis of *l*-cis-rose oxide has been achieved with a diastereomeric yield of 80% d.e. by cyclization of (3S)-3,7-dimethyl-6,7-octadien-1-ol (3S)-(22), which in turn had been obtained from the *l*-citronellol (3S)-(-)-(9) (98%e.e.), using a Pd-(R)-BINAP catalyst [21]. (Scheme 5)

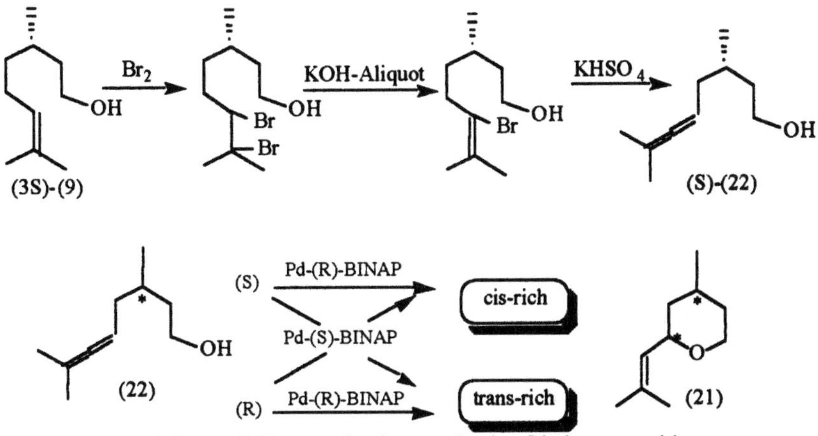

Scheme 5. Stereoselective synthesis of *l*-cis-rose oxide

Other differences were noticed in the chiral rose oxides apart from their odours. Interesting results were observed on their biodegradability and on the Contingent Negative Variation (CNV) of the stereoisomeric rose oxides (21). With respect to the biodegradability of rose oxide (21), after 4 weeks under the standard method the *l*-cis-isomer has completely vanished, the *l*-trans-isomer partly disappeared, and the *d*-isomers remained intact [10, 21].

CNV is known to transmit olfactory properties of odour molecules to the brain in terms of α-waves. An α-wave is reported to be changeable under some experimental conditions. The total CNV forms a complex curve. The CNV waves of the diastereomers of rose oxide (21) were compared, resulting in the discovery that the *l*-forms exhibit stimulative effects, while the amplitudes of the *d*-forms stayed at the control (neutral) position, without any move at all in either a stimulative or sedative direction [10,19]. It is interesting to note that the *l*-cis-isomer found most abundantly in nature exhibits better odour properties, higher biodegradability and stronger stimulative effect than the other diastereomers.

In the case of dihydroroseoxide (23) obtained through the hydrogenation of rose oxide (98% e.e.), the (4R)-cis-dihydroroseoxide (2R,4R)-(23) also showed the strongest and most characteristic odour properties out of the four diastereomers [21](Table 5).

2.1.6. Citrus-related compounds

Looking at the citrus-type flavour-fragrance raw materials, the olfactory properties of the optically active citronellal (5), citronellyl nitrile (24) and Bergamal™ (25) (α-methylene citronellal), prepared from the enantiomeric citronellal (5) or its diethylenamine (4) (98%e.e.) were examined (Table 6).

(4R)-form	Odour	Threshold (ppb)	(4S)-form	Odour	Threshold (ppb)
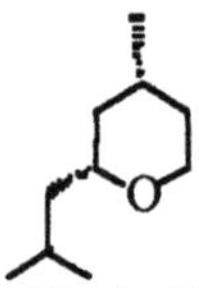 (4R)-cis-(23)	**Floral-Green** clean, light ripe fruit, herbal rosy, green, leafy	17	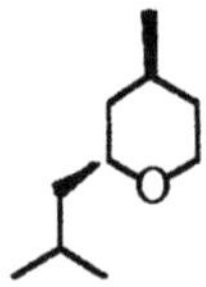 (4S)-cis-(23)	**Herbal-Green** herbal, leafy, green heavy	450
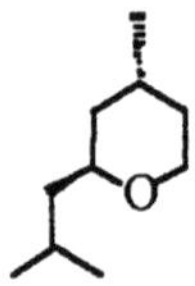 (4R)-trans-(23)	**Herbal-Floral** fruity, minty, dusty floral green	150	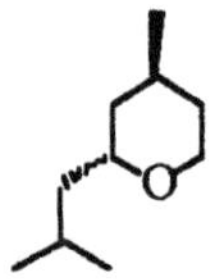 (4S)-trans-(23)	**Herbal-Green-Citrus** **fruity, herbal, fresh citrus (grapefruit)**	160

Table 5. Odour profiles of the diastereomeric dihydroroseoxide (23)

	(3S)-form		(3R)-form	
Compound	Odour	Threshold (ppb)	Odour	Threshold (ppb)
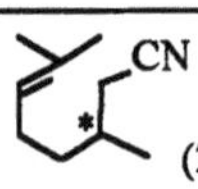 (24)	Strong citrus odour with an aldehydic note	75	Fresh mild citrus	100
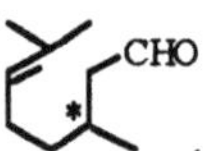 (5)	Powerful, fresh, bright clean, herbaceous-citrus	25	Powerful, fresh herbaceous-citrus	25
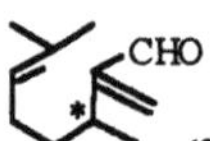 (25)	Clean citrus odour like bergamot	250	Citrus odour with citronellal-like side note	250

Table 6. Odour profiles of the enantiomeric citrus-related compounds (24), (5) and (25)

Optically active citronellal (5) is found in many plants, and exists in a wide range of enantiomer composition ratios. For example, *l*-citronellal (3S)-(-)-(5) is found in lemon peel oil (78 %e.e.), and *d*-citronellal (3R)-(+)-(5) is present in citronella oil Ceylon type (73 %e.e.) [3b].

As for its olfactory properties, the (3S)-(-)-(5) imparts a strong citrus odour with an aldehydic note and is useful in enhancing the feeling of fresh lemon peel.

Citronellyl nitrile (24) is one of the most important citrus compounds due to its stability in basic media such as soap and detergent. Although the odoriferous difference between the enantiomers of citronellyl nitrile (24) is small, the (3R)-form (fresh mild citrus, threshold; 0.1ppm) shows better odour quality than the antipode (strong citrus with aldehydic note, threshold; 0.075 ppm) [10].

In the case of methylene citronellal (25), the (3S)-form (clean bergamot-like odour) is much superior in olfactory quality to the (3R)-form (citrus odour with citronellal-like side note), although the threshold value (0.25ppm) is the same between the (3S)-form and (3R)-form [22, 23].

2.1.7.d-Norlimbanol™ and its homologues

Racemic 1-(2,2,6-trimethylcyclohexyl)hexan-3-ol (26) {Norlimbanol™ (Firmenich; trans-form rich) and Timberol™ (Dragoco; cis-form rich) } are commercially used as fragrance raw materials for woody-amber fragrance compounds. As for the odours of these geometrical isomers, the trans-isomer is known to be stronger and contributes more of an amber-like odour than the cis-form [24]. The synthesis routes of Timberol™ and Norlimbanol™ are shown in Scheme 6.

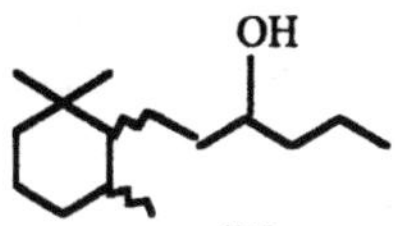

(26)

Racemic cis-rich mixture ; Timberol ™
Racemic trans-rich mixture ; Norlimbanol ™

Synthesis route of Timberol ™

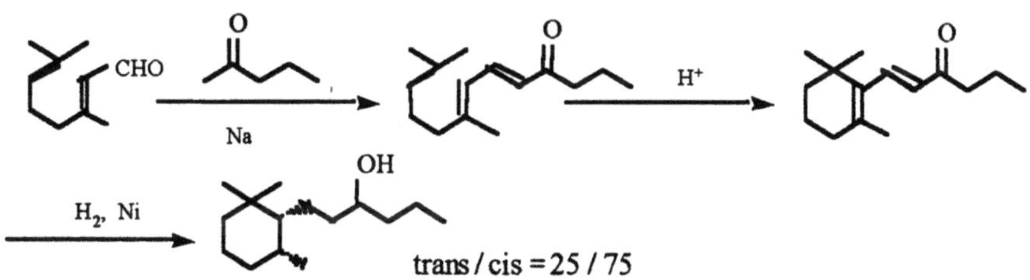

Synthesis route of Norlimbanol ™

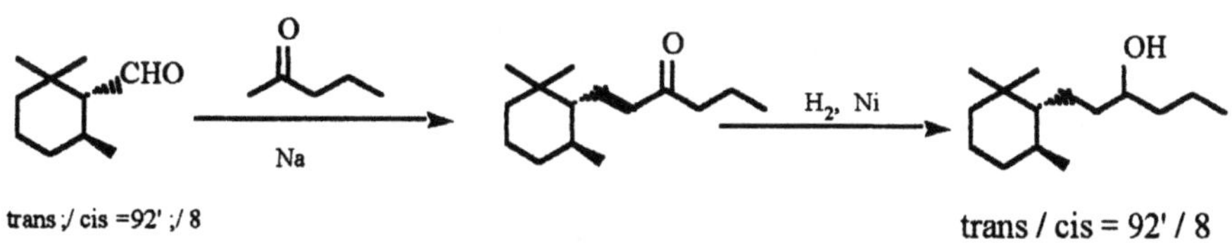

Scheme 6

The odour properties of diastereomeric trans-1-(2,2,6-trimethylcyclohexyl)hexan-3-ol (26) and its homologues were studied by two groups (Firmenich and Takasago) [25,26,27]. Diastereomeric trans-1-(2,2,6-trimethylcyclohexyl)hexan-3-ol (26) and its homologues (27)-(30) were synthesized from enantiomeric citronellal (98 %e.e.). From (3S)-citronellal, the (1R,6S)-(26) and its homologues (1R,6S)-(27)-(30) were synthesized as were the (1S,6R)-diastereomers from (3R)-citronellal [27] (Scheme 7).

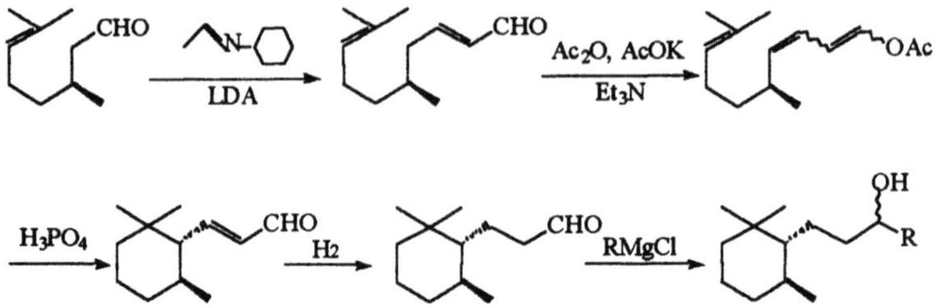

(1R,6S)-form

Scheme 7

As for their olfactory characteristics, significant differences between the diastereomers were observed, namely, all the (1R,6S)-forms (C3-epimer) display fundamentally strong characteristic sharp amber-like odours, whereas the (1S,6R)-forms (C3-epimer) show uncharacteristic weak odours. Particularly, the 1-(2,2,6-trimethylcyclohexyl)hexan-3-ol (26) showed the largest difference among the threshold values of the diastereomers. Its threshold value was found to be 4 ppb for the (1R,6S)- and 130 ppb for the (1S,6R)-(26) [26,27] (Table 7).

(1R,6S)-form	Odour (Threshold ; ppb)	(1S,6R)-form	Odour (Threshold ; ppb)
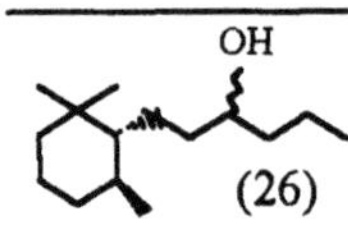 (26)	Very powerful diffusive sharp amber odour with slightly orris-like animal note (4)	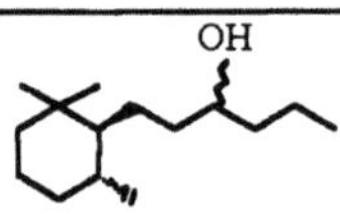	Faint vetiver-like woody odour with a slight mould-like side note (130)
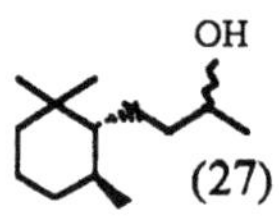 (27)	Diffusive amber odour with slightly floral, orris - like and earthy note (400)	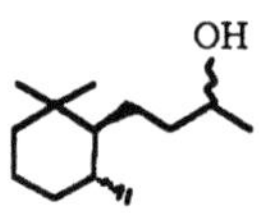	Faint vetiver - like woody odour with slight amber and mould notes (1000)
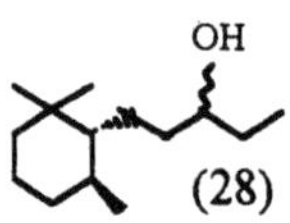 (28)	Very diffusive sharp amber odour with slightly camphoric side note (56)	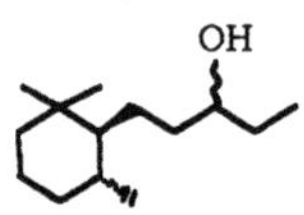	Faint vetiver -like woody odour with slight amber and mould notes (890)
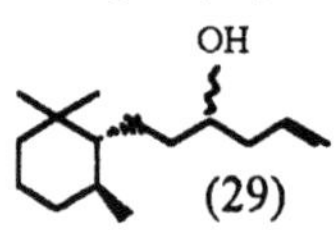 (29)	Powerful sweet amber odour with slightly musky note (10)	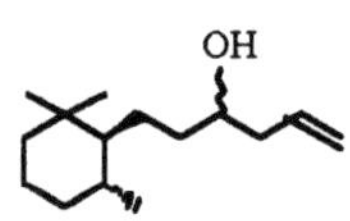	Phenolic woody odour with mould and vetiver-like notes (300)
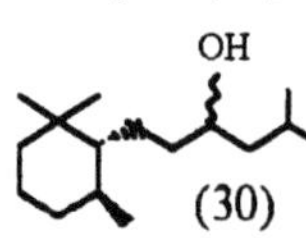 (30)	Highly diffusive amber odour with slightly camphoric side note (70)	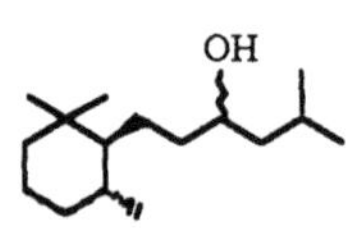	Faint woody odour with a slight mould-like side note (1000)

Table 7. Odour profiles of the diastereomeric 1-alkyl-3-(2,2,6-trimethylcyclohexyl)propan-1-ols (26)-(30)

The configuration of the hydroxyl group in the side chain of these compounds also influences their odours significantly. In the case of 1-(2,2,6-trimethylcyclohexyl)hexan-3-ol (26), the (1R,3'S,6S)-(26) gives the strongest amber-like odour among the four trans-diastereomers [25]. A similar observation was made on 1-(2,2,6-trimethylcyclohexyl)pentan-3-ol (28) [28]. Asymmetric alkylation of (1R,6S)-3-(2,2,6-

trimethylcyclohexyl)propanal with diethyl zinc, using the corresponding chiral titanium complex as a ligand, gives (1R,3'S,6S)-(28) or (1R,3'R,6S)-(28) with a diastereomeric yield of 94-96 % d.e. (Scheme 8). As for the olfactory properties, the (1R,3'S,6S)-(28) exhibits a powerful sharp clean amber odour and a threshold value eighteen times lower than the (1R,3'R,6S)-(28) (Table 8).

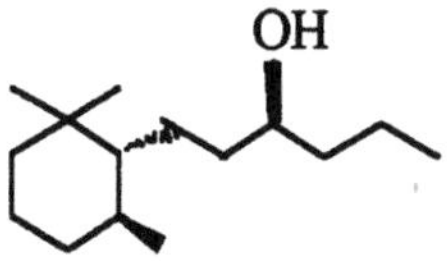

(1R,3'S,6S)-(26)

(28)	%d.e.	Threshold (ppb)	Odour
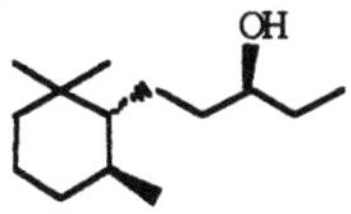 (1R,3'S,6S)-(28)	96	33	Powerful sharp clean amber odour
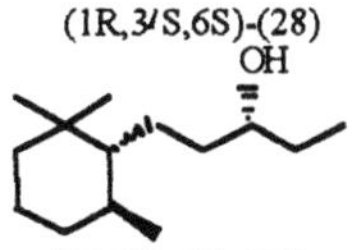 (1R,3'R,6S)-(28)	94	600	Weak uncharacteristic woody odour

Table 8. Odour profiles of the (1R,3'S,6S) and (1R,3'R,6S)-28

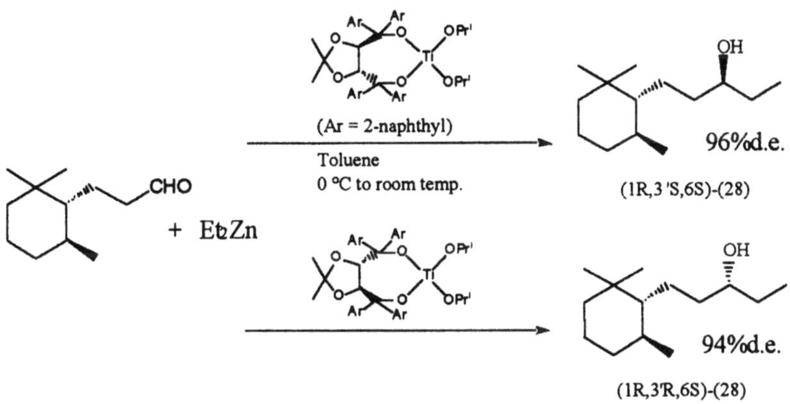

Scheme 8

2.2. ASYMMETRIC HYDROGENATION OF OLEFINS CATALYZED BY Ru-BINAP COMPLEXES

2.2.1. (3S)-(-)-(6E)-2,3-Dihydrofarnesol

2,3-Dihydrofarnesol (31) has been found in a few species of flowers, plants and insects. Especially, it has been revealed to be the main and key odorous component of white cyclamen (*Cyclamene purpurascens*). As another example of the Ru-BINAP-catalyzed asymmetric hydrogenation of allyl alcohols [11] as mentioned earlier in the synthesis of *l*-citronellol, chiral (6E)-2,3-dihydrofarnesol (6E)-(31) was synthesized with an optical yield of 89-90 %e.e., using (2E), (6E)-farnesol (32) as substrate. The odour profile of chiral (6E)-(31) corresponds to that of citronellol, which has the same skeleton as 2,3-dihydrofarnesol. That is to say, both (3S)-forms exhibit clean and fresh odours superior to those of their antipodes. The (3S)-(-)-(6E)-(31) (which is found enantiomerically pure in white cyclamen) has a clean and fresh floral-green odour reminiscent of white cyclamen and lily of the valley, whereas the antipode a weak floral green odour with a balsamic nuance (Scheme 9).

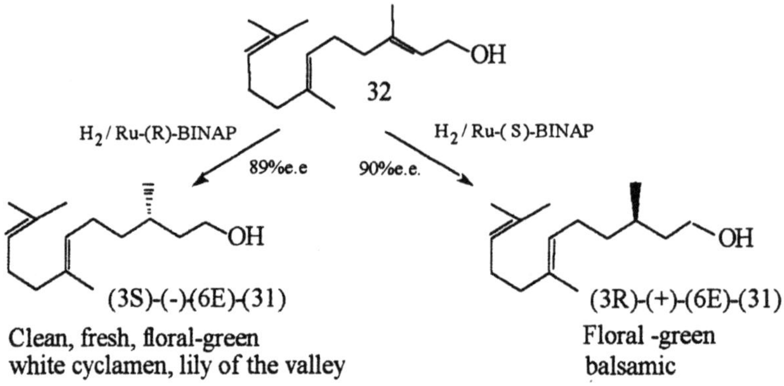

Scheme 9

Farnesol is known to possess some antibacterial activity against a wide range of microorganisms including those that are known to cause body malodour. However, it suffers from its sensitising potential. On the other hand, (6E)-2,3-dihydrofarnesol (6E)-(31) showed a higher antibacterial activity but less sensitising potential than farnesol. Furthermore, the (3S)-(-)-(6E)-(31) showed a slightly higher antimicrobial activity than the racemate (6E)-(31). The (3S)-(-)-(6E)-(31) is not as active as an antibacterial pharmaceutical such as Irgasan™, but the fragranced consumer products containing sufficient amounts of the (3S)-(-)-(6E)-(31) are active enough to inhibit the growth of bacteria that are responsible for body malodour [29,30].

2.2.2. l-Muscone

Muscone (33) has greatly motivated organic synthesis chemists because this natural musk is more expensive than gold and its structure is a curious fifteen-membered monocyclic ring as established by Ruzika in 1929. The chirality of muscone (33) obtained from male musk deer is laevorotatory and the odour of *l*-muscone (R)-(-)-(33) is reported to be stronger than, and much superior to that of the (S)-(+)-(33). The natural (R)-(-)-(33) exhibits a very nice rich and powerful musky note, and its threshold value (determined in water) is 61 ppb, whereas the (S)-(+) is olfactorily inferior to and weaker in strength than the former and the threshold value is 223 ppm[15]. Numerous attempts to synthesize *l*-muscone have been reported. Generally speaking, however, the syntheses require many steps for completion. To date, there seems to be no practicable synthesis method. To examine the scope of Ru-BINAP chemistry, asymmetric hydrogenation of 3-methyl-2-cyclopentadecenone (34), synthesized by Tsuzi *et al.* [31] was studied. When the double bond of the substrate (34) is fixed to the (Z)- or (E)-form, the optical purity of (R)-(-)- or (S)-(+)-muscone reaches 99%e.e. by the asymmetric hydrogenation of the (Z)-or (E)-(33) with a corresponding chiral Ru-BINAP catalyst [32,33] (Scheme 10).

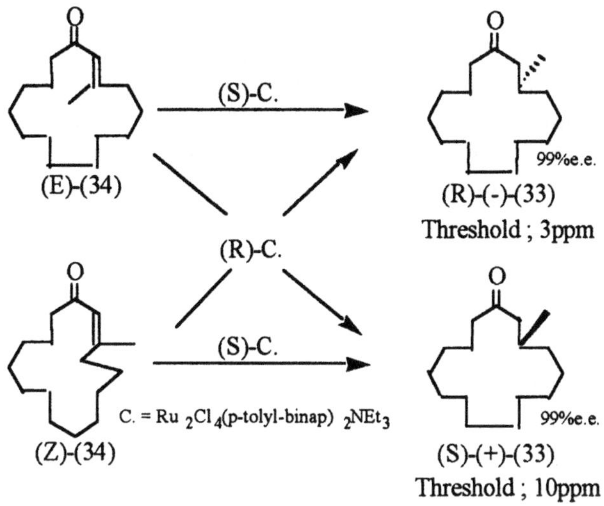

Scheme 10. The stereochemical dependency of muscone produced, on the catalyst configuration, and also on 3-methyl-2-cyclopentadecenone

This result indicates that the key to a commercial production of *l*-muscone is to develop a method to prepare the substrate (34). The odour threshold values, determined in aqueous-ethanol solution of the enantiomerically pure (99 %e.e.) muscone (33) were 3 ppm for the (R)-(-)-form and 10 ppm for the antipode [33].

Considering that natural musk is said to be the sex pheromone of musk deer, it is interesting to note that the CNV of *l*-muscone (R)-(-)-(33) demonstrated different results between men and women, namely a sedative effect for women and a stimulative one for men [34]. The configuration of the methyl group of muscone significantly influences the musk odour modality and a similar correlation has also been observed in the case of ethyl citronellyl oxalate (35). Looking at the relationship between chemical structure and musk odour, Beets [35] described an interesting example of the principle that a non-cyclic, flexible molecule may alter its conformation, mimicking the oriented profile of a macrocyclic molecule in its interaction with the collection of epithelial features. Ethyl citronellyl oxalate (35) has a clearly discernible musk odour, presumably because its molecules can assume in their contacts with the olfactory epithelium a cyclic conformation comparable in shape with that of, for example, the monocyclic musk dodecamethylene oxalate. The odour property of enantiomeric ethyl citronellyl oxalate (35), synthesized from enantiomeric citronellol (9) (98 %e.e.), is shown in Table 9 [33].

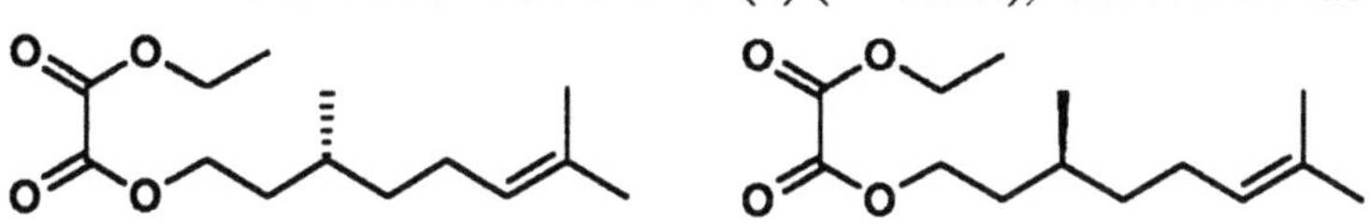

(R)-(+)-35	(S)-(-)-35
Threshold; 0.22ppm	Threshold; 0.45ppm
Floral Musk	Rose-like floral

Table 9. Odour profiles of the enantiomeric ethyl citronellyl oxalate (35)

2.2.3. Alkylcyclopentanones and δ-Lactones

The alkyl cyclopentanones are important ingredients in creating or reconstituting floral odours. Among these, methyl jasmonate and methyl dihydrojasmonate are frequently used in fragrance compounds. Regarding the odour of enantiomeric methyl jasmonate (36), only the (1R, 2S)-isomer has a very strong jasmine odour while the other three isomers exhibit weak notes or are rather odourless in principle [12] (Table 10).

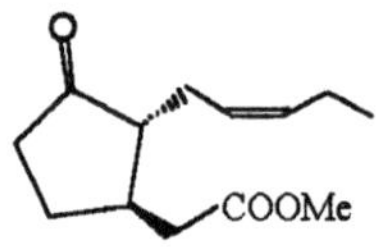

Weak odour (!R, 2R)

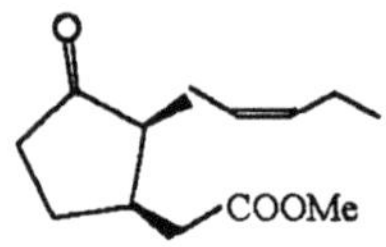

Strong jasmnine odour (1R, 2S)

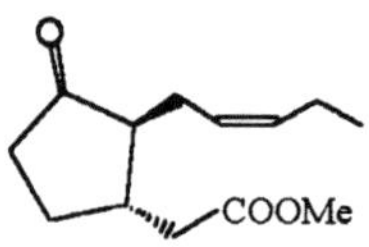

Odourless (1S, 2S)

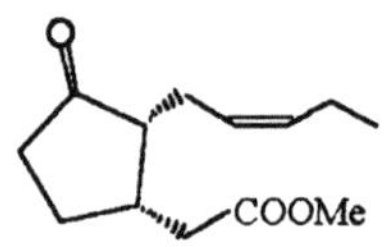

Odourless (1S, 2R)

Table 10. Odour profiles of enantiomeric methyl jasmonate

Racemic 2-n-hexylcyclopentanone (37) and 2-n-heptylcyclopentanone (38) are used as raw materials for jasmine-floral fragrance compounds. These enantiomeric 2-alkylcyclopentanones (37) and (38) were prepared by the asymmetric hydrogenation of the readily available 2-hexylidenecyclopentanone (39) and 2-hepthylidenecyclopentanone (40), with optical yields of 95-97 %e.e. [36] (Scheme 11).

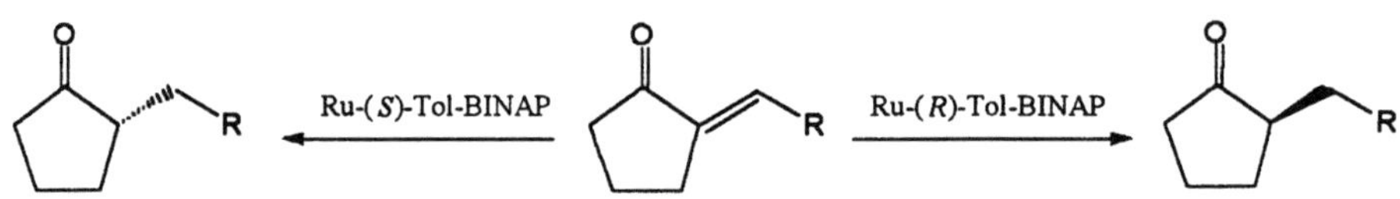

(R)-(-)-(37), (38) (39), (40) (S)-(+)-(37), (38)

Compound	R	(R)-(-)-form %e.e.	(S)-(+)-form %e.e.
(37)	n-C_5H_{11}	95.5	95.3
(38)	n-C_6H_{13}	97.1	96.7

Scheme 11. Synthesis of enantiomeric 2-alkylcyclopentanones (37) and (38)

Odour differences between the enantiomers of (37) and (38) are not so significant as in the case of the corresponding δ-lactones, and the threshold values are almost the same between the pairs. As for their olfactory qualities, fundamentally they showed a jasmine-floral modality [33] (Table 11).

Compound	Odour	Odour threshold (ppb)
(R)-(-)-(37) (% e.e. 95.5)	Powerful diffusive sweet fruity, fatty somewhat jasmine -like floral odour with slightly oily minty citrus note	70
(S)-(+)-(37) (% e.e. 95.3)	Powerful diffusive warm jasmine-like floral odour with coconut-like fruity and slightly herbaceous note	70
(R)-(-)-(38) (% e.e. 97.1)	Powerful diffusive jasmine-like warm floral odour with somewhat mandarin-like citrus side note and olfactorily more tenacious than the (S)-(38)	10
(S)-(+)-(38) (% e.e. 96.7)	Heavy, coconut -like oily fruity and jasmine -like floral odour with somewhat herbaceous side note	10

Table 11. Odour profiles of the enantiomeric 2-*n*-hexyl- and 2-*n*-heptycyclopentanones (37) and (38)

δ-Lactones are found widely in many different kinds of fruits and play very important roles in flavours. The comparison of odour properties of the enantiomers of δ-lactones has been reported by Mosandl using the samples prepared by HPLC separation [37]. One of the most convenient synthesis methods for preparing chiral δ-lactones is the Baeyer-Villiger oxidation. Applied to the chiral 2-alkylcyclopentanones (39) and (40) yields the corresponding enantiomeric δ-lactones 41 and 42 whilst retaining their chirality (Scheme 12).

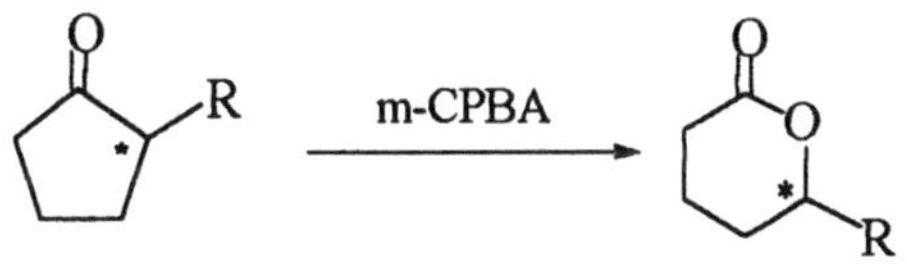

R = n-hexyl (39)
n-hepthyl (40)

R = n-hexyl (41)
n-hepthyl (42)

Scheme 12

Olfactory qualities of the chiral δ-undecalactone (41) and dodecalactone (42), synthesized from the enantiomeric 2-alkylcyclopentanones 39 and 40, are cited in Table 12.

Compound	R	Config.	%e.e.	Threshold value (ppm)	O dour
Undecalactone (41)	C_5H_{11}	*l* (S)	88.9	0.03	Fruity-sweet, creamy
		d(R)	88.8	0.10	Fruity-sweet, milky
Dodecalactone (42)	C_7H_{15}	*l* (S)	92.6	0.05	Fruity-sweet, apricot
		d(R)	92.2	0.50	Fruity-sweet

Table 12. Odour profiles of the enantiomeric δ-lactones (41) and (42)

As in the case of δ-lactones, it has also been reported that (R)-δ-lactones exhibit more potent odour characteristics than their corresponding antipodes [37B]. Evaluation of milk flavour with the chiral δ-lactones added to it indicated that (S)-δ-lactones show better sensory properties and a more enhanced feel of fresh milk than the antipodes [38].

2.2.4. Brahmanol ™

Recently, the odour properties of chiral sandalwood-type aromatic chemicals, synthesized from campholenaldehyde (44), such as Sandalore™ [39]', 4-(2,2,3-trimethylcyclopenten-3-yl) -2-methyl-4-methylen-butan-1-ol [40] , Bacdanol™ [41]., etc., have been reported.

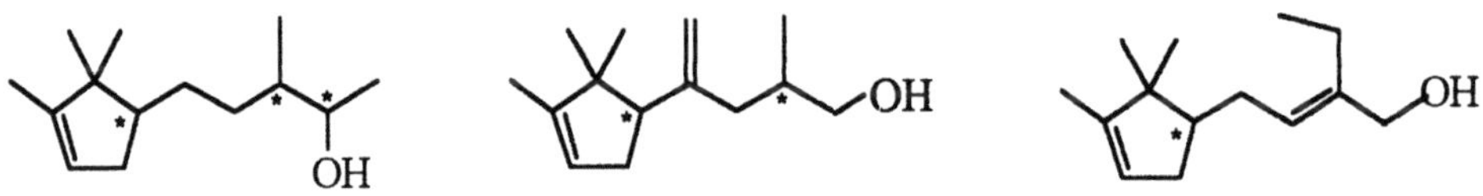

Sandalore™ 4-(2,2,3-trimethylcyclopent-3-enyl) -2-methy-4-methylen-butan-1-ol Bacdanol™

In these compounds, the stereochemistry is very important for their odour profiles. Brahmanol ™ (43), one of the commercially used sandalwood-type aromatic chemicals derived from campholenaldehyde (44), has two asymmetric carbons and consists of four diastereomers. Each diastereomer was synthesized with a diastereomeric yield of 73-78 % d.e. by asymmetric hydrogenation of chiral 2-methyl-4-(2,2,3-trimethyl-3-cyclopenten-1-yl)-2-buten-1-ol (45) which was derived from the optically active campholenaldehyde (97 %e.e.) as shown in Scheme 13.

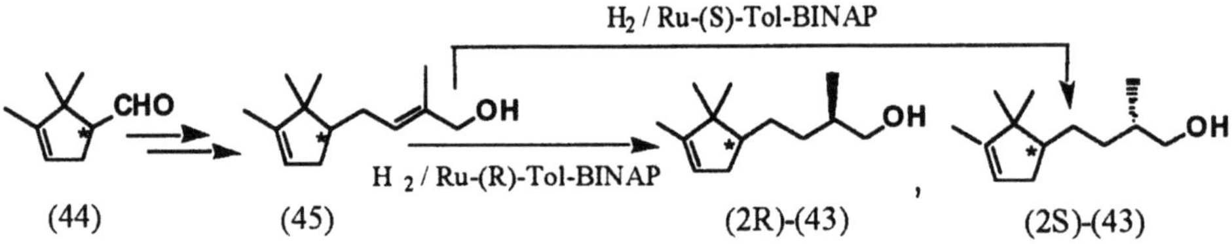

Scheme 13

As shown in Table 13, the configuration of the methyl group in the side chain of the (43) significantly influences the odour properties of Brahmanol™ and the (2R)-configuration has an important role to play in exhibiting its sandalwood-odour modality. The (1'S, 2R)- and the (1'R, 2R)-(43) show strong and surpassing odour characteristics much superior to those of the corresponding (1'S, 2S)- and (1'R, 2S)-(43). In the case of Bacdanol™, the configuration of the asymmetric carbon in the ring significantly influences its odour characteristics, whereas this influence in Brahmanol™ is very small. Among the four diastereomers, the (1S,2R)-(43) showed the strongest and best odour quality [42].

(43)	Odour	Threshold (ppb)
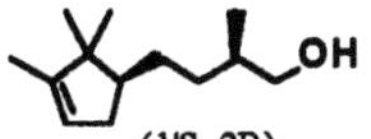 (1'S, 2R)	Strong sandalwood-type odour with characteristic clean bright elegant floral, fruity side note	14.5
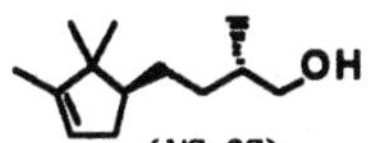 (1'S, 2S)	Weak slightly dusty, oily sandalwood-type odour	166.0
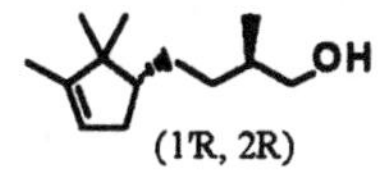 (1'R, 2R)	Strong sandalwood-type odour with characteristic clean bright elegant floral woody note	18.7
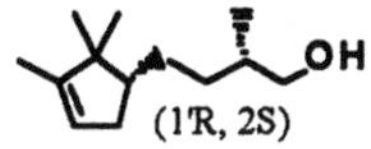 (1'R, 2S)	Weak slightly dusty oily and metallic sandalwoodtype odour	176.5

Table 13. Odour profiles of the diastereomeric Brahmanol ™ (43)

2.2.5. Lilial™- related compounds

Lilial ™ (44) is one of the most widely used lily of the valley-type aroma chemicals. The odour property of optically active Lilial ™ (44) was examined by Enders *et al.* [43], who stated that, the (S)-(-)-isomer has a stronger lily of the valley-lilac note than its antipode. By utilising the asymmetric hydrogenation of unsaturated carboxylic acids [44], the attempted asymmetric synthesis of α-methyl-p-alkyldihydrocinnamic aldehydes, Lilial (44), cyclamen aldehyde (45), and Suzaral™ (46) was examined, using 3-p-*tert* butylphenyl-, 3-p-isopropylphenyl- and 3-p-isobutylphenyl-α-acrylic acids (47) as the starting materials. The Ru-(S)-BINAP-catalyzed asymmetric hydrogenation of substrates (47) afforded the corresponding (S)-(-)-forms (44), (45) and (46) with optical yields of 71-77 %e.e., whilst the same hydrogenation using Ru-(R)-BINAP as catalyst resulted in the corresponding (R)-(+)-forms with optical yields of 72-80 %e.e. (Scheme 14).

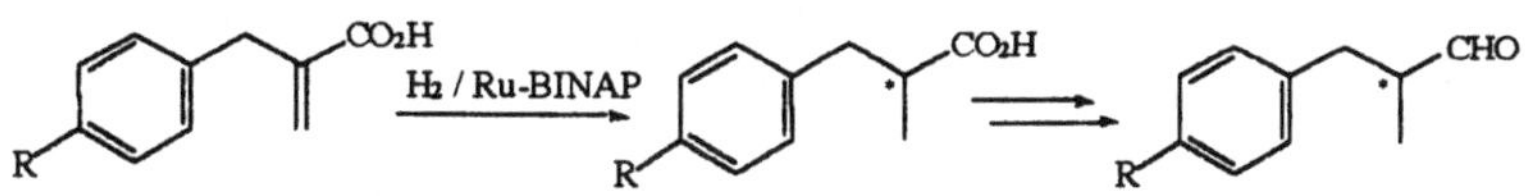

(47)

R= t-butyl (44)
i-propyl (45)
i-butyl (46)

Scheme 14

Compound	R	Config.	%e.e.	Odour
Lilial™ (44)	t-C_4H_9	*l (S)*	71	Floral, oily, lily of the valley
		d(R)	80	Aldehydic, chemical
Cyclamen aldehyde (45)	i-C_3H_7	*l(S)*	77	Green-floral, cyclamen
		d(R)	72	Ozone-like, aldehydic, chemical
Suzaral™ (46)	i-C_4H_9	*l(S)*	70	Lily of the valley, green-floral
		d(R)	75	Ozone-like, clean

Table 14. Odour profiles of the enantiomeric Lilial ™ (44), and its homologues (45) and (46)

The odour profiles of the enantiomers obtained are shown in Table 14. From evaluation of the results, it can be construed that the *l*-isomers are responsible for the lily of the valley-type floral notes [45].

2.2.6. 2-Methylbutyric acid and its ethyl esters

2-Methylbutyric acid (48) and its esters are found widely in various kinds of natural products. Especially, many of these esters are found abundantly in fruits. (S)-2-methylbutyric acid (S)-(+)-(48) exists profusely in different fruits (strawberry, apple, etc.), whereas the (R)-form is present in cacao beans. As for its esters, generally the (S)-forms are more plentifully present than the antipodes. These flavouring materials are essential for imparting a fermentation aroma and palatable sensation of ripe fruits to foods. Enantiomeric 2-methylbutyric acid is synthesized with an optical yield of 91-92 %e.e. from the asymmetric hydrogenation of tigric acid with Ru-BINAP catalyst [46] (Scheme 15).

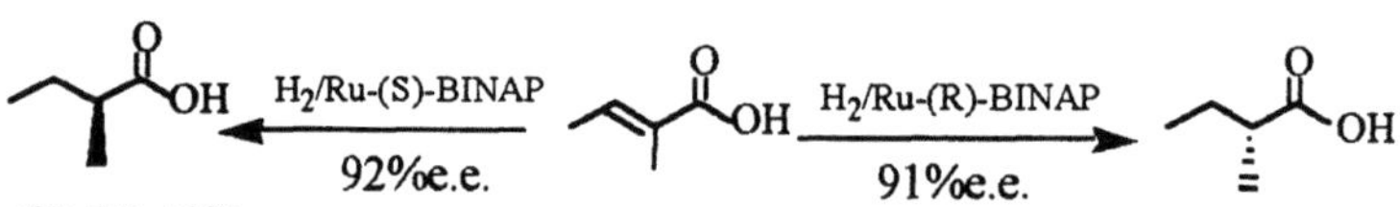

(S)-(+)-(48) (R)-(-)-(48)

Scheme 15

The odour properties of optically active 2-methylbutyric acid (48) and ethyl 2-methylbutyrate (49), reported by Rettinger, are shown in Table 15 [47].

Compound	Config.	Odour
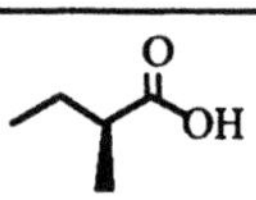	(S)-(+)-(48)	Pleasant sweet fine fruity note
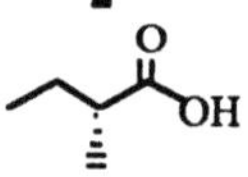	(R)-(-)-(48)	Penetrating cheesy, sweaty note
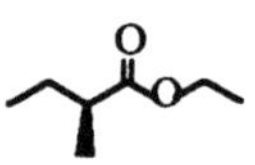	(S)-(+)-(49)	Etherial, sweet, unspecific pleasant apple note at extreme dilution
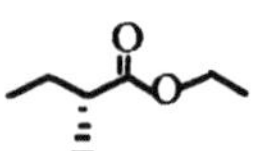	(R)-(-)-(49)	Initially medicinal-phenolic then fruity, sweet but unspecific

Table 15. Odour profiles of enantiomeric 2-methylbutyric acid (48) and ethyl-2-methylbutylate (49)

The aqueous solution threshold values for the enantiomeric 2-methylbutyric acid (48) were found to be 50 ppb for the (R)-(-)- and 100 ppb for the (S)-(+)-(48). Evaluation of strawberry flavour, with the enantiomeric 2-methylbutyric acid (48) added to it, resulted in the (R)-(-)-(48) enhancing the flavour of ripe fruit, and the (S)-(+)-(48) boosting the unripe fresh fruity note. Similarly, evaluation of strawberry flavour, with the enantiomeric ethyl 2-methylbutyrate (49) added to it, resulted in (S)-(49) boosting a muscat odour and improving the top note of the flavour, and (R)-(49) imparting an oily, heavy and medicinal nuance to the flavour [38].

2.3. ASYMMETRIC HYDROGENATION OF KETONES CATALYZED BY Ru-BINAP COMPLEXES

2.3.1. (R)-Styrallyl acetate

Racemic styrallyl acetate (50) is widely used as a floral green aromatic raw material which has an odour reminiscent of gardenia or strawberry. Noyori *et al.* [48] have reported that the asymmetric hydrogenation of acetophenone (51) with the Ru-BINAP-diamine-isopropanol system gave the chiral styrallyl alcohol (52) with an optical yield of 87-94 %e.e. (Scheme 16).

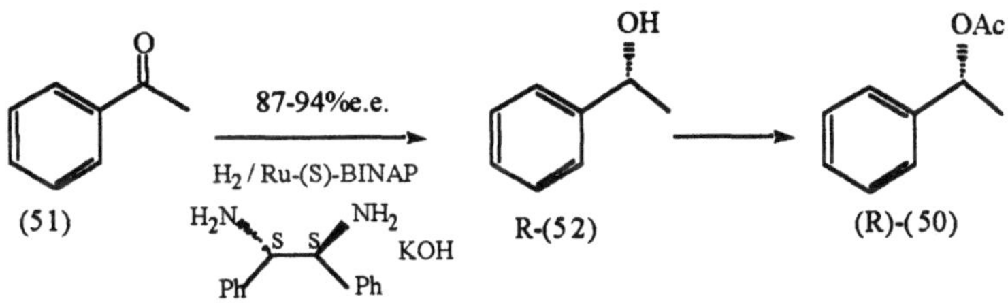

Scheme 16

Esterification of the chiral styrallyl alcohol (52) yielded the chiral styrallyl acetate (50). As for the odour properties, the enantiomeric pair (50) show fundamentally a floral-green note, reminiscent of gardenia, fresh and herbal profiles. However, the (R)-form gives a pronounced fruity-green note, whilst the (S)-form gives a sharp metallic note.

The odour properties of the enantiomeric analogues of styrallyl acetate (53), synthesized by the same catalyst system, are shown in Table 16.

The differences in olfactory characteristics among the enantiomers of the analogues of styrallyl acetate (53) are not great. For instance, the odours of their enantiomeric pairs (the dextrorotatory and laevorotatory) belong to the same odour classification categories, but differ from each other in specific olfactory characteristics such as tones and nuances. Generally speaking, the odours of the (R)-forms resemble those of the racemates but with more fruity, cleaner and lighter nuances. On the other hand, the (S)-forms have greener, sharper and more metallic nuances than the racemates [49].

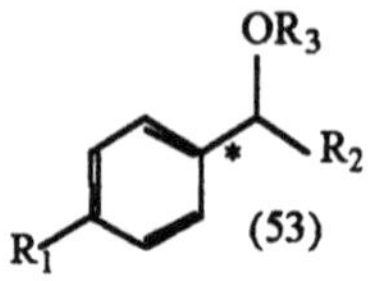

(53)

R_1	R_2	R_3	Comprehensive odour description	Difference in olfactory tone of enantiomeric pair (R)	(S)
H	Me	Et	Floral-green reminiscent of gardenia, strawberry, apple	Clean, green	Thin, light metallic
H	Me	n-Pr	Floral, fruity, apple pineapple, strawberry	Fruity	Fatty, green
H	Me	i-Pr	Floral-green reminiscent of strawberry, pineapple, herb	Light, fruity	Green
H	Me	n-Bu	Floral, fruity, strawberry	Heavy	Fruity, fatty
H	Me	i-Bu	Floral, fruity reminiscent of strawberry drop	Fruity	Sharp green
H	Et	Me	Floral (rose) with green (twig)	Metallic, bitter	Flat
Me	Me	Me	Floral (rose) with fresh green	Light, fresh fruity	Metallic
Me	Me	Et	Floral (rose) with green, fruity	Pineapple	Peach
Me	Me	n-Pr	Floral (rose) with strawberry-pear character	Honey	Fatty
Me	Me	i-Pr	Floral (rose) with strawberry pineapple, green	Fruity	Balsamic, dirty
MeO	Me	Me	Floral (rose, lilac) with balsam	Flat	Flat

Table 16. Odour profiles of the enantiomeric analogues of styrallyl acetate (53)

2.3.2. l-Matsutakeol

Matsutake mushroom, (*Tricholoma matsutake*), is very expensive in Japan. The Japanese are very fond of it not only for its taste but also its odour. Matsutakeol (1-octen-3-ol) (54) is a key odoriferous component of this mushroom. The enantiomeric mixture of (R)-(-)- and (S)-(+)-(54) has been identified in numerous species of mushrooms and each of them has a characteristic distribution of each of the enantiomers respectively. The optical distribution ratio of (R)-(-)- and (S)-(+)-(54) in the matsutake is 94 : 6, while it is 96 : 4 in the well-known truffle (*Tuber melanosporum*) [50]. The odours of the enantiomers of matsutakeol (54) have been compared by Mosandl, who separated the racemic mixture into the corresponding enantiomers, namely, the dextrorotatory and laevorotatory respectively, by HPLC. Mosandl showed that the (R)-(-)-(54) exhibited a strong mushroom-like odour whereas the antipode had a herbaceous-green-musty odour [51].

As in the example of the Ru-BINAP-catalyzed asymmetric hydrogenation of β-ketoesters [52], (R)-(-)-matsutakeol was synthesized with an optical yield of 99 %e.e. by the Ru-(R)-BINAP-catalyzed asymmetric hydrogenation of 3-ketooctanoate. This was the key step for introducing chirality into the molecule. Similarly, the (S)-(+)-(54) was synthesized with an optical yield of 99 %e.e. using the Ru-(S)-BINAP catalyst (Scheme 17).

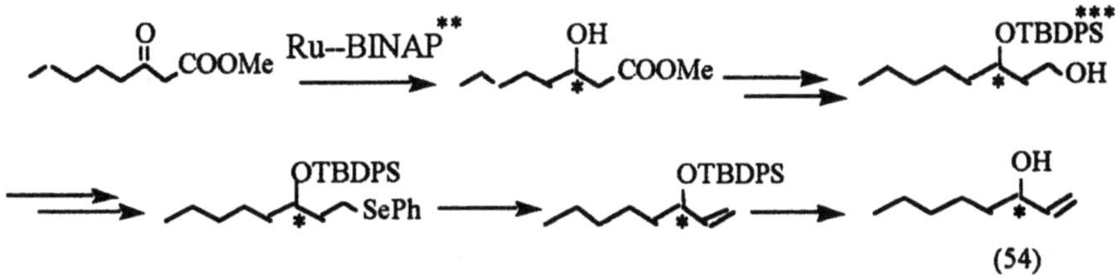

(54)

Total yield: 19.8%Th

** : Ru_2Cl_4[T-BINAP$]_2$$NEt_3$ (0.1mol%), in MeOH, 80 atm, r.t., 40 hs.

*** TBDPS ; t-butyldiphenylsilyl

Scheme 17. Asymmetric synthesis of 1-octen-3-ol (54)

The difference in the odour threshold values between the enantiomeric pair of matsutakeol (99 %e.e.) is fairly large. The threshold values measured in aqueous-ethanol solution were found to be 10 ppb for the (R)-(54), and 100 ppb for the (S)-(54) [10,33].

2.3.3. γ-Lactones

Enantiomeric mixtures of γ-lactones are found in numerous kinds of fruits, examples being peach, apricot, mango, passion-fruit and strawberry. Each fruit has a characteristic distribution of enantiomers [53] with the (R)-isomers always existing more abundantly. Looking at the sensory properties of these enantiomeric γ-lactones, it has been reported that the (R)-γ-lactones exhibit stronger sensory characteristics than the corresponding (S)-forms and are responsible for the pleasant and natural fruity aromas in the preliminary odour evaluation [37]. Four optically active γ-lactones have been synthesized with an optical yield of 98 %e.e. via the corresponding chiral γ-hydroxyesters which in their turn were prepared by Ru-BINAP-catalyzed asymmetric hydrogenation of the corresponding γ-ketoesters as shown in Scheme 18 [54].

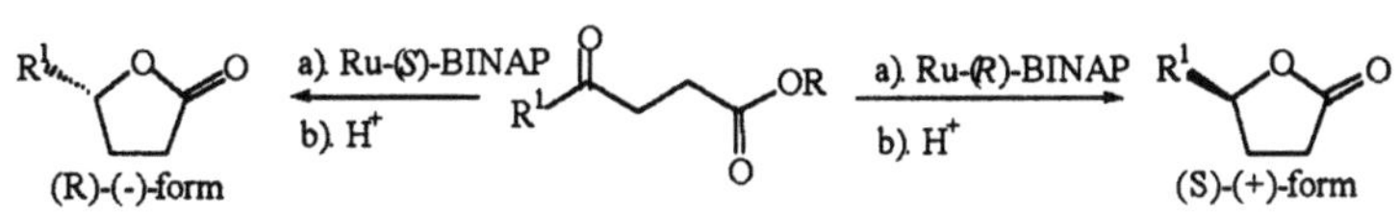

Scheme 18. Asymmetric hydrogenation of γ-ketoesters

The odour properties of the enantiomeric pairs of γ-hepta-, γ-octa-, γ-nona-, and γ-undecalactones (55-58) synthesized by this method are shown in Table 18 [10,33].

In the applied evaluation of peach flavour containing the enantiomeric pair of γ-undecalactone (58), the flavour with the (S)-(58) added showed an excellent light, fresh taste and a sweet top note, whereas the flavour with the (R)-(58) in it left an unpleasant taste with heavy and oily tones [38].

Compound	R	Config.	Threshold value (ppm)	odour
Heptalactone (55)	C_3H_7	*l (S)*	0.89	Fatty, coconut
		d(R)	0.70	Sweet, spicy
Octalactone (56)	C_4H_9	*l (S)*	0.13	Fatty, coconut
		d(R)	0.30	Spicy, green
Nonalactone (57)	C_5H_{11}	*l (S)*	0.15	Fatty, coconut
		d(R)	0.22	Sweet, coconut
Undecalactone (58)	C_7H_{15}	*l (S)*	0.012	Sweet, coconut
		d(R)	0.004	Strong, sweet, coconut

Table 18. Odour profiles of the enantiomeric-γ-lactones

3. Summary

The present metal (Rh, Ru)-BINAP catalyst has enabled us to prepare a number of chiral aromatic molecules for flavours and fragrances. As suggested previously in various reports, the results of the sensory evaluation of the chiral compounds prepared through the metal-BINAP-catalyzed asymmetric synthesis support the concept that such chiral molecules exhibit sensory properties superior to the corresponding antipodes and the racemates. They seem destined to become coveted superb tools to meticulously reproduce Nature's tastes and smells and thus to create new flavours and fragrances. Furthermore, some chiral molecules are superior to the antipodes and the racemates in other areas and functions such as safety, biodegradability (an environmental issue), antibacterial activity, psychological effect, olfactory tenacity, etc. Consequently, these proactively functional characteristics seem to point to the potential importance of chiral molecules for next-generation flavours and fragrances. Further improvement in production technology and progress in the chemistry of asymmetric synthesis such as enantioselectivity, catalytic activity and new catalytic systems will hasten this development.

4. Acknowledgment

I thank Dr. H. Tsuruta, Dr. H. Kumobayashi, Dr. T. Kanisawa, and Dr. S. Akutagawa for their encouragement and Mr. T. Yamada, Dr A. Lupo and Mr M.A. Murray-Pearce for their proof-reading of the manuscript.

5. References

1. H. Kumobayashi, N. Sayo, S. Akutagawa, T. Sakaguchi, H. Tsuruta, *Nippon Kagakukai Shi*,**1997**, 835.
2. R. Noyori *et al.*, *Chemistry and Chemical Industry*, **1998**, **51**, 713.
3. a) A. Mosandl, *J. Chromatogr.*, 1992, **624**, 267. b) P. Werkhoff, S. Brennecke, W. Bretschneider, M. Guntert, R. Hopp and H. Surburg, Proceedings of the 12th International Congress of Flavours, Fragrances and Essential Oils, (Post-congress volume), October 4-8, 1992, Vienna, Austria, p.64-128.
4. a) R. Noyori, *Asymmetric Catalysis in Organic Synthesis,* John Wiley & Sons, New York, 1994. b) B. M. Trost, *Angew. Chem., Int. Ed. Engl.*, 1995, **34**, 259. c) G. Wilkinson, F.G.A. Stone, E.W. Abel, L.S. Hegedus, *Comprehensive Organometallic Chemistry II*, Pergamon, New York,1995. d) I. Ojima, *Catalytic Asymmetric Synthesis,* VCH Publishers, New York, 1998.
5. R. Enberger and R. Hopp, in *Topics in Flavour Research*, R.G. Berger, S. Nits and P. Schreier, eds, Marzling-Hangenham ; H. Echhon, 1985, p.201-218.
6. G. S. Clark, *Perfumer & Flavorist*, 1998, **23** (5), 33.
7. a) K. Tani, T. Yamagata, S. Otsuka, S. Akutagawa, H. Kumobayashi, T. Taketomi, H. Takaya, A. Miyashita, and R. Noyori, *J. Chem. Soc. Chem. Commun.*, 1982, 600. b) K. Tani, T. Yamagata, S. Akutagawa, H. Kumobayashi, T. Taketomi, H. Takaya, A. Miyashita, R. Noyori and S. Otsuka, *J. Am. Chem. Soc.*, 1984, *106*, 5208.
8. S. Akutagawa, *Organic Synthesis in Japan*, R. Noyori (Ed), Tokyo Kagaku Dojin, 1992, p.75-82.
9. S. Otsuka, *Acta Chemica Scandinavia*, 1996, **50**, 353-360.
10. a) H. Tsuruta, the 12th International Congress of Flavours, Fragrances and Essential Oils, Vienna, Austria, October 4-8, 1992, Invited lecture. b)T. Yamamoto, H. Matsuda, T. Ohmoto, A. Shimada and T. Sato, Proceedings of the 12th International Congress of Flavours, Fragrances and Essential Oils, Vienna, Austria, October 4-8, 1992, p.442-455. c) T. Yamamoto, *Koryo*, 1994, **184**, 57.
11. H. Takaya, H. Ohta, N. Sayo, H. Kumobayashi, S. Akutagawa, S. Inoue, I. Kusahara and R. Noyori, *J. Am. Chem. Soc.*, 1987, **109**, 1596.
12. T. E. Acree, R. Nishida and H. Fukami, *J. Agric. Food Chem.* 1985, **33**, 425.
13. S. Watanabe, A. Kinosaki, M. Kawasaki, T. Yoshida and K. Kaidbey, Proceedings of the 10th International Congress of Essential Oils, Fragrances and Flavors, November 16-20, 1986, Washington, DC, U.S.A., p.1029-1038.
14. T. Yamamoto, T. Sakurai, S. Watanabe and A. Kinosaki, JP (B) 1905757 (1995).
15. W. Pickenhagen, in *Flavor Chemistry*, Trends and Development, R. Teranishi, RG Buttery and F. Shahid, eds, ACS Symp Series 388, Washington, DC, ACS, 1989 p.151-157.
16. A. Shimada, T. Ohmoto and T. Yamamoto, JP (B) 2748184 (1998).
17. A. Shimada, M. Ogura, H. Matsuda and T. Yamamoto, JP (A) 5-112495 (1993).
18. D. Rlenk de Haan and D.K. Kettenes, US 3956392 (1976).
19. T. Yamamoto, *The Takasago Times*,1993, **111**, 23.
20. G. Ohloff, in *Olfaction and Taste IV*, D. Schneider, ed., Stuttgart ; Information Retrieval Ltd.,1980, p.3-11.
21. a) H. Matsuda and T. Yamamoto, JP (A) 9-169993 (1993). b) H. Matsuda, T. Yamamoto, and T. Kanisawa, Proceedings of the 13th International Congress of Flavours, Fragrances and Essential Oils, October 15-19 1995, Istanbul, Turkey, p.13-26.
22. A. Shimada and T. Yamamoto, JP (B) 2759546 (1998).
23. A. Shimada, T. Ohmoto, T. Yamamoto and T. Sato, Proceedings of the 25th JASTS Annual Meeting, 1991, p.57-59.
24. K.H. Schulte-Elte, G. Ohloff, B.L. Muller and W.K. Giersch, US 4623750 (1986).
25. K.H. Schulte-Elte, M. Christian, C. Christian and S. Dana P., EP-457022 A1.(1991).
26. A. Shimada, T. Ohmoto and T. Yamamoto, EP-456932 A1 (1991).
27. a) T. Ohmoto and T. Yamamoto, JP (A) 5-286893 (1993).b) T. Ohmoto and T. Yamamoto, JP (A) 5-286876 (1993).
28. T. Yamamoto, *The Takasago Times*, *110*, 33 (1993).
29. M. Harada, H. Matsuda and T. Yamamoto, JP (A) 8-245979 (1995).
30. T. Yamamoto, *Koryo*,1997, **196**, 91.
31. J. Tsuji, T. Yamada, M. Kaito and T. Mandai, *Bull. Chem. Soc. Japan*, 1980, **53**, 1417.
32. M. Ogura and T. Yamamoto, JP (A) 6-192161 (1994).
33. T. Yamamoto, Proceedings of the 37th Symposium on Terpenes, Essential Oils and Aromatic Chemicals (TEAC), Japan, 1993, p.1-4.
34. Y. Okazaki, M. Koma, Y. Takashima and T. Kanisawa, Proceedings of the 13th International Congress of

Flavours, Fragrances and Essential Oils, Istanbul, Turkey, October 15-19, 1995, p.385-391.
35. M. G. J. Beets, The proceedings of International Congress of Food Science and Technology, Washington, DC.,1971, p.379-394.
36. T. Ohta, T. Miyake, N. Seido, H. Kumobaysshi, S. Akutagawa and H. Takaya, *Tetrahedron Lett.*, 1992, *33*, 635.
37. a) A. Mosandl and M. Gessor, *Z.Levensm Unter-Forsch,* 1988,**40**, 197. b) A.Mosandl and Gunther, *J. Agric. Food Chem.*, 1990, **37**, 413.
38. A. Amano, *The Takasago Times*,1993, **112**, 31.
39. R.E. Nalpawer and Wallington N.J., US 4696766 (1987).
40. C. Chapyui and P. Alan Rlan, JP (A) 8-67661 (1996).
41. M. Harada, T. Aida, A. Amamo, T. Yamazaki, T. Yamamoto and H. Iwai, JP (A) 10-147347 (1998).
42. T. Aida, H. Matsuda and T. Yamamoto, JP (A) 8-268940 (1996).
43. D. Enders and H. Dykers, *Liebigs Ann. Chem.*, 1990, 1107.
44. T. Ohta, H. Takaya, M. Kitamura, K. Nagai and R. Noyori, *J. Org. Chem.*,1987, **52**. 3178.
45. T. Sotoguchi, T. Yamamoto and T. Sato, Proceedings of the 33rd Symposium on Terpenes, Essential Oils and Aromatic Chemicals (TEAC), Japan, 1989, p.1163-1165.
46. a) X. Zhang, T. Uemura, N. Sayo, H. Kumobayashi and H. Takaya, *Synlett*, 1994, 501.
47. K. Rettinger, C. Burschka, P. Scheeben, H. Fuchs and A.Mosandl, *Tetrahedron : Asymmetry*,1991, **2**, 965.
48. T. Ohkuma, H. Ooka, S. Hashiguchi, T. Ikariya and R. Noyori, *J. Am. Chem. Soc.* 1995, **117**, 2675.
49. T.Yamasaki, The proceedings of the 1997 RSC / SCI International Conference on Flavours and Fragrances held on 30 April to 2 May 1997 in Warwick, UK., ed., Karl A. D. Swift, 1997, p.185-195.
50. Unpublished data.
51. A. Mosandl, G. Heusinger and M. Gessner, *J. Agric. Food Chem.*, 1986, **34**, 119.
52. R. Noyori, T. Ohkuma, M. Kitamura, H. Takaya, N. Sayo, H. Kumobayashi, S. Akutagawa, *J. Am. Chem. Soc.*,1987, **109**, 5856.
53. A. Mosandl and C. Gunther, *J. Agric. Food Chem.*, 1989, **37**, 413.
54. R. Noyori, M. Kitamura, T. Ohkuma, N. Sayo and H. Kumobayashi, JP (A) 4-108782 (1992).

TOWARDS ENVIRONMENTALLY FRIENDLY CHEMICAL PROCESSES

ROGER A. SHELDON
Delft University of Technology
The Netherlands

1. Introduction

Increasingly stringent environmental regulations have stimulated chemical manufacturers to develop alternative technologies that produce a minimum amount (preferably zero) of waste and avoid, as much as possible, the use of toxic and/or hazardous reagents and solvents [1-4]. Emphasis is clearly on the reduction of waste at source - primary pollution prevention - rather than incremental end-of-pipe solutions. Sustainable development and benign by design are the catch phrases that paraphrase this trend towards 'green chemistry' [5, 6]. Consequently, traditional concepts of process efficiency are changing from an exclusive focus on chemical yield to one that assigns economic value to eliminating waste.

2. The E-Factor and Atom Efficiency

The urgency of the need for waste reduction is readily apparent from a consideration of the amount of waste produced per kg of product, what we have designated as the E factor [1-4], in various segments of the chemical industry (Table 1).

Table 1. The E- factor

Industry segment	Product tonnage	kg waste/kg product
Oil refining	10^6-10^8	< 0.1
Bulk chemicals	10^4-10^6	< 1 - 5
Fine chemicals	10^2-10^4	5 - > 50
Pharmaceutical	10-10^3	25 - > 100

In this context waste is defined as everything that is formed in the process except the desired product. The waste consists primarily of inorganic salts, e.g. sodium chloride, sodium sulphate and ammonium sulphate, formed in the reaction or subsequent neutralisation steps or derived from stoichiometric inorganic reagents, e.g. heavy metal salts. The E factor increases substantially on going downstream from bulk to fine

K.A.D. Swift (ed.), Current Topics in Flavours and Fragrances, 59–78.

chemicals and specialities such as pharmaceuticals and flavours and fragrances. This is partly owing to the fact that the production of fine chemicals and specialities involves multi-step syntheses but is also a reflection of the widespread use of stoichiometric inorganic and organic reagents rather than catalytic methodologies.

Obviously the key to waste minimisation is selectivity in organic synthesis. Unfortunately, attention is generally focused on the selectivity (efficiency) of the conversion of the key raw material to the desired product, which provides little or no indication of the total amount of waste generated in the process. Hence, most processes were originally developed primarily with emphasis on product yield with little regard for the production of (inorganic) waste. Put in another way, the E factor constitutes the ratio of 'kilos in to kilos out', i.e. it is the mass balance or the weight ratio of all the raw materials, including solvent losses, to the amount of desired product. Hence, it can be readily calculated for a particular product line, production site or even a whole company. It is quite amazing, therefore, that so few companies are aware of what their E factor is.

The atom utilisation or atom efficiency concept [7-9] is a uscful tool for rapid evaluation of the amount of waste generated by alternative routes to a particular product. It is calculated by dividing the molecular weight of the desired product by the sum total of the molecular weights of all substances produced in the stoichiometric equation for the reaction(s) involved. The comparison is made on a theoretical basis (i.e. 100% chemical yield). The reciprocal of the atom efficiency is the theoretical E factor, i.e. assuming 100% yield, using equivalent amounts of reagent(s), with no subsequent neutralisation steps and no solvent losses. In practice the E factor is, of course, significantly higher. A further refinement should also take the energy consumption of alternative routes into consideration, as this also results in waste generation (mostly as CO_2). Furthermore, an evaluation of the environmental impact of alternative routes must also take the nature of the waste into account. After all, one kilo of NaCl or one kilo of a chromium salt do not have the same environmental impact. To this end we introduced [4] the term environmental quotient, EQ, which is the product of the E factor and an arbitrarily assigned unfriendliness quotient, Q. For example, NaCl could be assigned a Q value of 1 and heavy metal salts, say 100-1000, depending on their toxicity, ease of recycling, etc. Although the magnitude of Q values assigned to waste streams is obviously debatable it is nevertheless possible to quantitatively evaluate the environmental impact of processes on this basis. Further refinement leads to the concept of environmental profile analysis, in which processes are assessed on the basis of a weighting of various factors: raw materials and energy consumption, waste generation and safety aspects.

3. The Role of Catalysis

As noted above a prime cause of high E factors is the use of stoichiometric inorganic reagents. Fine chemicals and specialities manufacture is rampant with classical 'stoichiometric' technologies, many of which date from the last century. Examples which readily come to mind are stoichiometric reductions with metals (Na, Mg, Fe, Zn, etc.) and metal hydrides ($LiAlH_4$, $NaBH_4$) and stoichiometric oxidations with permanganate or chromium(VI) reagents. Hence, the solution is simple: replacement of antiquated stoichiometric methodologies with cleaner catalytic alternatives. For example, the atom efficiencies of the stoichiometric oxidation of an alcohol, to the corresponding ketone, with chromium trioxide in sulphuric acid and catalytic oxidation with O_2 are compared in Figure 1.

Stoichiometric:

$$3\,PhCH(OH)CH_3 + 2CrO_3 + 3\,H_2SO_4 \longrightarrow$$

$$3\,PhCOCH_3 + Cr_2(SO_4)_3 + 6\,H_2O$$

Atom efficiency = 360/860 = 42%

Catalytic:

$$PhCH(OH)CH_3 + 1/2\,O_2 \xrightarrow{\text{catalyst}} PhCOCH_3 + H_2O$$

Atom efficiency = 120/138 = 87%

Figure 1. Atom efficiency: stoichiometric vs. Catalytic oxidation

The longer term trend in fine chemicals manufacture is towards the use of the simplest raw materials - H_2, O_2, H_2O, H_2O_2, NH_3, CO and CO_2 - in catalytic, low-salt processes [10]. Catalytic hydrogenation and carbonylation are typical examples of 100% atom efficient processes (Reactions 1 and 2).

$$RCOCH_3 + H_2 \xrightarrow{\text{catalyst}} RCH(OH)CH_3 \qquad (1)$$

$$RCH(OH)CH_3 + CO \xrightarrow{\text{catalyst}} RCH(CH_3)CO_2H \qquad (2)$$

Organic chemists are generally quite familiar with catalytic hydrogenation, although they often choose to employ stoichiometric metal hydride reagents, such as $LiAlH_4$ or $NaBH_4$. Catalytic carbonylation and oxidation, in contrast, find only sporadic use which is partly owing to unfamiliarity and partly to the lack of efficient generic methodologies, e.g. for oxidative transformations of various functional groups.

Another major cause of waste is the use of mineral (H_2SO_4, H_3PO_4, etc.) and Lewis acids ($AlCl_3$), often in stoichiometric amounts, which cannot easily be recovered and recycled. Hence, there is a discernible trend towards their replacement by solid, recyclable Brønsted and Lewis acids, e.g. zeolites, acidic clays, etc. (see later).

4. Catalytic Reduction

Catalytic hydrogenation is undoubtedly the work-horse of catalytic organic synthesis. A wide variety of functional groups can be efficiently hydrogenated, often with high degrees of chemo-, regio- and stereoselectivity. Although it is a well-established catalytic methodology new advances continue to appear. For example, Rhône Poulenc scientists have reported [11] the direct hydrogenation of aromatic and aliphatic carboxylic acids to the corresponding aldehydes, a notoriously difficult transformation to perform. The reaction is carried out in the vapour phase over a supported ruthenium/tin alloy catalyst at temperatures between 250 and 300 °C and 1 bar H_2 pressure. For example, 1-dodecanoic acid and trifluoroacetic acid afford 1-dodecanal and fluoral, respectively, and an α,β-unsaturated acid is reduced to the corresponding α,β-unsaturated aldehyde (Reaction 3) in 91% yield.

(3)

Similarly, Mitsubishi scientists have developed commercial processes for the hydrogenation of aromatic, aliphatic and unsaturated carboxylic acids, to the corresponding aldehydes, in the vapour phase over zirconia or chromia catalysts [12].

The Meerwein-Ponndorf-Verley (MPV) reduction of aldehydes and ketones is a well-established methodology in organic synthesis. It involves the reaction of the substrate with a hydrogen donor, usually isopropanol, in the presence of an aluminium alkoxide, e.g. $Al(OPr^i)_3$. Although the latter is, in principle, a catalyst MPV reductions often employ stoichiometric amounts owing to the low rate of exchange of the alkoxy group in aluminium alkoxides. Recently, van Bekkum and co-workers have reported the zeolite beta-catalysed MPV reduction of cyclohexanones [13]. A major advantage of this method is that it is truly catalytic and the catalyst can be readily separated, by simple filtration, and recycled. An additional benefit is that reduction of 4-tert-butylcyclohexanone affords the thermodynamically less stable cis-alcohol, an important fragrance chemical intermediate, in high (> 95%) selectivity. In contrast, classical MPV reduction of this ketone affords mainly the trans- isomer. Preferential formation of the cis-isomer was attributed to transition state selectivity imposed by confinement of the substrate in the zeolite pores (Figure 2).

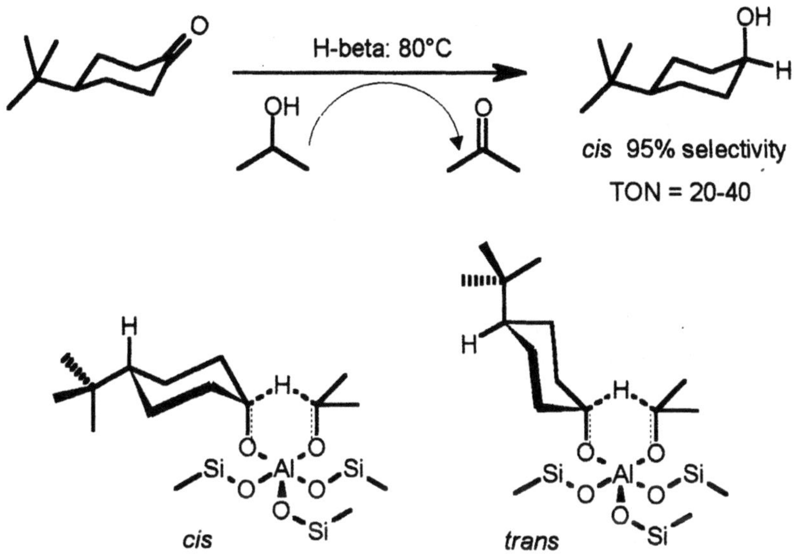

Figure 2. Zeolite-catalysed MPV reduction

Catalytic hydrogenation can sometimes provide benefits in unexpected places. For example, the Williamson ether synthesis (Reaction 4) has been widely used by organic chemists for more than a century. Although it is a well-tried and convenient methodology it, unfortunately, produces stoichiometric amounts of sodium chloride. Recently, a catalytic alternative, involving reductive O-alkylation (Reaction 5) has been described [14]. Similarly, reductive N-alkylation of amides (Reaction 6) is a low-salt alternative to classical alkylation procedures.

$$R^1CH_2Cl + R^2ONa \longrightarrow R^1CH_2OR^2 + NaCl \quad (4)$$

$$R^1CHO + R^2OH + H_2 \xrightarrow{Pd/C} R^1CH_2OR^2 + H_2O \quad (5)$$

$$R^1CONHR^2 + R^3CHO + H_2 \xrightarrow{Pd/C} R^1CONR^2R^3 + H_2O \quad (6)$$

5. Catalytic Oxidation

Catalytic oxidation with O_2 is widely used in the manufacture of commodity chemicals [15]. In some cases such technologies can be directly applied to the synthesis of fine chemicals. An elegant example of an atom efficient, low-salt process, in which the key step is a catalytic oxidation in the vapour phase, is the BASF process (Figure 3) for the manufacture of citral [16], a key intermediate in the synthesis of vitamin A and fragrance chemicals. The supported silver catalyst is essentially the same as that used in the manufacture of formaldehyde from methanol. The original route to citral started from beta-pinene and involved, inter alia, a stoichiometric oxidation with manganese dioxide. The same vapour phase oxidation technology is also commercially applied in the oxidation of straight-chain alcohols to the corresponding aldehydes, e.g. octanal from 1-octanol.

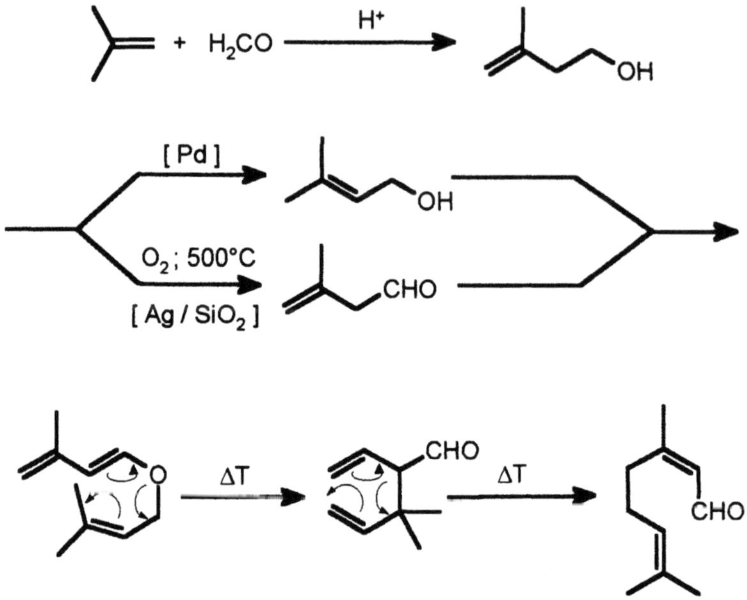

Citral

Figure 3. BASF process for citral

Another example of an atom inefficient process that is widely used in organic synthesis is the Beckmann rearrangement of oximes. Substantial amounts of salts are formed both in the formation of the oxime, by reaction of a ketone with a hydroxylamine salt, and in the rearrangement which employs stoichiometric quantities of, e.g. sulphuric acid. In the manufacture of caprolactam from cyclohexanone, for example, about 4.5 kg of ammonium sulphate are produced per kg of caprolactam. In an alternative process, developed by Enichem [17] cyclohexanone oxime is produced, in a salt-free process, by ammoximation of cyclohexanone with a mixture of ammonia and hydrogen peroxide. The process employs a solid catalyst, titanium silicalite, the prototype of a new generation of solid catalysts for liquid phase oxidations, so-called redox molecular sieves [18]. The ammoximation proceeds via in situ formation of hydroxylamine by titanium-catalysed oxidation of ammonia with hydrogen peroxide. Subsequent reaction of hydroxylamine with the ketone occurs in the bulk solution and, hence, the methodology is applicable to other ketones (and aldehydes), e.g. p-hydroxyacetophenone was converted to the corresponding oxime [19]. Beckmann rearrangement of the latter affords the analgesic, paracetamol (Figure 4).

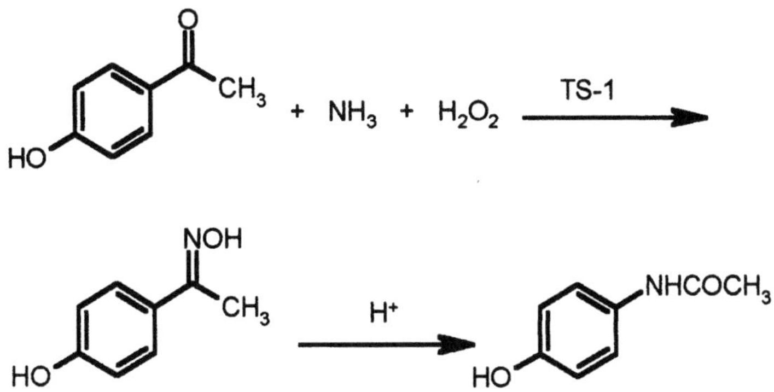

Paracetamol

Figure 4. TS-1 catalysed ammoximation

Substitution of classical oximation by the ammoximation procedure does not entirely solve the problem of salt generation as more than half of the ammonium sulphate production derives from the Beckmann rearrangement. However, this problem also has a solution: Sumitomo workers have reported [20] the vapour phase rearrangement of cyclohexanone oxime to caprolactam, in 95% selectivity at 100% conversion, over a high-silica ZSM-5 (Si/Al > 1000) zeolite as a solid catalyst.

6. Catalytic C-C Bond Formation

An important catalytic methodology for generating C-C bonds is carbonylation. A prominent example from the bulk chemicals arena is provided by the carbonylation of methanol (Reaction 7) catalysed by rhodium or, more recently, iridium complexes [21]. This process accounts for more than half of the five million tons of acetic acid produced annually on a world-wide basis.

$$CH_3OH + CO \xrightarrow[I^-]{Rh^I \text{ or } Ir^I} CH_3CO_2H \qquad (7)$$

The process is 100% atom efficient, highly selective (> 99%) and proceeds in one step from inexpensive, basic feedstocks. Extrapolation of this methodology to more complex substrates can afford atom efficient routes to fine chemicals and specialities. An elegant example is the Hoechst-Celanese process (Figure 5) for the manufacture of ibuprofen [22], an over-the-counter analgesic with an annual production in excess of 8000 tons. In this process ibuprofen is produced in two catalytic steps (hydrogenation and carbonylation), with 100% atom efficiency, from p-isobutylacetophenone. The latter is produced by Friedel-Crafts acylation of isobutyl-benzene in liquid hydrogen fluoride. Although the hydrogen fluoride can be completely recycled, thus circumventing the waste generation associated with traditional aluminium chloride-mediated acylations, there is still a definite need for truly catalytic procedures, e.g.

using solid acids, that are both effective and broadly applicable (see later).

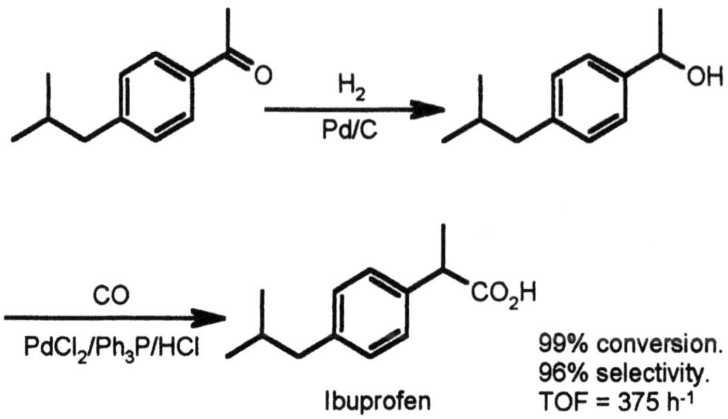

Figure 5. Hoechst-Celanese process for ibuprofen

In the Hoffmann La Roche process [23] for the anti-Parkinson drug, lazabemide, palladium-catalysed amidocarbonylation of 2,5-dichloropyridine replaced an original synthesis which involved 8 steps starting from 2-methyl-5-ethylpyridine and an overall yield of 8%. The new process affords lazabemide hydrochloride in 65% yield in one step with 100% atom efficiency (see Figure 6).

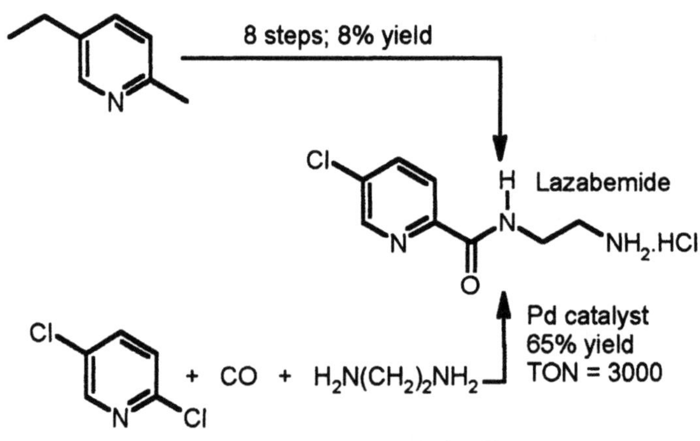

Figure 6. Two routes to Lazabemide

Palladium-catalysed amidocarbonylation has also been used [24] for the one-step, atom efficient synthesis of amino acid derivatives from olefins, CO and an amide (Reaction 8).

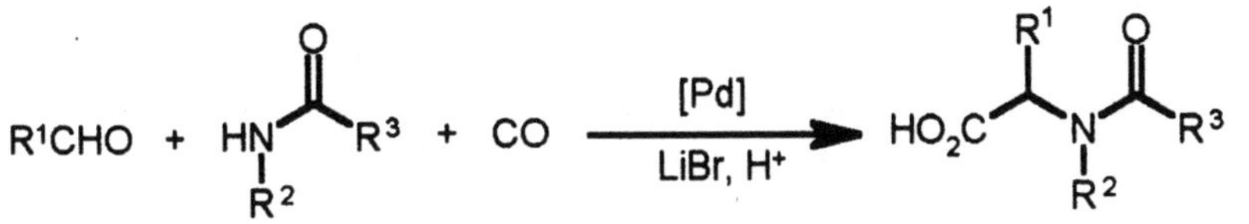

(8)

[Pd] = $PdBr_2$ or $(Ph_3P)_2PdBr_2$

Another catalytic methodology for C-C bond formation that has recently attracted attention is the Heck reaction [25,26]. The Heck reaction involves the palladium-catalysed arylation of olefinic double bonds (Reaction 9) and provides an alternative to Friedel-Crafts procedures for attaching carbon fragments to aromatic rings. The reaction is broadly applicable but has its limitations: the arylating agent is largely limited to relatively expensive aryl bromides and iodides and the process generates one equivalent of halide ion. Of interest in this context, therefore, is the recent report of a halide-free Heck reaction by DSM scientists [27]. In this procedure (Reaction 10) an aromatic carboxylic anhydride is used as the arylating agent. No addition of base or phosphine ligands is needed and fast reactions (< 3 hrs for completion) were observed at substrate/catalyst ratios of 1000. The carboxylic acid co-product can be readily recycled to its anhydride.

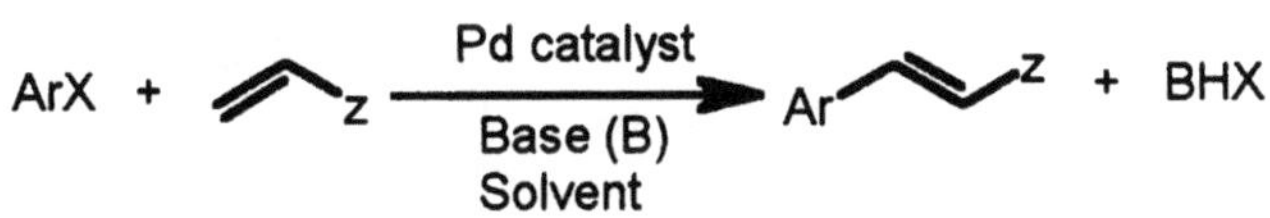

(9)

X = Br, I, OTf, COCl, SO_2Cl.
Z = H, alkyl, Ar, CN, CO_2R, etc.

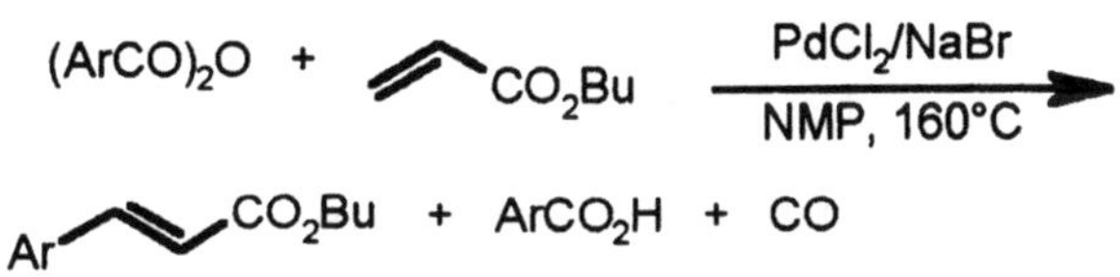

(10)

80-85% yield

The reaction can be considered as an alternative for Friedel-Crafts methodologies as the Heck products can be hydrogenated to alkyl aromatics or oxidised to aromatic ketones.

7. Catalysis with Solid Acids and Bases

Many standard reactions in organic synthesis employ strong mineral or Lewis acids, such as sulphuric acid and aluminium chloride, often in stoichiometric quantities. This generates waste streams containing large amounts of spent acid, which cannot easily be recovered and recycled. Replacement of soluble mineral and Lewis acids by recyclable solid acids, such as zeolites and clays, would constitute a substantial improvement, especially if they function in catalytic quantities. Hence, there is currently much interest in the application of solid acids in the synthesis of fine chemicals [28].

A case in point is Friedel-Crafts acylation, which is widely used in fine chemicals manufacture. In contrast to Friedel-Crafts alkylations, which proceed readily in the presence of catalytic amounts of mineral or Lewis acids, acylations require more than one equivalent of, for example, $AlCl_3$ or BF_3. This is owing to strong complexation of the Lewis acid by the ketone product. Although zeolites have already widely replaced mineral and Lewis acids in Friedel-Crafts alkylations, the more difficult acylations have proven recalcitrant. Nevertheless, Rhône Poulenc scientists recently reported [11] the first example of the commercial application of a zeolite-catalysed aromatic acylation: the acetylation of anisole over zeolite beta in the liquid phase (Figure 7).

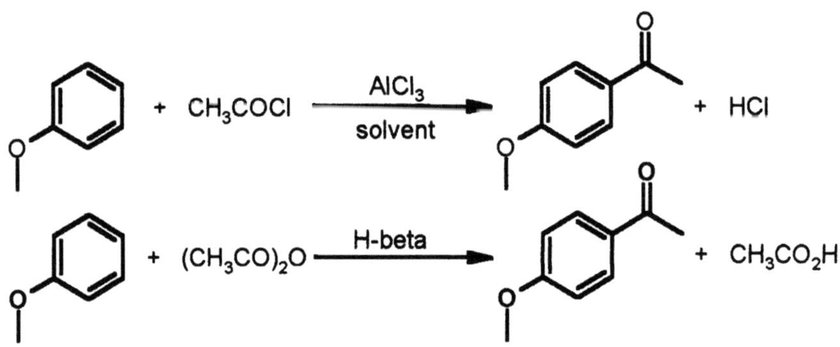

Homogeneous	Heterogeneous
$AlCl_3$ > 1 equivalent	H-beta, catalytic & regenerable
Solvent	No solvent
Hydrolysis of products	No water necessary
Phase separation	-
Distillation, organic phase	Distillation, organic phase
Solvent recycle	-
85-95% yield	> 95% yield / higher purity
4.5kg aqueous effluent per kg	0.035kg aqueous effluent per kg

Figure 7. Classical vs. Zeolite-catalysed Friedel-Crafts acylation

The zeolite-catalysed acylation employs acetic anhydride as the acylating agent, in contrast with the classical process which uses acetyl chloride in combination with $AlCl_3$ (use of the anhydride would require > 2 equivs. of $AlCl_3$). This circumvents the generation of HCl in both the acylation and the synthesis of the acetyl chloride. The original process generated 4.5 kg of aqueous effluent, containing $AlCl_3$, HCl, chlorinated hydrocarbon solvent residues and acetic acid, per kg of p-methoxyacetophenone. The catalytic alternative generates 0.035 kg aqueous effluent, i.e. > 100 times less, consisting of 99% H_2O, 0.8% CH_3CO_2H and < 0.2% other organics, per kg of product and requires no solvent. Moreover, a higher product yield > 95% vs. 85-95%) is achieved, the catalyst is recyclable, and the number of unit operations is reduced from twelve to two.

The above process involves the acylation of a relatively reactive aromatic substrate. For broad application in fine chemicals synthesis such a catalytic methodology should also be applicable to less reactive substrates. Previous attempts to affect the solid acid-catalysed acylation of less reactive aromatics, such as benzene, met with little success, which we attributed to an adsorption imbalance, i.e. the more polar acylating agent is selectively adsorbed and the aromatic substrate does not reach the active site [28]. This problem is reduced when the reaction is performed in the vapour phase and the successful acetylation of benzene with acetic acid over H-ZSM-5 (Si/Al = 41), at 250 °C and 1 bar in a tubular reactor, has recently been reported [29].

Zeolite beta also catalyses the synthesis of coumarin derivatives via the Pechmann condensation, e.g. condensation of resorcinol with ethyl acetoacetate over H-beta in refluxing toluene affords methylumbelliferone (Figure 8), a perfumery ingredient, in 70-80% yield [30].

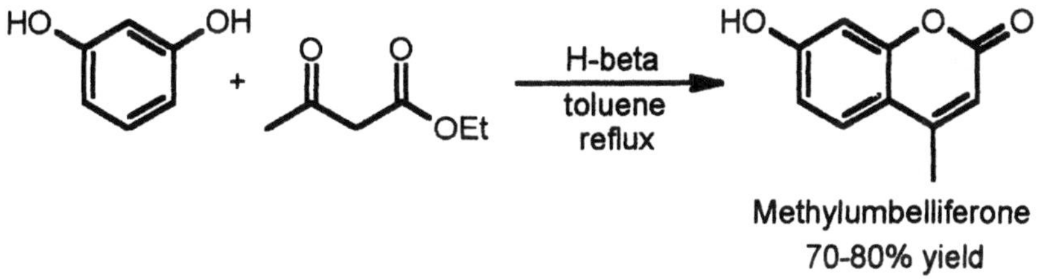

Figure 8. Zeolite-catalysed Pechmann condensation

The examples discussed above represent the proverbial tip of the iceberg; zeolites are potentially interesting recyclable catalysts for a wide variety of acid-mediated reactions in organic synthesis, e.g. epoxide rearrangements [31, 32], Claisen rearrangement [33], aromatic nitrations and hydroxymethylations [11], to name but a few. Hence, we expect that catalysis by zeolites or related solid acids will be widely applied in the synthesis of fine chemicals and specialities in the future.

Similarly, one can expect the application of recyclable solid bases, such as anionic clays, as substitutes for organic and inorganic bases in a variety of reactions, e.g. aldol condensations, Michael additions, and related reactions. An interesting development in this context is the use of guanidines anchored to silica, as recyclable solid catalysts in Michael additions [34].

8. Catalytic Conversions In Non-Conventional Media

A characteristic feature of homogeneous (transition metal) catalysts is their superior activity and/or selectivity compared to their heterogeneous counterparts. A serious shortcoming of homogeneous catalysts, however, is the often cumbersome separation of the expensive catalyst from the reaction products and its subsequent recycling. A novel approach to solving this problem is to perform the reaction in a liquid/liquid biphasic system, whereby the catalyst is dissolved in one phase (e.g. water) and the product in the other (organic) phase [35-37]. This concept also touches on another environmental issue: solvents. The use of chlorinated hydrocarbon solvents has been severely curtailed. In fact, so many of the solvents favoured by organic chemists are now on the black list, that the whole question of solvents in organic synthesis requires rethinking. The best solvent is no solvent and if a solvent (diluent) is needed it should preferably be water [38]. Moreover, there is no fundamental law that says that reactants have to be completely dissolved in order to react. There is a definite need, therefore for more chemistry in aqua and the use of water-soluble catalysts that function in biphasic systems could be the answer in some cases. Water is non-toxic, non-inflammable, inexpensive and abundantly available. Furthermore, owing to its highly polar character one can expect novel reactivities and selectivities for organometallic catalysis in water.

A large scale application of biphasic organometallic catalysis is the Ruhrchemie/Rhône Poulenc process for the hydroformylation of propylene to n-butanal (Reaction 11) which employs a water-soluble rhodium(I) complex of trisulphonated triphenylphosphine (tppts) as the catalyst [39].

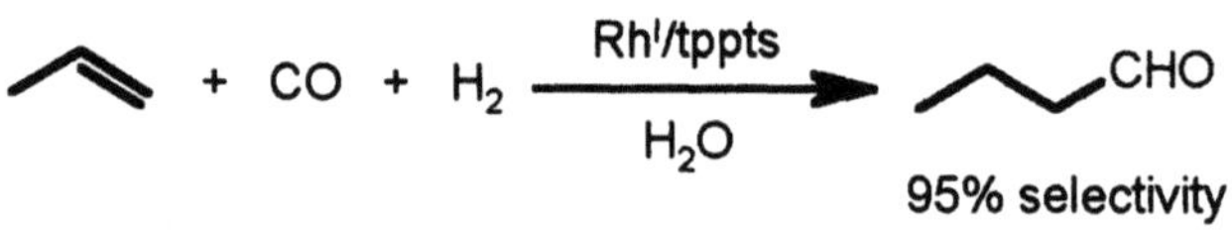

(11)

The same complex is used as the catalyst in the Rhône Poulenc process [40] for the manufacture of geranyl acetone via reaction of myrcene with methylacetoacetate (Figure 9) in a biphasic system.

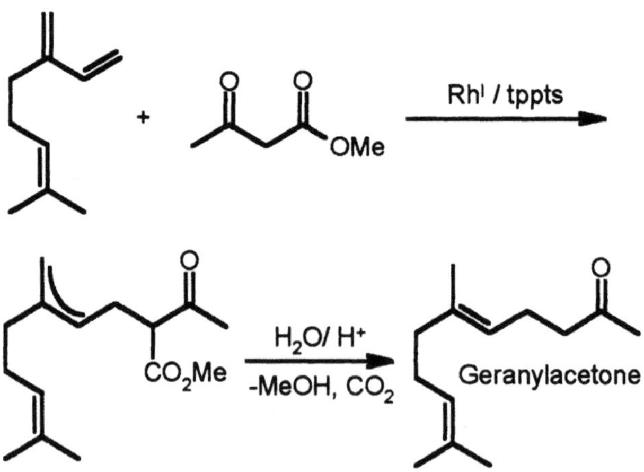

Figure 9. Rhône Poulenc process for geranylacetone

We found that the water soluble palladium(0) complex, $Pd(tppts)_3$, catalyses the selective biphasic carbonylation of benzylic alcohols in an acidic aqueous medium [41]. We subsequently applied [42] this methodology to the synthesis of ibuprofen (see earlier) via biphasic carbonylation of 1-(4-isobutylphenyl)ethanol (Reaction 12).

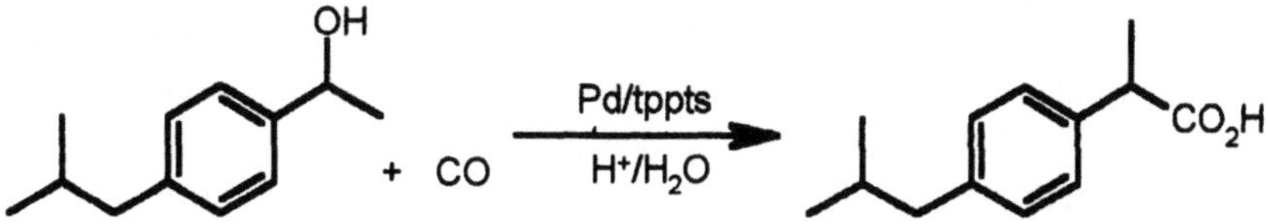

(12)

83% conversion
82% selectivity

Subsequently, we [43] and others [44] have reported the Pd(tppts)$_3$-catalysed biphasic hydrocarboxylation of olefins (Reaction 13). With propylene exceptionally high turnover frequencies (TOF > 2500 h^{-1}) were observed at 110-120 °C and 50 bar [43].

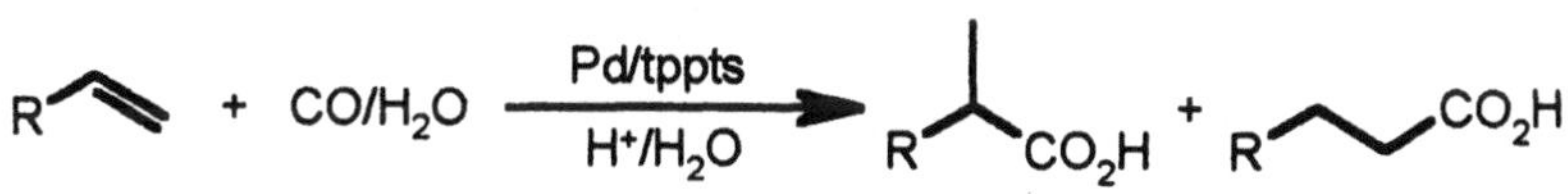

(13)

R	Selectivity (%)	
	iso-	*n-*
CH_3	43	57
C_6H_5	56	33
4-*i*-BuC_6H_4	82	18

Other non-conventional reaction media have also attracted attention recently, e.g. fluorous biphasic systems developed by Horvath and co-workers [45]. In this concept the catalyst is contained in a perfluorohydrocarbon phase, by employing perfluorinated ligands, while substrate and product reside in a second organic phase. In many cases raising the temperature produces a single phase which separates into two phases again on cooling.

The use of supercritical carbon dioxide as a solvent for catalytic conversions, e.g. hydrogenation [46] is also gaining ground and the use of ionic liquids as reaction solvents [47, 48] is expected to contribute significantly to the development of clean technologies.

9. Biocatalysis

Biocatalytic methodologies have several advantages in the context of fine chemicals and specialities manufacture. They are generally performed in water and often involve fewer steps than conventional procedures by virtue of the fact that functional group protection and deprotection steps are not necessary. For example, enzymatic hydrolysis of Penicillin G to 6-APA (Figure 10) proceeds in one step in water at room temperature while chemical deacylation requires 3 steps, and a temperature of -40 °C, and various reagents resulting in a high E factor. However, the main stimulus for switching from chemical to enzymatic deacylation was to avoid the use of dichloromethane as solvent [49].

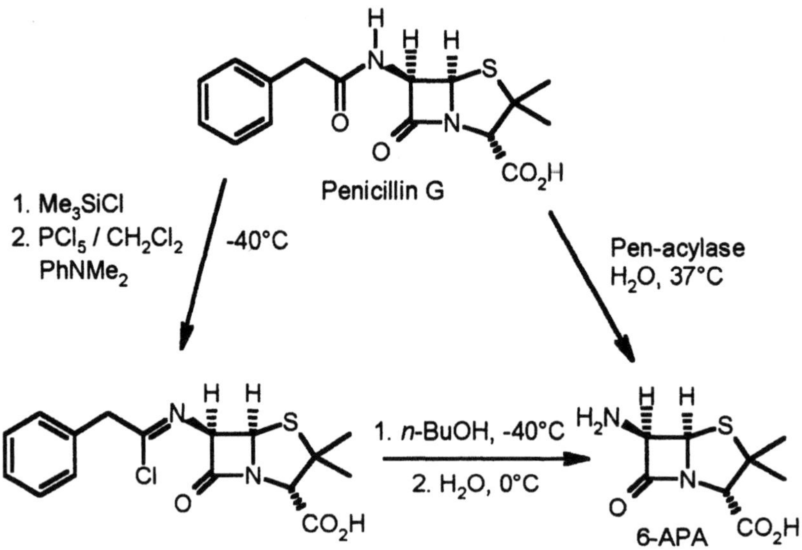

Process	Chemical	Enzymatic
Reagents	Me_3SiCl, PCl_5 $PhNMe_2$, *n*-BuOH	Pen-acylase (catalyst)
Solvent	CH_2Cl_2	H_2O
Temperature	-40°C	37°C
Atom efficiency	Low	High

Figure 10. Chemical vs. Enzymatic deacylation of Penicillin G to 6-APA

Another advantage of biocatalytic conversion is that chemo-, regio- and stereoselectivities can be attained which are difficult or impossible to achieve by chemical means. For example, Lonza [50] has developed commercial processes for the microbial ring-hydroxylation and side-chain oxidation of heteroaromatics. Some examples are shown in Figure 11. Such conversions would not be possible by conventional chemical means.

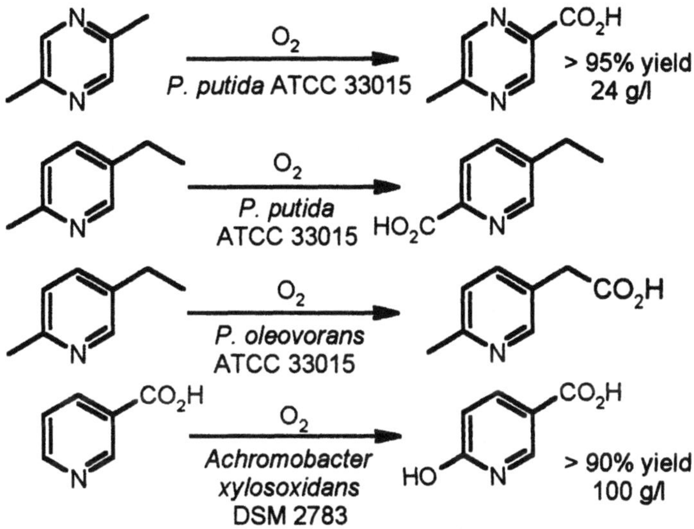

Figure 11. Regioselective microbial oxidation of heteroaromatics

Biocatalytic methods are particularly useful for the synthesis of enantiomerically pure compounds, which is often needed in the manufacture of pharmaceuticals, agrochemicals and flavours and fragrances [51]. In the case of flavour and fragrance chemicals there is an additional benefit. The use of natural raw materials and biocatalytic methods qualifies the product for labelling as natural which commands a substantially higher price than the corresponding synthetic material prepared by chemical means.

10. Asymmetric Catalysis

Another major trend in the chemical and allied industries is the development of products that are more targeted in their action with less undesirable side-effects. In the case of chiral molecules that exhibit biological activity the desired bioactivity almost always resides in only one of the enantiomers. The other enantiomer constitutes isomeric ballast that does not contribute to the desired activity and may even exhibit undesirable effects. Hence, there is a marked trend, in pharmaceuticals, agrochemicals and flavours and fragrances, towards the synthesis of pure enantiomers. In the case of flavours and, to a lesser extent, fragrances the issue is complicated by the consumer-driven trend towards products that are "natural".

The same reasoning applies to the synthesis of pure enantiomers as to organic syntheses in general: the processes should be atom efficient, i.e. involve catalytic methodologies. Hence, in recent years attention has been increasingly focused on asymmetric catalysis, using either enzymes or chiral metal complexes.

An elegant example of a highly efficient catalytic asymmetric synthesis is the Takasago process [52, 53] for the manufacture of l-menthol, an important product in the flavours and fragrances industry. The key step is a Rh-Binap catalysed isomerisation of a prochiral enamine to a chiral imine (Figure 12). The latter is produced in 99% ee at a substrate/catalyst ratio of 8000 and recycling of the catalyst leads to turnover numbers up to 400,000. The Takasago process accounts for about half of the world production (ca. 4,500 tons) of l-menthol.

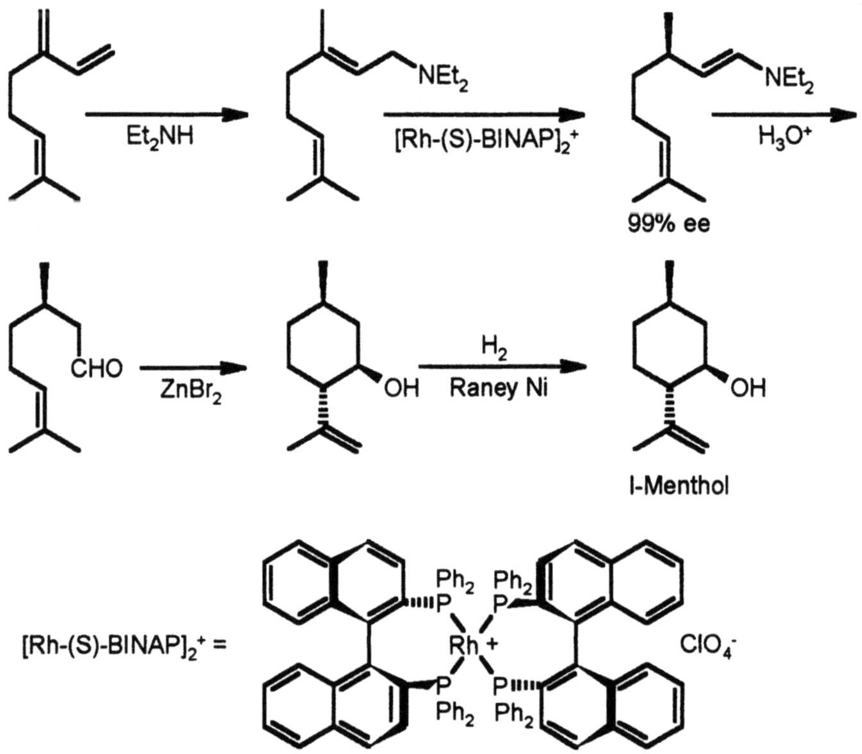

Figure 12. Takasago l-menthol process

An even more impressive example of catalytic efficiency, this time in an asymmetric hydrogenation, is provided by the Novartis process for the synthesis of the optically active herbicide, (S)-metolachlor [54]. The key step (Figure 13) involves the asymmetric hydrogenation of a prochiral imine catalysed by a ferrocenyldiphosphine complex of iridium(I). The substrate/catalyst ratio in this step is 75,000 and 1 million turnovers are achieved in 6 hrs. The product has an ee of 80%. Higher ee's can be obtained, at the expense of activity, but are actually not necessary for this product.

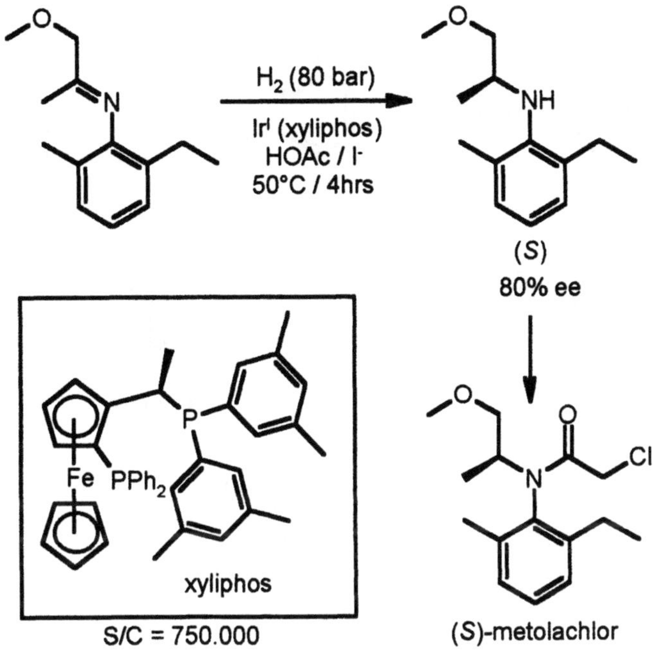

Figure 13. Novartis process for (*S*)-metolachlor

11. Concluding Remarks

The widespread application of atom efficient, catalytic processes in the manufacture of fine chemicals and specialities, such as flavours and fragrances, pharmaceuticals, etc., provides the key to environmentally friendly processes. The first hurdle to be taken is promoting an awareness among fine chemical manufacturers that this is so. It is gratifying, therefore, to note that a recent patent application, assigned to Quest International [55], compares the atom efficiencies of different routes for preparing dihydromyrcenol, a well-known fragrance material.

The enormous potential of catalysis is well illustrated by the synthesis of the flavour chemical, vanillin, recently reported by Rhône Poulenc scientists [11]. The process involves four catalytic steps, all involving a heterogeneous catalyst, from phenol (Figure 14).

Hence, we conclude that a growing awareness of the importance of E factors of chemical processes is stimulating the development of atom efficient catalytic methodologies for fine chemicals manufacture. Seen in this light, elegance in organic synthesis takes on a whole new meaning.

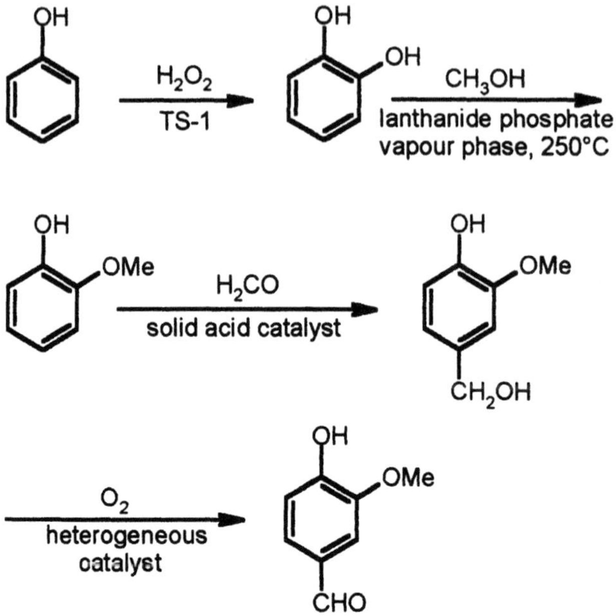

Figure 14. Rhône Poulenc process for vanillin

12. References

1. Sheldon, R.A. (1997) Catalysis: The key to waste minimization. *J. Chem. Techol. Biotechnol.*, **68**, 381-388.
2. Sheldon, R.A. (1997) Catalysis and pollution prevention. *Chem. Ind. (London)*, 12-15.
3. Sheldon, R.A. (1996) Selective catalytic synthesis of fine chemicals: opportunities and trends. *J. Mol. Catal. A: Chemical*, **107**, 75-83.
4. Sheldon, R.A. (1994) Consider the environmental quotient. *CHEMTECH*, March, 38-47.
5. Anastas, P.T., Farris, C.A., Eds. (1994) *Benign by Design*, ACS Symp. Ser. 577, Am. Chem. Soc., Washington, DC.
6. Clark, J.H., Ed. (1995) *Chemistry of Waste Minimization*, Blackie, Glasgow.
7. Sheldon, R.A. (1992) Catalysis, the atom utilization concept and waste minimization. In *Industrial Environmental Chemistry*, Sawyer, D.T. and Martell, A.E., Eds., Plenum, New York, 99-119.
8. Sheldon, R.A. (1993) The role of catalysis in waste minimization. In *Precision Process Technology*, Weijnen, M.P.C. and Drinkenburg, A.A.H., Eds., Kluwer, Amsterdam, 125-138.
9. Trost, B.M. (1995) Atom economy - a challenge for organic synthesis: homogeneous catalysis leads the way. *Angew. Chem. Int. Ed. Engl.*, **34**, 259-281.
10. Sheldon, R.A. (1998) Organic reactions and catalysis: a marriage of convenience. *Organic Reactions and Catalysis*, in press.
11. Ratton, S. (1998) Heterogeneous catalysis in the fine chemicals industry: from dream to reality. *Chem. Today*, March/April, 33-37.
12. Yokoyama, T., Setoyama, T., Fujita, N., Nakajima, M., Maki, T. (1992) Novel direct hydrogenation of aromatic carboxylic acids to the corresponding aldehydes with zirconia catalyst. *Appl. Catal. A: General*, **88**, 149-161.
13. Creyghton, E.J., Ganeshie, S.D., Downing, R.S., van Bekkum, H. (1997) Stereoselective Meerwein-Ponndorf-Verley and Oppenauer reactions catalysed by zeolite BEA. *J. Mol. Catal. A: Chemical*, **115**, 457-472.
14. Fache, F., Bethmont, V., Jaquot, L., Lemaire, M. (1996) Reductive O- and N-alkylation. Alternative

catalytic methods to nucleophilic substitution. *Recl. Trav. Chim. Pays-Bas*, **115**, 231-238.
15. Sheldon, R.A., Kochi, J.K. (1981) *Metal Catalyzed Oxidations of Organic Compounds*, Academic Press, New York.
16. Hölderich, W.F. (1993) New reactions in various fields and production of specialty chemicals. In *New Frontiers in Catalysis*, Guczi, L., Solymosi, F., Tetenyi, P., Eds., Elsevier, Amsterdam 127-163.
17. Roffia, P., Leofanti, G., Cesana, A., Mantegazza, M., Padovan, M., Petrini, G., Tontí, S., Gervasutti, P. (1990) Cyclohexanone ammoximation: a breakthrough in the 6-caprolactam production. *Stud. Surf. Sci. Catal.*, **55**, 43-52.
18. Arends, I.W.C.E., Sheldon, R.A., Wallau, M., Schuchardt, U. (1997) Oxidative transformations of organic compounds mediated by redox molecular sieves. *Angew. Chem. Int. Ed. Engl.*, **36**, 1190-1211.
19. le Bars, J., Dakka, J., Sheldon, R.A. (1996) Ammoximation of cyclohexanone and hydroxyaromatic ketones over titanium molecular sieves. *Appl. Catal. A: General*, **136**, 69-80.
20. Sato, H., Hirose, K., Kitamura, M., Nakamura, Y. (1989) A vapor phase Beckmann rearrangement over high-silicious ZSM-5. *Stud. Surf. Sci. Catal.*, **49**, 1213-1222.
21. Maitlis, P.M., Haynes, A., Sunley, G.J., Howard, M.J. (1996) Methanol carbonylation revisited: thirty years on. *J. Chem. Soc. Dalton Trans.*, 2187-2196.
22. Elango, V., Murphy, M.A., Smith, B.L., Davenport, K.G., Mott, G.N., Moss, G.L. (1991) Process for producing ibuprofen. US Pat. 4981995 to Hoechst Celanese Corp.
23. Roessler, F. (1996) Catalysis in the industrial production of pharmaceuticals and fine chemicals. *Chimia*, **50**, 106-109.
24. Beller, M., Eckert, M., Vollmüller, F., Bogdanovic, S., Geissler, H. (1997) Palladium-catalyzed amidocarbonylation - A new, efficient synthesis of N-acyl amino acids. *Angew. Chem. Int. Ed. Engl.*, **36**, 1494-1496.
25. Heck, R.F. (1985) *Palladium Reagents in Organic Syntheses*, Academic Press, New York.
26. Tsuji, J. (1995) *Palladium Reagents and Catalysts: Innovations in Organic Synthesis*, Wiley, New York.
27. Stephan, M.S., Teunissen, A.J.J.M., Verzijl, G.K.M., de Vries, J.G. (1998) Heck reactions without salt formation: aromatic carboxylic anhydrides as arylating agents. *Angew. Chem. Int. Ed. Engl.*, **37**, 662-664.
28. Downing, R.S., van Bekkum, H., Sheldon, R.A. (1997) Zeolites and related materials for the solid-acid catalyzed production of fine chemicals: a view from Delft. *Cattech*, **2**, 95-109.
29. Singh, A.P., Pandey, A.K. (1997) Acetylation of benzene to acetophenone over zeolite catalysts. *J. Mol. Catal. A: Chemical*, **123**, 141-147.
30. Gunnewegh, E.A., Hoefnagel, A.J., Downing, R.S., van Bekkum, H. (1996) Environmentally friendly synthesis of coumarin derivatives employing heterogeneous catalysis. *Recl. Trav. Chim. Pays-Bas*, **115**, 226-230.
31. Elings, J.A., Lempers, H.E.B., Sheldon, R.A. (1997) Zeolite-catalysed rearrangement of isophorone oxide. *Stud. Surf. Sci. Catal.*, **105**, 1165-1172.
32. Liebens, A.T., Mahaim, C., Hölderich, W.F. (1977) Selective isomerisation of α-pinene oxide with heterogeneous catalysts. *Stud. Surf. Sci. Catal.*, **108**, 587-594.
33. Elings, J.A., Downing, R.S., Sheldon, R.A. (1995) Zeolite-catalysed Claisen rearrangement of allyl aryl ethers. *Stud. Surf. Sci. Catal.*, **94**, 487-494.
34. Subba Rao, Y.V., De Vos, D.E., Jacobs, P.A. (1997) Silica-supported 1,5,7-triazabicyclodecene as a catalyst for Michael additions. In *Supported Reagents and Catalysts in Chemistry*, Hodnett, B.K., Kybett, A.P., Clark, J.H., Smith, K., Eds., The Royal Society of Chemistry, Cambridge, 110-115.
35. Papadogianakis, G., Sheldon, R.A. (1997) Catalytic Conversions in Water. An Environmentally Benign Concept for Heterogenisation of Homogeneous Catalysis. In *Catalysis*, Vol. 13, Specialist Periodical Report, Royal Society of Chemistry, Cambridge, 114-193.
36. Papadogianakis, G., Sheldon, R.A. (1996) Catalytic conversions in water: environmentally attractive processes employing water soluble transition metal complexes. *New. J. Chem.*, **20**, 175-185.
37. Cornils, B., Herrmann, W.A., Eds. (1998) *Aqueous-Phase Organometallic Catalysis. Concepts and Applications*, Wiley-VCH, Weinheim.
38. Grieco, P.A., Ed. (1998) *Organic Synthesis in Water*, Blackie, Glasgow.
39. Cornils, B., Wiebus, E. (1996) Virtually no environmental impact: the biphasic oxo process. *Recl. Trav.*

Chim. Pays-Bas, **115**, 211-215.

40. Mercier, C., Chabardes, P. (1994) Organometallic chemistry in industrial vitamin A and vitamin E synthesis. *Pure Appl. Chem.*, **66**, 1509-1518.
41. Papadogianakis, G., Maat, L., Sheldon, R.A. (1997) Catalytic conversions in water. Part 4. Carbonylation of 5-hydroxymethylfurfural (HMF) and benzyl alcohol catalysed by palladium trisulfonated triphenylphosphine complexes. *J. Mol. Catal. A: Chemical*, **116**, 179-190.
42. Papadogianakis, G., Maat, L., Sheldon, R.A. (1997) Catalytic conversions in water: Part 5. Carbonylation of 1-(4-isobutylphenyl)ethanol to ibuprofen catalysed by water-soluble palladium-phosphine complexes in a two-phase system, *J. Chem. Technol. Biotechnol.,*. **70**, 83-91.
43. Papadogianakis, G., Verspui, G., Maat, L., Sheldon, R.A. (1997) Catalytic conversions in water. Part 6. A novel biphasic hydrocarboxylation of olefins catalyzed by palladium TPPTS complexes (TPPTS = $P(C_6H_4$-m-$SO_3Na)_3$). *Catal. Lett.*, **47**, 43-46.
44. Tilloy, S., Monflier, E., Bertoux, F., Castagnet, Y., Mortreux, A. (1997) A new fruitful development in biphasic catalysis: the palladium-catalyzed hydrocarboxylation of alkenes. *New. J. Chem.*, **21**, 529-531.
45. Horvath, I.T., Rabai, J. (1994) Facile catalyst separation without water: fluorous biphase hydroformylation of olefins. *Science*, **266**, 72-75.
46. Hitzler, M.G., Smail, F.R., Ross, S.K., Poliakoff, M. (1998) Selective catalytic hydrogenation of organic compounds in supercritical fluids as a continuous process. *Org. Proc. Res. Dev.*, **2**, 137-146.
47. Seddon, K.R. (1997) Ionic liquids for clean technology. *J. Chem. Technol. Biotechnol.*, **68**, 347-356.
48. Chauvin, Y., Olivier-Bourbigou, H.O. (1995) Non aqueous ionic liquids as reaction solvent. *CHEMTECH*, September, 26-30.
49. Bruggink, A., Roos, E.C., de Vroom, E. (1998) Penicillin acylase in the industrial production of β-lactam antibiotics. *Org. Proc. Res. Dev.*, **2**, 128-133.
50. Kiener, A. (1995) Biosynthesis of functionalized aromatic N-heterocycles. *CHEMTECH*, September, 31-35.
51. Sheldon, R.A. (1993) *Chirotechnology: the Industrial Synthesis of Optically Active Compounds*, Marcel Dekker, New York.
52. Akutagawa, S. (1992) A Practical Synthesis of (-)-Menthol with the Rh-BINAP Catalyst. In *Chirality in Industry*, Collins, A.N., Sheldrake, G.N., Crosby, J., Eds., Wiley, New York, 313-323.
53. Kumoboyashi, H. (1996) Industrial application of asymmetric reactions catalyzed by BINAP-metal complexes. *Recl. Trav. Chim. Pays-Bas*, **115**, 201-210.
54. Blaser, H.-U., Spindler, F. (1998) Enantioselective catalysis for agrochemicals. The case histories of (S)-metolachlor, (R)-metalaxyl and clozylcan. *Topics in Catalysis*, **5**, 275-284.
55. Davey, P.N., Richardson, C.D., Newman, C.P., Hart, B.R. (1997) Transesterification process for the preparation of dihydromyrcenol and myrcenol. Eur. Pat. Appl. 0784043 A1 to Quest International.

ADVANCES IN THE INDUSTRIAL SYNTHESIS OF MEDIUM TO LARGE RING MOLECULES

Simon Ellwood, Tom Haines
Quest International
Ashford, Kent, TN24 0LT
United Kingdom

1. Introduction

The synthetic chemist's interest in materials with musk odours began when it was realised that crude natural musks, such as those isolated in limited amounts from the musk deer and civet cat, were prized as indispensable fragrance ingredients in the 1800's[1]. Over a period of time the active musks in these crude materials were identified as macrocyclic compounds such as methyl substituted cyclopentadecanones and cycloheptadecenones[2]. Similar musk materials of vegetable origin were also found for example, ambrettolide which can be isolated from ambrette seed oil[3]. Since the availability of these musks from their natural source is limited, they command a very high price and their use in perfumery is relatively very small.

In 1888 Baur found that 2,4,6-trinitro-3-methyl-tert-butylbenzene possess a strong musk odour and, as a result a number of musk materials based upon substituted nitrobenzenes were synthesised [4]. The discovery of these materials provided a relatively cheap supply of musk odorants for use in fragrances. There are five commercially important nitro musk materials, Musk Xylene, Musk Ambrette, Musk Tibetene, Musk Ketone and Moskene, however, their use has recently diminished in response to toxicity and environmental issues [5].

It was not until the year 1948 that a non-nitrobenzene based musk material was found when Ambral a benzaldehyde derivative was synthesised by Carpenter [6]. With this discovery effort began to focus on alternative families of synthetic musk materials. The first commercially important material to come from these efforts was Phantolide [7]. This and other indane based structures were quickly followed by tetralin [8] and tricyclic [9] based structures. The most commercially significant examples of the latter types of materials are Tonalid and Galaxolide respectively.

K.A.D. Swift (ed.), Current Topics in Flavours and Fragrances, 79–95.

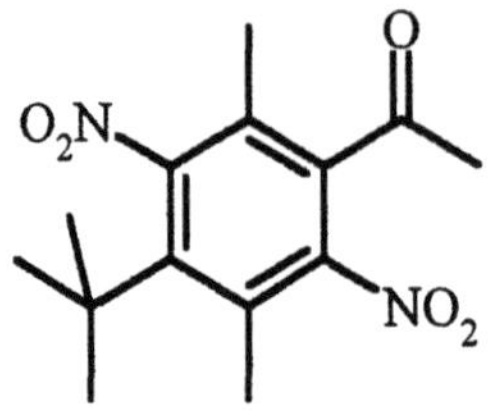

Musk Ketone

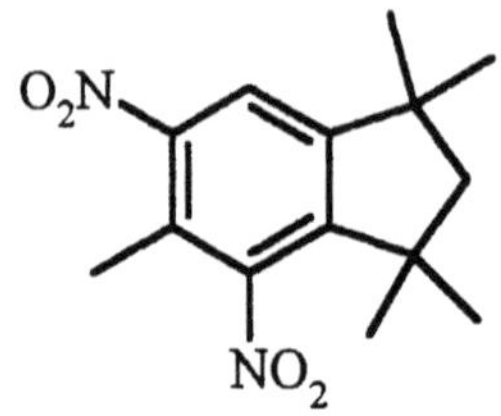

Moskene

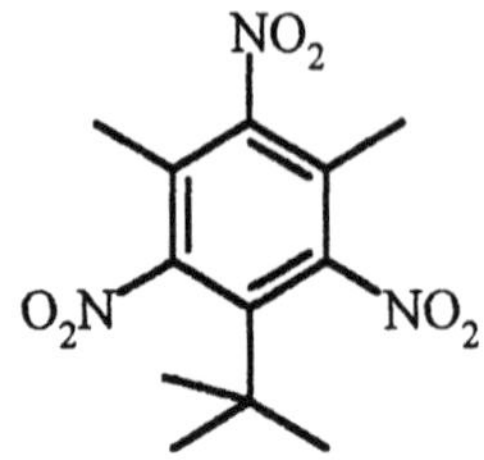

Musk Xylene

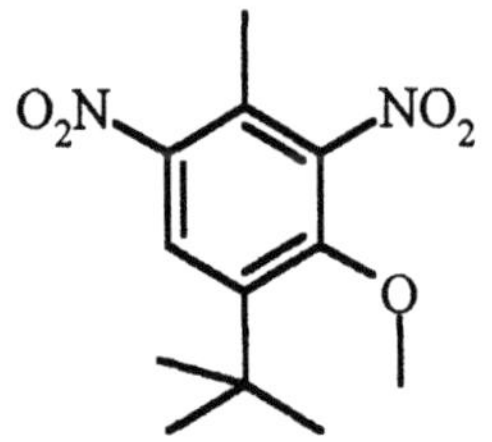

Musk Ambrette

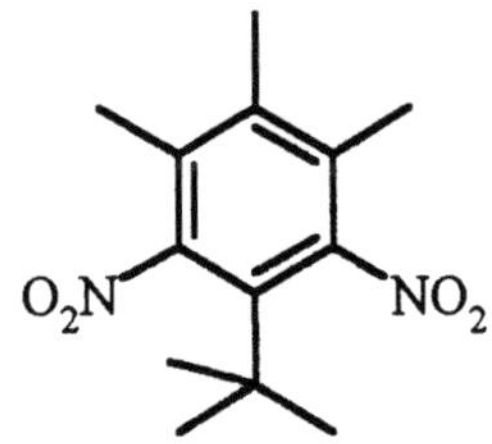

Musk Tibetene

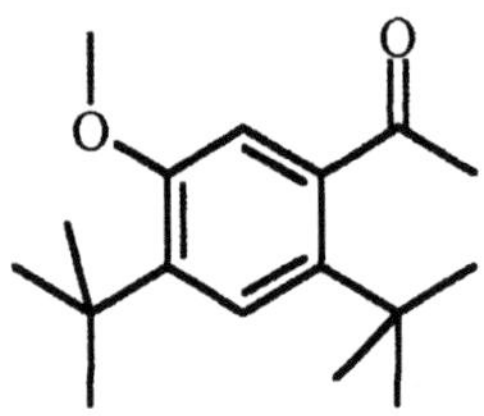

Ambral

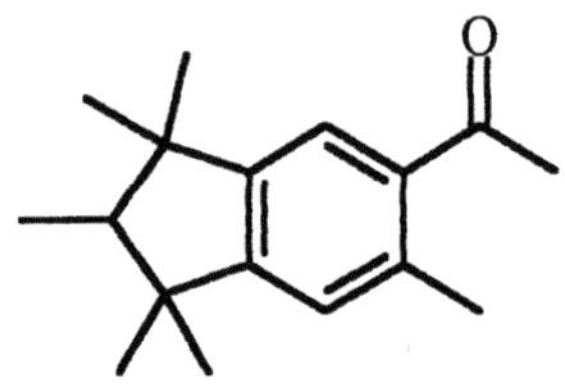

Phantolide

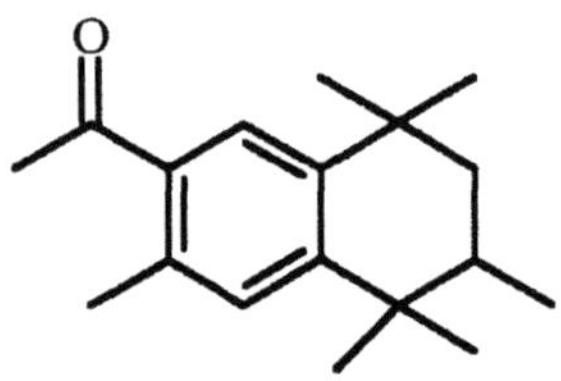

Tonalid

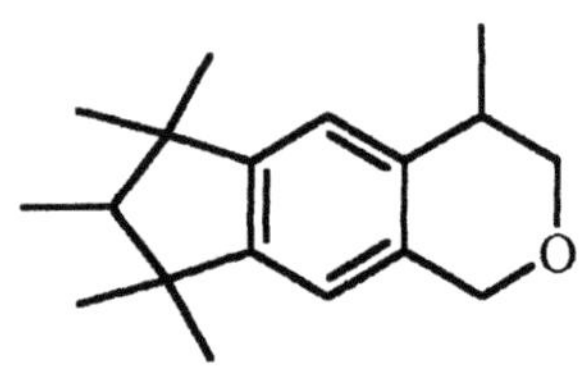

Galaxolide

With the wider/greater availability and lower cost provided by these synthetic musk materials, their use in areas other than fine fragrances grew dramatically e.g. in detergent and cosmetic formulations and they are thus an extremely important ingredient in the perfumer's palette.

In recent years, however, effort towards the synthesis of macrocyclic musks and understanding their structure odour relationships [10] has increased within the fragrance

industry - a result of their uniqueness in the musk odour area and their good biodegradability.[11,12]' There have also been reports that there is an increase in commercial interest in existing macrocyclic musks. Examples include the development of a new production facility for the generation of musk ambrettolide (~3000kg),[13] Soda Aromatic Co. doubling its macrocyclic musk capacity,[14] Japan Energy Corp. increasing its brassylic acid capacity by 320 tpa (ultimately used to make the macrocyclic musk ethylene brassylate)[15] and Synarome introducing cis-isoambrettolide[16].
This review intends to highlight some of the growing interest in macrocyclic musk synthesis, from the aspect of potentially commercial application. The following three key approaches to make macrocyclic compounds considered here are:

i. The synthesis of α,ω-bifunctional chains followed by macrocyclisation.
ii. Ring expansion reactions *via* either one, two, three or four atom expansions.
iii. Formation of macrocyclic rings directly *via* oligomerisation reactions.

2. Synthesis and Macrocyclisation of Bifunctional Chains

This approach requires a bifunctionalised long chain molecule and the key to its success is the availability/ease of synthesis of these starting materials, and how reactor efficient the thermodynamically disfavoured macrocyclisation can be made.

2.1 SYNTHESIS OF BIFUNCTIONAL CHAINS

Some of the most common bifunctional chains for macrocyclic musk synthesis are the α,ω-dicarboxylic acids. These materials are versatile starting materials for the synthesis of a number of macrocyclic musks (Scheme 1). 1,13-Tridecane dicarboxylic acid (brassylic acid) is a key raw material for producing one of the largest volume macrocyclic musks, ethylene brassylate (see scheme 1). A number of procedures are known for producing brassylic acid, one popular example of a commercial route being *via* ozonolysis of erucic acid (present in rape seed oil)[17].

Other interesting alternative routes to brassylic acid include:

- An electrochemical route based upon the crossed Kolbe reaction between monomethyl adipate and monomethyl azelate (obtained from castor oil). The procedure suffers from homo-coupling reactions but selectivities to dimethyl brassylate of 71% (based on monomethyl azelate) can be obtained by using a 10 mole excess of monomethyl adipate [18].

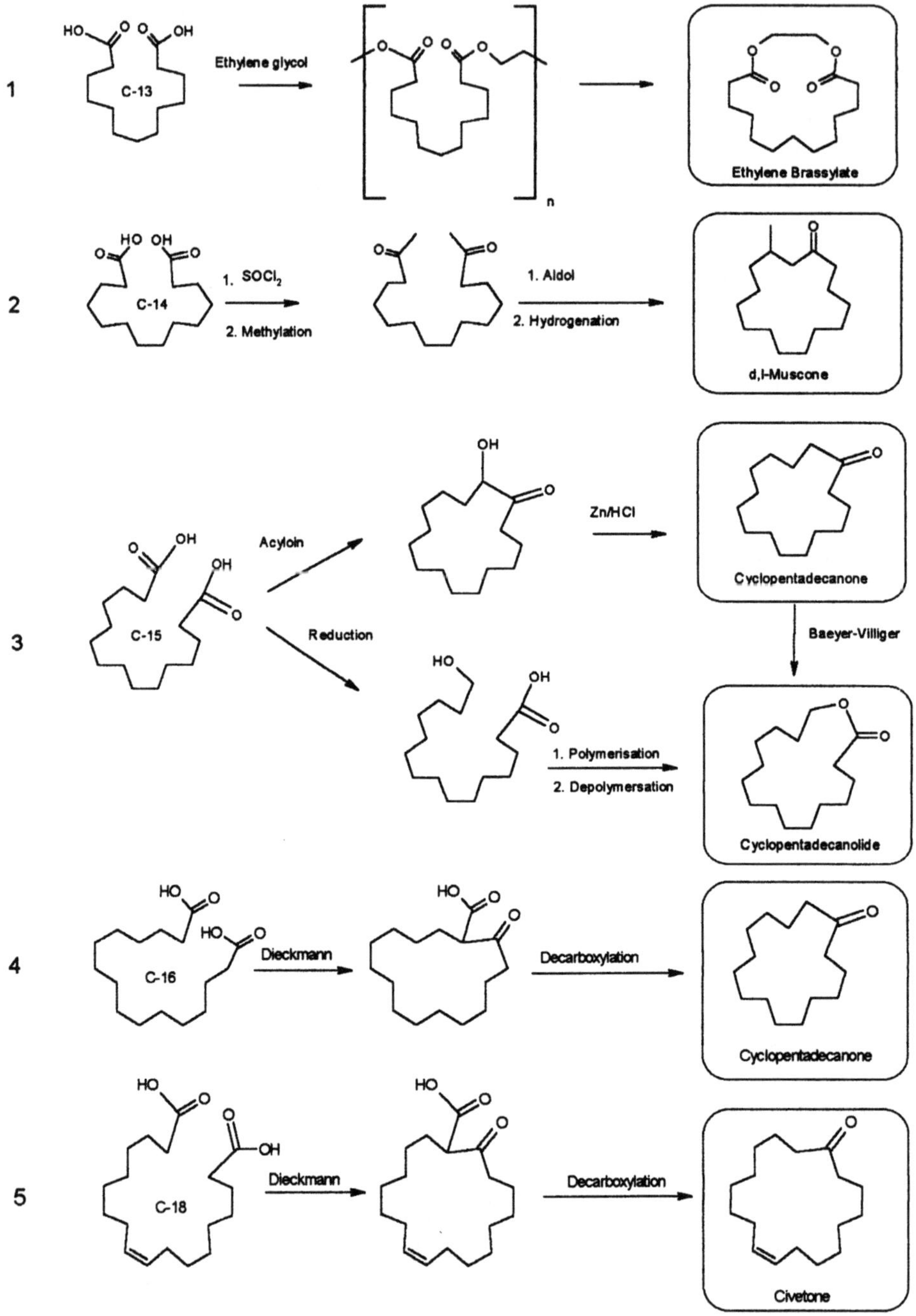
1
HO
O
O
OH
C-13
Ethylene glycol
n
Ethylene Brassylate
2
HO
OH
C-14
1. SOCl₂
2. Methylation
1. Aldol
2. Hydrogenation
d,l-Muscone
3
C-15
Acyloin
Reduction
Zn/HCl
Cyclopentadecanone
Baeyer-Villiger
1. Polymerisation
2. Depolymersation
Cyclopentadecanolide
4
C-16
Dieckmann
Decarboxylation
Cyclopentadecanone
5
C-18
Dieckmann
Decarboxylation
Civetone

Scheme 1

- A base catalysed condensation between formaldehyde and two molecules of cyclohexanone, followed by a double Baeyer-Villiger reaction and hydrogenation in the presence of a Raney nickel/Amberlyst mixed catalyst (Scheme 2)[19].

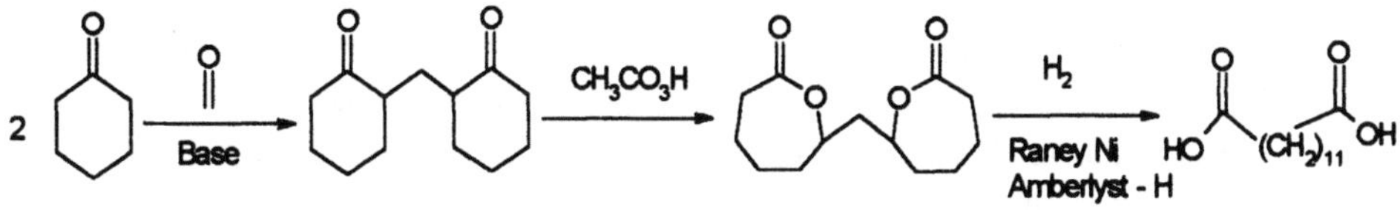

Scheme 2

- Very recently, Creanova Spezialchemie have developed a process to brassylic acid which provides a non-biochemical route using cheap petrochemical feedstocks [20].

Biochemical routes to 1,15-pentadecane dicarboxylic acid and 1,16-hexadecane dicarboxylic acid are also now well founded (both are precursors to cyclopentadecanone and cyclopentadecanolide). Both oleochemical [21] and petrochemical [22] feedstocks can be used to produce these materials. Several strains of yeast are known to excrete alpha, omega-dicarboxylic acids as a by-product when cultured on alkanes or fatty acids as the carbon source. In particular yeast belonging to the genus *Candida* such as *Candida cloacae*, *Candida lipolytica* and *Candida albicans* are known to produce dicarboxylic acids [23]. Attempts have also been made to increase the omega-hydroxlase activity of these yeast by amplifying one or more of the P450 genes in *Candida tropicalis* [24].

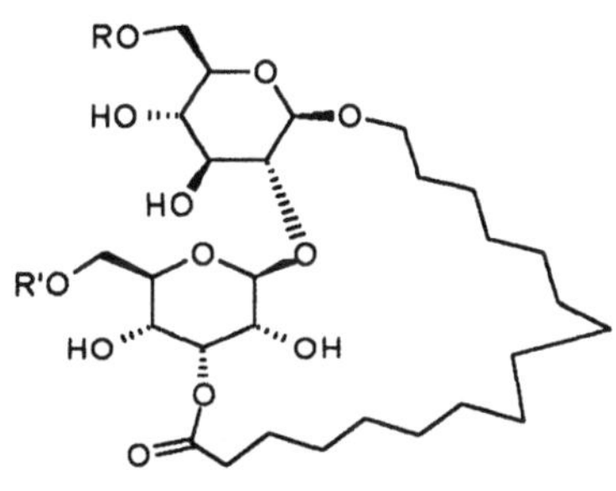

Figure 1

It has also been known for several decades that sophorose lipids (Figure 1) can be produced using the yeast *Candida bombicola*. The glycolipid separates from the culture medium as a thick syrup and the terminally hydroxylated fatty acid is obtained *via* hydrolysis.

The characteristics of the fatty acid moiety are dependent upon the hydrophobic feedstock. It has been found that C-16 through to C-18 fatty acids can be used as feedstocks, whilst longer chain length material *e.g.* C-20 and 22 tend to be incorporated only after being reduced in chain length during the fermentation to give sophorolipids containing fatty acids with chain lengths of 16 or 18. Shorter chain substrates are poorly incorporated [25]. Thus, by using palmitic acid (or its ester) as a feedstock it is possible to produce ω-hydroxy palmitic acid *en route* to hexadecanolide[14]. It should be noted that in addition to the desired product 15-hydroxypalmitic acid is also obtained as a result of ω-1 oxidation. Cyclisation of this material leads to 15-methyl cyclopentadecanolide found naturally as a constituent of Galbanum absolute and as a sex pheromone of the stink bug, *Piezodorus hybneri* [26]. Despite the activity in this area, the number of commercial processes based on these approaches currently remains small (for a commercial route to cyclopentadecanone *via* fermentation see ref. 17).

Returning to synthetic approaches, a fairly efficient route to ω-hydroxypentadecanoic acid is to react an excess of 1,12-dodecane diol (obtained from cyclododecene) with methyl acrylate using di-t-butyl peroxide as initiator [27]. The product actually obtained is the lactonised material (scheme 3). Yields of 80% based upon methyl acrylate are possible. The butyrolactone product is converted to the desired hydroxy acid by catalytic hydrogenation using zeolite HY containing ion exchanged-deposited palladium. The catalyst possesses acidic and hydrogenation catalytic sites which enable the direct conversion of the lactone to the saturated carboxylic acid (Scheme 3).

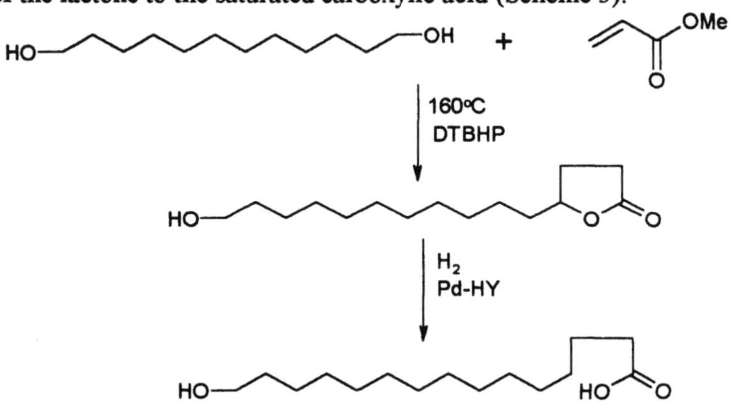

Scheme 3

An interesting chemical approach which represents the synthetic chemist's attempt to perform the same transformation as described above for the biochemical/fermentation is the ω- to ω-2 chlorination of fatty acids adsorbed onto alumina [28]. For hexdecanoic acid, selectivity towards the ω, ω-1 and ω-2 chlorinated material was found to be as high as 50% relative to 27% for non-adsorbed material using chlorine as the chlorinating agent.

Along with the development of ever more robust and efficient ring closing metathesis catalysts (see below) there is an increasing interest in unsaturated alcohol, acids *etc.* (especially terminally unsaturated materials). Some of the most commonly encountered examples are:

- 10-undecylenic acid derivatives and the corresponding alcohol (obtained from castor oil)[29].
- 9-decenol (obtainable from 1,9-decadiene, in turn obtained from metathesis between cyclooctene and ethylene).
- 4-pentenol derivatives [30]
- 5-hexenol derivatives [23,31]
- Oleic acid/oleoyl alcohol[22,32]

Finally, an interesting approach, based upon fermentation, to ω-hydroxy aldehydes has been developed *en route* to macrocyclic musks (in particular 7-hexadecenolide)[33]. The appeal of this approach is that the feedstock, a long chain α,ω-diol is directly mono-oxidised selectively without protection of the remaining hydroxyl group.

2.2 MACROCYCLISATION OF BIFUNCTIONLISED CHAINS

A study of the literature indicates the difficulty of cyclising suitably bifunctionalised chains to 15,16 and 17 membered macrocycles. Interestingly, it was believed that large-ring compounds could not be synthesised or isolated (von Baeyer strain theory[34]) until Ruzicka's structural elucidation of the macrocyclic musk materials civetone and muscone [35]. The majority of procedures require dilutions of ≤ 10gdm^{-3}. This approach is of value when highly functionalised macrolide antibiotic materials are required and a number of methods have been developed to do just that [36]. However, for relatively high volume materials, such as macrocyclic musks, these low concentrations are not acceptable with respect to reactor utilisation. For a typical 20h high dilution macrocyclisation reaction at 10gdm^{-3} a dedicated fully occupied reactor with a volume of ~30m^{-3} would be required to satisfy a product volume of 100tpa. Conversely, a cyclisation that can occur at 100gdm^{-3} will only require a 2.5m^{-3} reactor. This doesn't of course apply to irreversible techniques where the product is stable to the reaction conditions, such as alkyne coupling reactions, although these procedures still suffer from the prerequisite slow reagent additions.

Fortunately, there are "tricks" that can be applied such as the polymerisation/depolymerisation approach for macrolactonisation of α,ω-hydroxy acids originally developed by Carruthers in the 1930's [37,38]. This procedure is used commercially to produce a number of macrocyclic musks, including ethylene brassylate, and hexadecanolide.

Numerous catalyst systems have been applied to the synthesis of macrocyclic lactones by lactonisation e.g. hydrous zirconium (IV) oxide [39], tin (IV) compounds [40] and zeolites [41]. In selected cases, the technique of templating can increase the selectivity of cyclisation *vs.* polymerisation. A good example of this is the formation of macrocyclic dilactones from diacid anhydrides and poly-glycols using tris-(2-methoxyphenyl)bismuthane as a catalyst (Scheme 4). Here it was possible to obtain yields of macrocyclic material in the range of 56-83% at substrate concentrations of 0.5M.* This approach, compared to the uncatalysed reaction is significantly more efficient with respect to reactor volume and selectivity, but of course, is specific to polyether materials capable of templating around the metal centre [42].

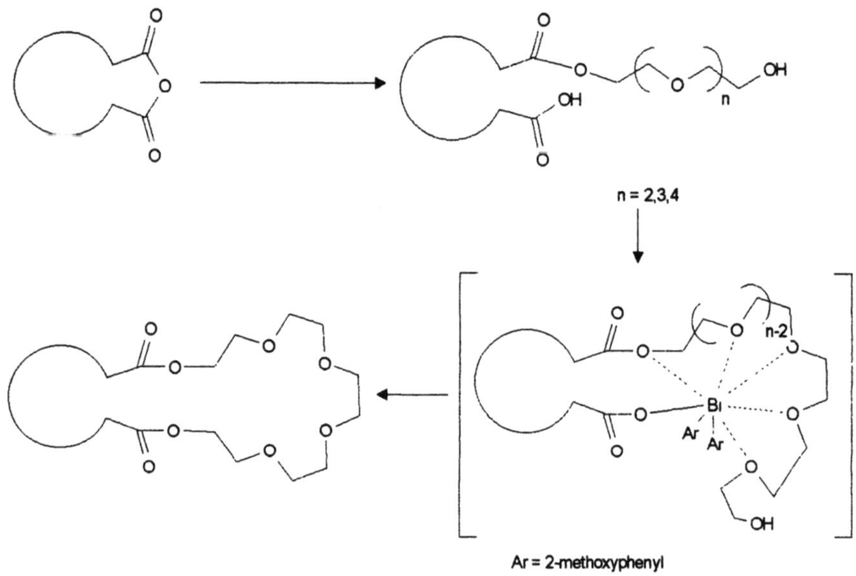

Scheme 4

Macrocyclic lactones can also be synthesised from an ω-hydroxy carboxylic acid using lipases as catalysts. These enzymes are very useful to the synthetic chemist, although their primary function *in-vivo* is to *hydrolyse* ester functionalities, by using sufficiently anhydrous conditions they can be used to prepare esters from the corresponding acid

* Note the template effect does not always mean a higher substrate concentration can be tolerated (see SD Burke *et al.*, *J. Org. Chem.*, 1998, **63**, 2715).

derivative and alcohol. In addition, since the enzyme's natural substrates tend to be long chain/hydrophobic glycerides, long chain ω-hydroxy carboxylic acids can act as ideal feedstocks. Finally, lipase enzymes don't require co-factors and thus isolated enzymes are suitable for performing the desired reaction. The challenge of favouring intramolecular esterification [43] (to lactones) over intermolecular esterification [44] still remains. A study on the effect of chain length of the hydroxy acid in *Pseudomonas* species lipase catalysed lactonisations indicated that diolide formation becomes significant for methyl 13-hydroxytridecanoate (25%) to the detriment of monolide formation (38%) and reached a minimum at or above methyl 16-hydroxyhexadecanoate (diolide 3% cf. monolide at 80%). However, as with other macrolactonisations the optimum substrate concentration remains low at $0.3gdm^{-3}$ (mM level) with a reaction time of 4h *i.e.* reactor volume for 100tpa would be $170m^{-3}$[45].

The acyloin reaction has been successfully developed for the synthesis of macrocyclic ketones since the procedure allows relatively high concentrations for the cyclisation of the lower alkyl esters of the dicarboxylic acid [46]. The intermediate cyclic 2-hydroxy ketone can be reduced to the ketone by using Clemmensen type reaction conditions [47]. These reactions have been commercially developed to produce cyclopentadecanone from the diester of 1,15-pentadecane dicarboxylic acid (obtained *via* a fermentation process - see above).

The carbonyl coupling reaction catalysed by low valent titanium, made popular by McMurry, provides a potentially useful route to macrocyclic musks (Scheme 5). The reaction uses $TiCl_3$ or $TiCl_4$ and a suitable reductant such as Zn, Mg, Zn-Cu, or $LiAlH_4$ and proceeds *via* a pinacol intermediate which is deoxygenated by the Ti(0). Alternatively, the pinacol product can, however, be isolated in high yields by maintaining a low reaction temperature [48]. The reaction has been used extensively to prepare macrocyclic materials from long chain α,ω-dialdehydes although high dilution techniques are still required (a typical procedure uses a syringe pump and addition over 30-40hrs) [49].

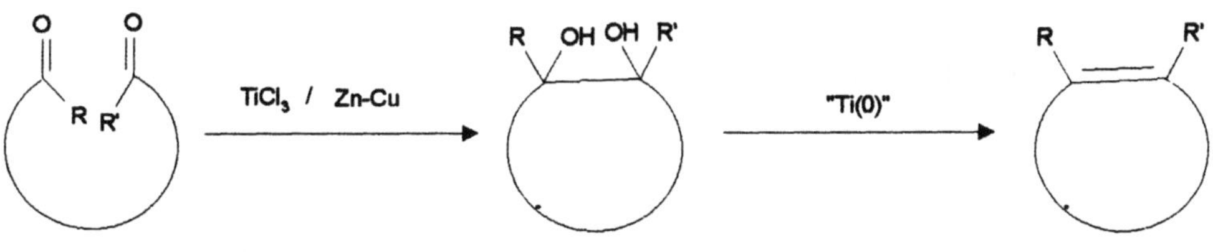

Scheme 5

As stated earlier, olefin metathesis has evolved significantly as a synthetic tool due to the development of new generations of well defined catalysts with high tolerances towards functional groups. Ring closing metathesis in particular, is gaining popularity as an approach to macrocyclic molecules (not only in the fragrance chemistry field [50] but also in the area of macrolide antibiotics)[51,52,53,54]. An early example of the successful use of metathesis to prepare civetone was provided by Mol and Plugge [55]. Here the catalytic entity was generated *in-situ* from $Re_2O_7/SiO_2.Al_2O_3$ and tetra-n-butyl tin. This procedure highlighted some of the problems associated with the metathesis approach e.g.:

- The reaction is reversible (*i.e.* thermodynamically controlled) and must be performed at high dilution;
- There is little or no selectivity with respect to the geometry of the resultant alkene.

A more recent example which, in part, overcomes the reversible nature of the metathesis reaction is the synthesis of cyclopentadecenolides (*en route* to cyclopentadecanolide) from either 5-hexenyl 10-undecenoate or 10-undecenyl 5-hexenoate [56]. In this reaction the driving force for the reaction is provided by the evaporation of the ethene gas that is generated (*c.f.* Mol's example above where the involatile 9-octadecene is produced as the co-product). However, again the geometry of the alkene from the reaction is still relatively non-selective. This problem was approached by the first synthesis of a musk material using alkyne metathesis [57]. The main interest in this route is the synthetic utility of the resulting macrocyclic alkyne *i.e.* hydrogenation of the alkyne functionality using a Lindlar catalyst yielded exclusively a Z-configured macrocyclic alkene which is not available *via* normal RCM. As with other alkene based RCM approaches to macrocyclic compounds, high dilutions, ≤0.02M, were required.

For an example of the use of the aldol reaction towards a musk, Muscone has been prepared *via* a multistep route involving the transient intermediate 1,8-nonadiene generated by the metathesis reaction between cycloheptene and ethylene, which undergoes further reaction with itself to ultimately yield 1,8,15-hexatriene. This can then be oxidised selectivity under Wacker conditions to the diketone which is cyclised *via* an aldol reaction then hydrogenated to the target molecule [58].

3. Macrocycles *via* Ring Expansions

This approach is dominated by the availability of the initial cyclic starting material. The most common ring sizes available for starting materials are 5 and 6 membered, although these are not ideally suited since the formation of thermodynamically disfavoured 9-12 membered second rings/transition states would be required. Fortunately, analogues of cyclododecane are commercially available *via* the cyclo-oligomerisation of 1,4-butadiene. These materials can then be used to make 15-17 membered macrocycles *via* 3,4 and 5 carbon ring expansions and there are numerous examples of processes that use this approach, some of which are detailed below.

One particular approach involves the preparation of a bicyclic enol ether from the addition of a 3-functionalised propanol followed by an acid catalysed intramolecular condensation [59]. Numerous methods have been used for the cleavage of the alkene, the product usually being a ketolactone, the keto group being removed by Wolf-Kishner or Clemmensen reduction. A recent example of the use of this approach to produce enantiomerically pure muscone and muscolide is shown in Scheme 6 [60].

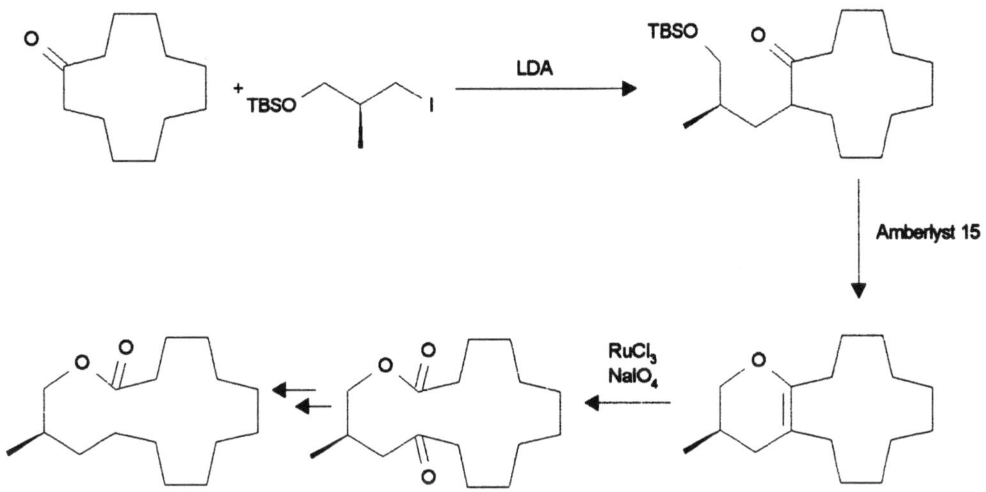

Scheme 6

Although significant differences in musk type odour were not found, the S-enantiomer of muscolide was found to have greater animalic notes. Enantioselective synthesis in the macrocyclic musk area provides an intriguing challenge to the development chemist. One approach towards this goal based upon asymmetric hydrogenation of Z-3-methyl-2-cyclopentadecene-1-one has been patented using a homogeneous optically active ruthenium catalyst [61].

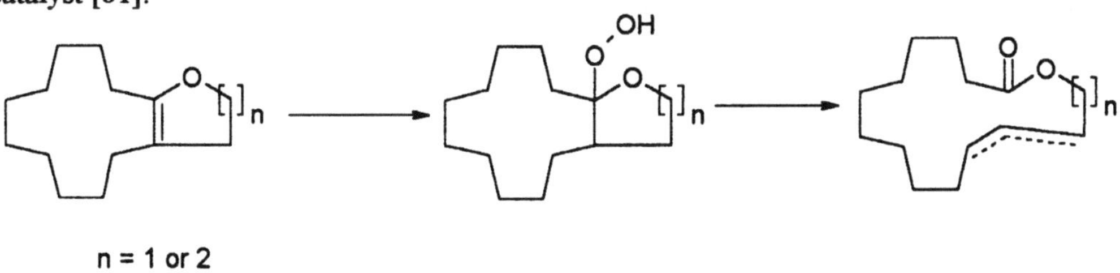

Scheme 7

An alternative, more direct, route involves converting the bicyclic lactol to the hydroperoxide (as generically shown in Scheme 7) by treatment with hydrogen peroxide and acidic conditions.

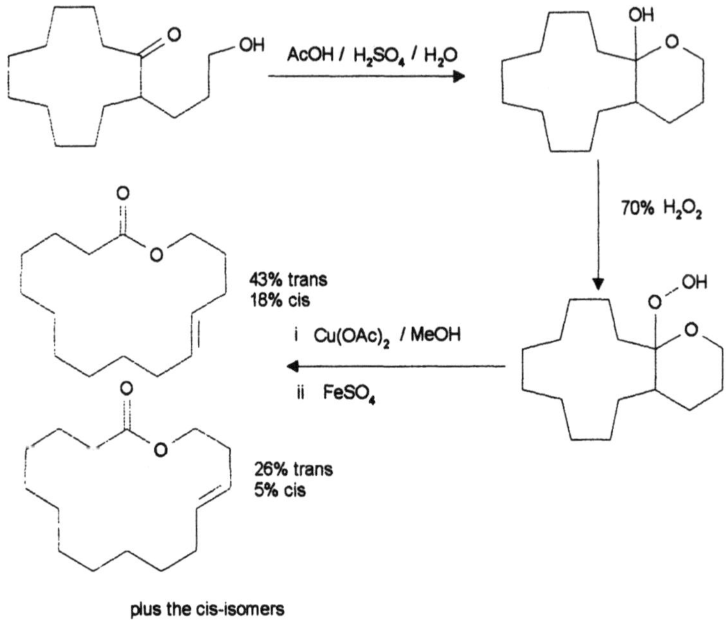

Scheme 8

This approach has been used to prepare unsaturated macrocyclic mixtures, previously seen as intermediates towards cyclotetradecanolide and cyclopentadecanolide, and which possess notable musk odours in their own right. New routes to these materials have been patented recently (see Scheme 8 for cyclopentadecenolide) [62].

An alternative procedure for the breakdown of the hydroperoxide intermediate has also been patented [63]. This approach is dependent upon a relatively dilute thermolysis at high temperature (140°C) and yields the saturated cyclopentadecanolide along with ~10% of the unsaturated cyclopentadec-11(12)-enolide.

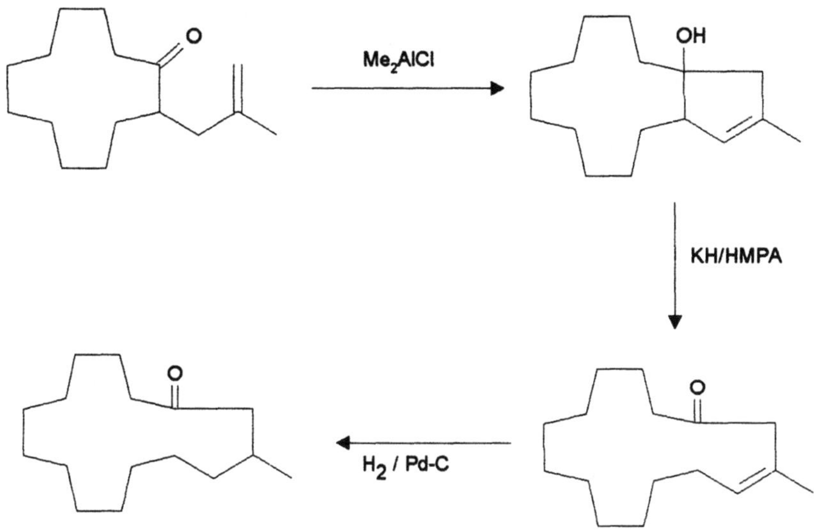

Scheme 9

Finally, β-Cleavage of a homoallylic alkoxide to yield an unsaturated ketone provides another approach from cyclododecanone which has been used to yield muscone in only three synthetic steps from the readily available 2-(2-methylprop-2-enyl)cyclododecanone (Scheme 9)[64].

Macrocyclic lactones are also readily obtained from the corresponding ketone *via* the Baeyer-Villiger (BV) reaction although the usual synthetic approach using percarboxylic acids is prohibited by the use of the relatively expensive peracid. However, recently developed methodologies which involve the *in-situ* production of peroxy species suitable for the Baeyer-Villiger reaction using molecular oxygen as the oxidant and aldehydes as precursors to peracids may increase the interest in this approach [65].

4. Macrocycles *via* Cyclooligomerisation

This approach to macrocyclic musks is particularly attractive as the macrocycle is formed directly from the precursors thus avoiding the necessity for high dilution reactions. Another valuable aspect of this approach is that readily available, cheap materials can be used.

Metathesis can be used to produce macrocyclic dimeric dienes and trimeric trienes from smaller unsaturated cyclic molecules. Examples of this approach toward musks include:

- The dimerisation of cyclooctene to give 1,9-cyclohexadiene which has been converted to 8-cyclohexadecen-1-one *via* selective hydroboration and oxidation or by hydroperoxide rearrangement [66].

- Cyclooctadiene (obtained *via* butadiene dimerisation) can undergo a metathesis reaction over rhenium on alumina to produce cyclopolyenes having the general formula 8+4n carbon atoms. One of these products, cyclohexa-1,5,9,13-tetraene has been converted to the tris-unsaturated ketone as shown in Scheme 10 [67].

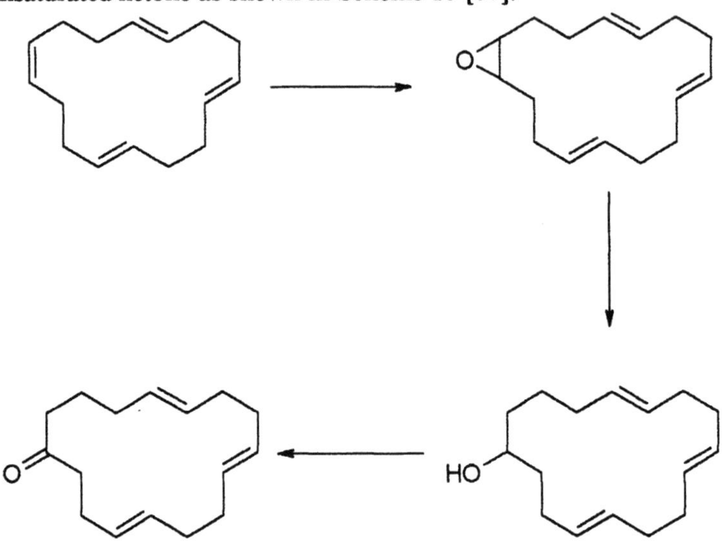

Scheme 10

- Attempts have also been made to create a 15 carbon macrocycle using cyclopentene as the precursor [68].

The greatest hurdle to overcome with this approach is the low yields provided by the oligomerisation. This is due to further reaction of the desired products with themselves and/or further monomer to ultimately yield polymeric material (in other words Ring-Opening Metathesis Polymerisation - ROMP). Thus, to make commercial processes *via* this chemistry will require either shape selectivity which limits the size of the molecules that are generated, or the removal of the desired product *in-situ* using some physical type of separation.

An attractive route to muscone has been developed which involves the reaction of the *bis*-π-allylnickel complex with allene to give a further *bis*-π-allylnickel complex which can be reacted with carbon monoxide to give the unsaturated muscone [69]. Hydrogenation gives the desired compound. An improved procedure involves replacing carbon monoxide with an alkyl isocyanide for the insertion step (Scheme 11) [70].

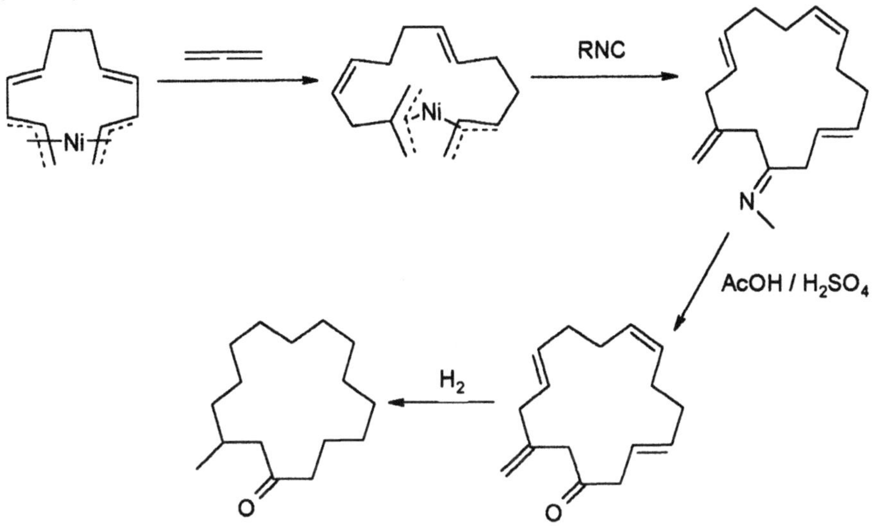

Scheme 11

As with the previous examples in this section, this approach has great potential as it is based upon relatively cheap raw materials (butadiene, allene and carbon monoxide or methyl isocyanide).

5. Summary

It's safe to say that there has been an increasing amount of activity in developing synthetic routes towards macrocyclic musks, some of which have been developed to a commercial scale. It does, however, appear that, to date, macrocyclic musks still do not compete with the polycyclic musks (such as those mentioned in the introduction). The routes developed suffer from either high raw material costs, high dilution reactions or a large number of synthetic steps. By analysing proposed synthetic routes using these factors, it should be possible to prioritise each proposal. Retrospectively, the approach which appears to suffer the least from these negatives are the ring expansion routes. The major raw materials, cyclododecane derivatives, are relatively cheap, readily available and the reaction steps don't suffer from requiring high dilution techniques. This is also backed up by the number of commercial routes developed using this chemistry. An area which may provide some serious competition for the polycyclic musks is the cyclooligomerisation of butadiene and its derivatives. Here the raw material is cheap and the problem of macrocyclisation is bypassed. The major effort will need to be focussed on developing efficient transformations of the resultant macrocycles to musk materials.

6. References

1. B.D. Mookherjee and RA Wilson, in *Fragrance Chemistry - The Science of the Sense of Smell* Ed ET Theimer, Academic Press 1982, 433.
2. H. Walbaum, *J. Prakt. Chem.*, 1906, **73**, 488., L. Ruzicka, *Helv. Chim. Acta.*, 1926, **9**, 715.
3. US 3,415,813 (1968)
4. A. Baur, *Ber. Dtsch. Chem. Ges.*, 1891, **24**, 2836 and 284.
5. T. Yamagishi *et al.*, Bull. Environ. Contam. Toxicol., 1981, **26**, 656.
6. US 2,450,878 (1948), M.S. Carpenter *et al.*, J. Org. Chem., 1951, **16**, 59.
7. A. Rebmann, C. Wauschkuhn, W. Waizenegger, *Dragoco Rep. (Engl Ed)*, 1998, (2), 49.
8. T.F. Wood *et al.*, J. Org. Chem., 1963, **28**, 2248.
9. US 3,360,530.
10. For a review see K. J. Rossiter, *Chem. Revs.*, 1996, **96**, 3201-3240.
11. D.P. Anonis, *Perfumer & Flavourist*, 1992, **17**, 23.
12. C. Allan *et al.*, Stud. Nat. Prod. Chem., 1991, 219.
13. *Chemical-Business-India*, 1996, **10**(2), 69.
14. *Japan Chemical Week*, 1998, **39**(1966), 3.
15. *Japan Chemical Week*, 1998, **39**(1979), 3.
16. Parfums-Cosmetiques-Aromes, 1998, **143**, 67.
17. J.L. Greene Jr., *et al.*, *J. Polymer Sci.* A-1, 1967, **5**, 391.
18. US Patent, 4,525,251 (1985).
19. US Patent, 4,310,685 (1982).
20. Chemical-Market-Reporter, 20 Apr 1998.
21. R. Jeffcoat and B.J. Willis in Flavors and Fragrances: A World Perspective. *Proceedings of the 10th International Congress Essential Oils, Fragrances and Flavors, Washington DC, USA*, 16-20 November 1986, pp743, Eds. B.M. Lawrence, B.D. Mookherjee and B.J. Willis.
22. H. Okino, A. Taoka and N. Uemura in Flavors and Fragrances: A World Perspective. *Proceedings of the 10th International Congress Essential Oils, Fragrances and Flavors, Washington DC, USA*, 16-20 November 1986, pp753, Eds. B.M. Lawrence, B.D. Mookherjee and B.J. Willis.
23. R. Uchio and I. Shiio, *Agr. Biol. Chem.*, 1972, **36**, 426.
24. US 5,648,247 (1997).
25. A. Tulloch, J. Spencer and P. Gorin, *Can. J. Chem.*, 1962, **40**, 1326, H. Asmer, S. Lang, F. Wagner and V. Wray, *J. Am. Oil Chem. Soc.*, 1988, **65**, 1460.
26. S. Kuwarhara *et al.*, *Biosci. Biotechnol. Biochem.*, 1998, **68**, 1261.
27. US Patent 4,246,182 (1981).
28. L.H. Hinkamp, H.J. Shafer, B. Wippich and H. Luftmann , *Liebigs Ann. Chem.* 1992, 559.
29. J. Tsuji and S. Hashigucji, *J. Organomet. Chem.*, 1981, **218**, 69.
30. D. Villemin, *Tet. Lett.*, 1980, 1715.
31. US Patent 4,559,168 (1985).
32. J. Tsuji and S. Hashiguchi, *Tet. Lett.*, 1980, 2955.
33. JP 08217716 (1996).
34. T.G. Back, *Tetrahedron*, 1977, **33**, 3041.
35. L Ruzicka, *Helv. Chim. Acta.*, 1926, **9**, 230, L Ruzicka, *Ibid*, 1926, **9**, 715, L Ruzicka, *Ibid*, 1926, **9**, 1008.
36. K.C. Nicolaou, *Tetrahedron*, 1977, **33**, 683, K.C. Nicolaou, Y. He, F. Roschangar, P.N. King, D. Vouloumis, T. Li, *Angew. Chem. Int. Ed.*, 1998, **37**, 84, A. Fuerstner, K. Langemann, *J. Am. Chem. Soc.*, 1997, **119**, 9130.
37. J.W. Hill and W.H. Carothers, *J. Am. Chem. Soc.*, 1933, **55**, 5031.
38. For recent examples of the polymerisation/depolymerisation procedure see: JP 7,267,945-A (1995), US Patent 4,709,058-A (1987).

39. H. Kuno et al., *Chemistry Letters*, 1992, 571-574.
40. J.D. White, N.J. Green, F.F. Fleming, *Tet Lett*, 1993, **34**, 3515, K. Steliou, S. Hannesian *et al.* , *J. Am. Chem. Soc.*, 1980, **102**, 7579.
41. T. Tatsumi, H. Sakashita, K. Asano, *J.Chem.Soc. Chem. Commun.*, 1993, 1264
42. T. Ogawa, A. Yoshikawa, H. Wada, C. Ogawa, N. Ono and H. Suzuki, *J Chem Soc Chem Commun., 1995,* 1407.
43. G.K. Robinson, M.J. Alston, C.J. Knowles, P.S.J. Cheetham, K.R. Motion, *Enzyme Microb. Technol.*, 1987, **16**, 855, C.S. Chen and C.J. Sih, Angew. Chem. Int. Ed. Engl., 1989, **28**, 695.
44. F. Binns, S.M. Roberts *et al.*, *J.Chem. Soc. Perkin Trans. I*, 1993, 899.
45. A. Makita, T. Nihira and Y. Yamada, *Tet. Lett.*, 1987, 805-808.
46. L. Bouvealt, R. Loquin, *Bull. Soc. Chim. France,* 1936, **35**, 629., K. Ruhlmann, *Synthesis*, 1971, 236.
47. JP 5155802 (1991).
48. J.E. McMurry and J.G. Rico, *Tet. Lett.*, 1989, 1169.
49. J.E. McMurry, *Chem. Rev.* 1989, **89**, 1513.
50. K.A.D. Swift, *Proceedings from Fragrances and Flavours Association of India XIV Biennial Seminar* "Fragrances and Flavours in the 21st Century", Jaipur, India, 1999.
51. S. Warwel *et al.*, *Seifen Oele Fette Wachse*, 1989, **115**, 538.
52. K.C. Nicolaou *et al.*, *Angew. Chem. Int. Ed. Engl.*, 1998, 37, 84-87.
53. K.E. Litinas, B.E. Salteris, *J. Chem. Soc. Perkin Trans. 1*, 1997, 2869.
54. S.J. Danishefsky *et al.*, *J. Am. Chem. Soc.*, 1997, **119**, 10073.
55. M.F.C. Plugge and J.C. Mol, *Synlett*, 1991, 507.
56. A. Furstner and K. Langemann, *J. Org. Chem.*, 1996, **61**, 3942.
57. A. Furstner and G. Seidel, *Angew. Chem. Int. Ed.*, 1998, **37**, 1734.
58. see G. Frater and D. Lamparsky in *Perfumes, Art, Science and Technology*, Elsevier Applies Science London & New York, ed. P.M. Muller and D. Lamparsky.
59. I.J. Borowitz, G. Gonis, *Tet Lett*, 1964, 1151, I.J. Borowitz *et al.*, *J. Org. Chem.*, 1966, **31**, 3032, I.J. Borowitz *et al.*, *J. Org. Chem.*, 1973, **38**, 1324, US 4,187,222 (1980).
60. W. Tochtermann *et al.*, *Tetrahedron*, 1996, **52**, 4581.
61. JP 7267968 (1995), JP 6192161 (1994).
62. EP 862911 (1998), US 5,266,559 (1993).
63. US 5,214,163 (1993).
64. V. Rautenstrauch *et al.*, *Helv. Chim. Acta.*, 1990, **73**, 896.
65. C. Bohm *etal.*, *Tet. Letts.* 1993, 3405, K. Kaneda *et al.*, *J. Chem. Soc. Chem. Commun.*, 1994, 797, K. Kaneda *et al.*, *J. Org. Chem.* 1994, **59**, 2915, F. Wang *et al.*, *Syn. Commun.*, 1996, **26**, 1613.
66. L. G. Wideman, *J. Org. Chem.*. 1968, **33**, 4541., S. Lee, R. Huff, *Q. Rev. Chem. Soc.*, 1967, **21**, 445.
67. GB 1,433,463 (1974).
68. US 4,668,836 (1987).
69. R. Baker *et al.*, *J. Chem. Soc. Chem. Commun.*, 1972, 802.
70. R. Baker *et al.*, *J. Chem. Soc. Chem. Commun.*, 1974, 515.

ANALYSIS TECHNOLOGY

ALAIN CHAINTREAU
Nestlé Research Center, Nestec Ltd
P.O. Box 44, Vers-chez-les-Blanc
CH 1000 Lausanne 26, Switzerland

1. Introduction

Analysis of flavours and fragrances is a hard task because of their complex composition and high dilution in a no less complex non-volatile matrix. This challenge has stimulated analysts to constantly develop and improve techniques yielding a more and more representative image of the volatile fraction composition. This subject has given rise to many reviews [1, 2, 3, 4, 5], therefore, rather than extensively reviewing and describing well-known techniques, such as classical application of GC or MS with which readers are certainly familiar, attention will be focused on new methods or on recent improvements of known methods. Due to their potential interest for flavours and fragrances, some references related to other analytical areas will also be cited. This paper will mainly deal with the analysis of volatiles as they represent all perfume ingredients and the majority of flavour constituents. However, techniques enabling the analysis of non-volatile taste compounds, or of odourant precursors will be discussed as well.

2. Sample Preparation

In flavour analysis, the objectives of sample preparation are: isolation of the flavouring volatiles from the matrix, and concentration of the extract to a level that is compatible with the detector's sensitivity.

Ideally, such an isolate should respect an additional criterion: its composition must be representative of the original flavour present in the matrix. In other terms, the isolation method should recover all components participating in the flavour of a given food, without discrimination. When ratios between constituents' concentrations are changed due to varying recoveries from one component to another, the original character of the aroma is also altered. In addition, care must be taken to avoid the formation of any artifacts during sample preparation.

K.A.D. Swift (ed.), Current Topics in Flavours and Fragrances, 97–122.

Sample preparation is a key step in the analytical procedure, as no analytical instrument, however sophisticated, will provide the analyst with the correct composition of a plant scent or of a food aroma if the extract is not representative.

2.1. METHODS BASED ON VOLATILITY

2.1.1. Distillation

Distillation as such is little used because flavours and fragrances normally occur at a too low concentration. However steam distillation is one of the oldest techniques to be applied to industrial preparation of essential oils. The old Marcuson's or Clevenger's device has been re-designed for micro-scale operations [6] and used for the preparation of Artemisia essential oil [7].

In the laboratory, steam distillation remains an isolation means that is only applicable to produce essential oil-like samples because it submits volatiles to the temperature of boiling water. Many artifacts may be generated under these conditions [8].

To avoid such drawbacks distillation of volatiles under high vacuum (10^{-1} down to 10^{-4} Torr) has been proposed for cheese aromas [9] and flower scents [10]. Under these conditions the water in the raw material presumably plays the role of steam generator, with reduced relative volatilities of organic compounds under a vacuum [11]. Non-thermally altered extracts are obtained after solvent extraction of the distillate and concentration, prior to GC analysis [12].

2.1.2. Headspace

Headspace sampling has recently gained popularity because it requires little sample preparation and automated injectors have been commercialised for static and purge-and-trap headspace (S-HS and P&T-HS, respectively). Therefore, manual injection with a gas-tight syringe is now declining. S-HS may be sampled either into the gas loop of an automated valve [13], or by pressurising the headspace of a closed vial containing the sample and subsequent release of the gas into the capillary column (Fig. 3-11 of [14]). In both cases, injection volumes are limited by the capacity of capillary column injectors. This technique is, therefore, dedicated to very volatile compounds. Readers interested in an extensive description of S-HS possibilities should refer to Kolb's book [14].

To overcome this lack of sensitivity for components of lower volatility, the aroma must be concentrated [15]. It may be stripped from a liquid sample using an inert gas, and trapped (P&T-HS) according to several possible ways. 1) Organic compounds of the gas flow can be dissolved in a solvent reflux, and GC injected [16]. 2) They may be cryo-focused, for instance on the top of the GC-column itself [17]. 3) Many automated systems trap them in an adsorbent (graphite, Tenax®, Porapak®, etc.), from which they are either thermally desorbed [15], or eluted with a solvent and GC injected [18]. With hydrophobic adsorbents, such as Tenax®, organic compounds can be enriched in the polymer without retaining water, which is always present in plants or food. Traps made of different adsorbent layers may be helpful for trace analysis at the ppt level [19, 20].

It must be recalled that a trapping step is only chromatography through a very retentive phase. Care must be taken to limit stripping gas volumes to less than the breakthrough volume of the target compound in a given adsorbent, otherwise losses must be expected. Breakthrough volumes of many volatiles in Tenax® have been published [21]. Very efficient adsorbents such as active charcoal may not quantitatively release trapped components [22]. The desorption step may give rise to artifact formation. For instance, microwave heating of graphite traps seems to systematically yield benzene derivatives (e.g. [23]). Therefore, blank experiments must be run, using the same system and, possibly, the same matrix (at least pure water for aqueous samples).

GC profiles obtained with P&T-HS are not representative of the real volatile composition of a plant or a food. The recovery ρ_i of a given compound i can be calculated:

$$\rho_i = 1 - \exp[-(V_g k_{gl})/ V_l] \quad \text{after [24]} \qquad (1)$$

(V_g: volume of stripping gas, V_l: volume of the solution, k_{gl}: gas-to-liquid partition coefficient).

For a given stripping gas volume, recoveries of volatiles will depend on individual partition coefficients.

Special attention has been paid to *in-situ* sampling of volatiles emitted by living plants because cutting them completely changes enzymatic processes responsible for the formation of the volatiles they emit. Sophisticated computer-controlled systems have been proposed [25]. The plant is put in a glass chamber under a flow of pure air, and the gas exiting the system passes through Tenax® traps. Switching the outlet from one trap to another yields a chromatographic image of the plant's fragrance as a function of time, sun intensity, etc. [26]. An alternative, proposed by Bicchi [27], used short capillary columns with a thick bonded phase (15μm) to trap volatiles from living plants in fields. Desorption is performed in the GC oven, by connecting the trap as the first column of a multidimensional GC system. This avoids artifacts inherent to adsorption processes, as mentioned above, because the trapping uses the same phenomenon as the chromatographic elution itself (solubilisation in the bonded liquid phase). S-HS and P&T-HS applications in essential oils have been reviewed [28], publications related to food aroma are myriad.

More recently, two techniques were proposed which combine S-HS with trapping of the headspace. Solid phase microextraction (SPME) was originally designed to extract solutes from a liquid, using a glass fiber covered by a film of a bonded phase (see 2.2.2). It may also be used to extract organic compounds from the headspace (SPME-HS) of a vial containing the liquid sample [29] (Fig. 1A). The recovery of a single compound is given by:

$$\rho_i = (V_f k_{fl})/(k_{fl} V_f + k_{gl} V_g + V_l) \qquad \text{after [30]} \qquad (2)$$

(k_{fl}: fiber-to-liquid partition coefficient, k_{gl}: gas-to-liquid partition coefficient, V_g: volume of headspace, V_l: volume of liquid, V_f : volume of phase on the fiber).

At equilibrium, the composition of volatiles adsorbed onto the fiber is similar to that from a direct extraction in the liquid alone when the HS volume can be neglected [30] as shown in Table 1. Therefore SPME-HS-GC profiles are not representative of HS composition. HS-SPME has been applied to the aroma of coffee [31], orange juice [32], truffles [33], cheese [34], etc.

With the second method [35], the aroma of a liquid or a solid sample is equilibrated with the headspace of the cell (Fig. 1B). When the piston is pressed, the air is evacuated through an adsorbent trap which is then thermally desorbed onto the GC column. Contrary to SPME-HS, the composition is representative of the equilibrium headspace and not of the solution. This static-and-trapped HS technique (S&T-HS) has been applied to yogurt [36] and coffee [37].

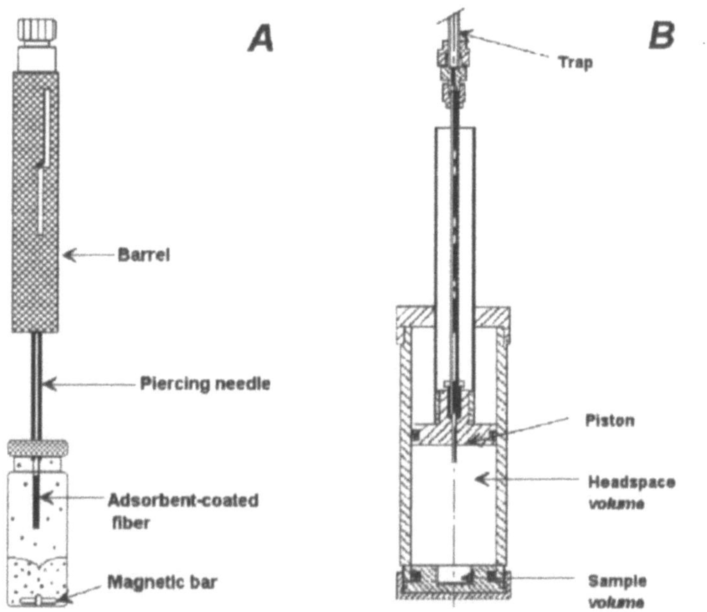

Figure 1A. SPME sampling in the headspace. **B.** static-and-trapped HS (reprinted with permission from [35], copyright 1995, American Chemical Society)

2.2. EXTRACTION METHODS

Extraction techniques are powerful tools to enrich flavours and fragrances, but non-volatile compounds are simultaneously extracted. They may pollute GC injectors and decompose into volatiles that are not present in the plant or in the food. In spite of this drawback, they do not alter original odorants as this isolation procedure does not involve any heating of the sample [38].

2.2.1. With a Fluid

Soxhlet extraction, and batch or continuous liquid/liquid extraction are well known methods. They remain irreplaceable techniques to recover very polar components from water with reasonable yields, such as furaneol [39, 40] or organic acids [41]. Laboratory-scale continuous extractors have also been designed [42].

To isolate volatiles from co-extracted non-volatiles the organic phase containing the volatiles is frozen in liquid nitrogen and "high vacuum sublimation" (HVS) is performed under low pressure (0.02 Pa) [43].

Supercritical fluid extraction (SFE), using carbon dioxide, may operate at nearly room temperature [44] with the advantage of a simple solvent elimination at atmospheric pressure. However, non-volatiles, specially fats, are co-extracted depending on the pressure applied to the supercritical fluid [45].

2.2.2. Solid Phase Extraction (SPE)

This technique involves adsorption onto an inorganic adsorbent (e.g. silica, activated carbon), a porous polymer (e.g. styrene-divinylbenzene resins) [46, 47], or extraction by a liquid phase chemically bound to a solid [47, 48, 49]. Good recoveries have also been observed with hydrophilic adsorbents which retain water, and from which organic compounds may be eluted with a solvent [50]. Like in HS trapping, most efficient adsorbents may not fully release organic compounds [51]. Either clean-up cartridges or HPLC columns themselves have been used to enrich flavour components [48, 52]. As many adsorbents are hydrophobic, the recovery of hydrophilic components may be improved by derivatising resins with polar groups [53]. In addition, the use of a column allows one to use an elution gradient to fractionate and trap volatiles. Fractions are further analysed by GC, GC/MS, etc. [54].

SPE application to UHT milk allowed lactone recovery at the ppb level [55]. Green tea SPE extracts were found to be more representative than SDE extracts because the aroma was isolated at room temperature [38].

Solid phase micro extraction (SPME), born in 1990, has had a lightning development in many areas as it greatly simplifies the SPE procedure and shortens sample preparation time. A liquid phase is bound to the surface of a glass fiber. This latter is first exposed to the liquid sample, then it is thermally desorbed into a GC injector. The method has been employed for many odorants (e.g. [56, 57]), and was recently reviewed by its author [58]. It is very applicable to flavours and fragrances, if its limitations are not forgotten. Like other batch extraction methods, partitioning occurs between the sample and the fiber-supported phase. The yield (ρ_i) depends on partition coefficients which differ from one compound to another:

$$\rho_i = V_f k_{fl} / (k_{fl} V_f + V_l) \text{ after [30]} \quad (3)$$

(V_l: volume of solution, V_f: volume of fiber-supported phase, k_{fl}: fiber-to-liquid partition coefficient).

Therefore, SPME-GC profiles are not directly representative of the solution composition, as with SDE extracts.

2.3. MEMBRANES

Membranes have not yet given rise to many applications in flavour and fragrance analysis. However they are mentioned here because of their increasing possibilities which are already exploited by analysts in other areas. For instance membrane techniques often allow on-line sample preparation and subsequent extract analysis [59].

2.3.1. Pervaporation

Pervaporation can be used to extract and concentrate organic compounds based on the solubility and diffusivity in the membrane [60]. A driving force is created either by applying a vacuum to one side followed by cryo-trapping of the permeate [61, 62], or by directly applying the vacuum to a mass spectrometer source [59], or by flushing the surface of the membrane with a pure gas [63]. Another alternative, also related to pervaporation, is MESI (membrane extraction with sorbent interface). A carrier gas flows through a hollow silicon fiber, whose external surface is in contact with the headspace to be analysed. Volatiles migrate through the polymer, and are conveyed to the analytical system [64]. Alternatively, the aqueous sample can be circulated in the hollow fiber whose external surface is swept by the carrier gas flow [63].

2.3.2. Pertraction and extraction across a membrane

Pertraction allows the extraction of solutes from the solution contacting one side of the membrane by a solvent flushing the other side [62]. Alternatively, a liquid/liquid extraction across a membrane may also be used such as for wine flavour [65]. It combines continuous solvent extraction of an aqueous solution with phase separation using a hydrophobic membrane (pore size: 0.45 μm). Stirring wine and solvent leads to an emulsion in which aqueous droplets are rejected from the fluorinated polymer. The aroma-containing solvent passes through the membrane, thus separating the emulsion.

2.3.3. Ultrafiltration and Nanofiltration

Ultrafiltration and nanofiltration are little used in volatile analyses as their selectivity is based on filtration of compounds through pores whose size is normally much larger than that of flavourings. However, nanofiltration has been successfully applied to the enrichment of non-volatile compounds responsible for taste [66].

2.4. COMBINED METHODS

2.4.1. Simultaneous Distillation-Extraction (SDE)

Simultaneous distillation-extraction, the most popular isolation technique in flavour and fragrance laboratories, combines volatility and extractability in a single step. The original device [67] has been modified by many scientists (e.g. [68]) to improve recoveries and to use small quantities of material. An aqueous solution or a slurry of the sample is steam distilled. Volatiles are continuously extracted from steam condensates by a solvent reflux (Fig. 2A). Theoretically, this apparatus should allow quantitative recoveries as it is based on a "double closed loop" (water and solvent are continuously recycled) which can be operated for hours. In practice, components with very low volatility such as furaneol would need an infinite isolation time. Two theoretical models have been published [69, 70]. The most recent one takes into account the extraction efficiency and a possible aroma redistillation from the solvent flask:

$$\rho_i = (k_{aw}SK/(k_{aw}SK+k_{as}W))[1-\exp\{-(k_{aw}Sk_{sw}+k_{as}W)F_sF_wt/WS(k_{aw}F_s+F_w)\}] \quad [70] \qquad (4)$$

(k_{aw}: air-to-water partition coefficient, k_{as}: air-to-solvent partition coefficient, S: volume of solvent, W: volume of water, k_{sw}: solvent-to-water partition coefficient, F_s: solvent flow, F_w: steam flow, t: duration of the isolation).

However, steam distillation at atmospheric pressure gives rise to formation of thermal artifacts [38]. Therefore composition of the extracts is close to that of an essential oil, but atmospheric pressure-SDE is not applicable to heat sensitive products. With this restriction, SDE gives the most representative GC-profiles over a wide range of volatility compared to other isolation means [71]. Applications are too numerous to be listed.

To avoid thermal modification of flavours and fragrances, a device working under vacuum (V-SDE), at nearly room temperature was proposed [72] (Fig. 2B). High recoveries were obtained without thermal artifacts, and a preparative version allows large scale operations [70,73].

2.4.2. Steam Distillation-Solid Phase Extraction

Steam distillation-solid phase extraction, to our knowledge, has only been used once in flavour and fragrance analysis, for the isolation of an essential oil [74].

2.5. REPRESENTATIVENESS

Researchers have often tried to check whether their isolation method is selective. Up to the 90's, this was estimated using analytical tools and model mixtures of volatiles that were submitted to a given isolation procedure [71, 72]. As theoretical models exist for many sample preparation methods, it is possible to calculate recoveries. In Table 1 yields after SPME-HS, liquid sampling-SPME, P&T-HS, and SDE are simulated using equations 2, 3, 4, and 5, respectively, for a mixture of 2-propanone, pyridine, 2-

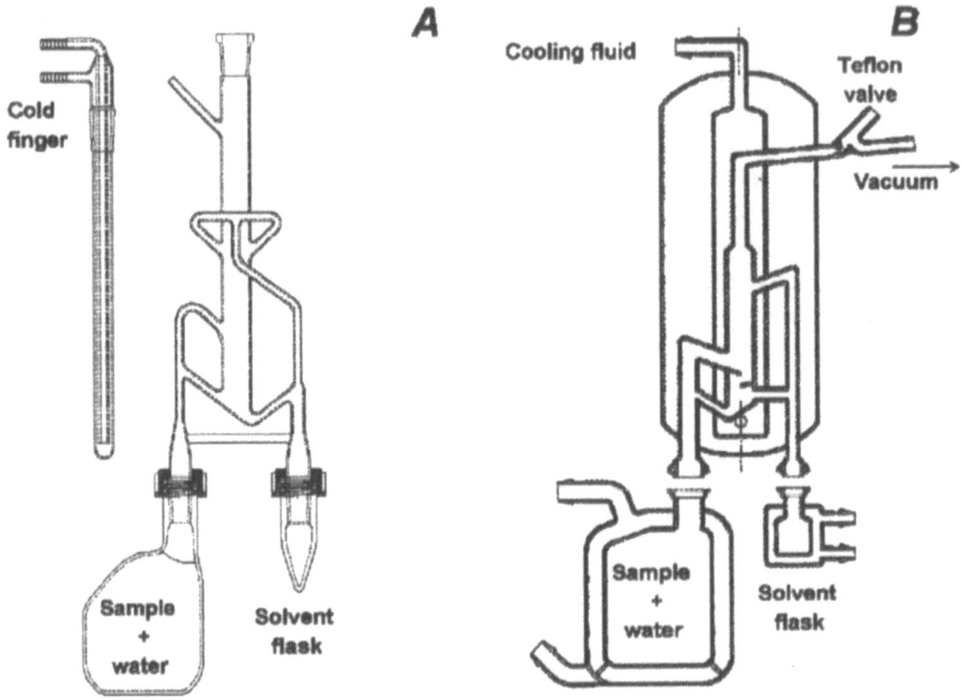

Figure 2A. Atmospheric pressure SDE. **B**. Vacuum SDE.
(Reprinted from [68] and [72], copyright 1981 & 1992, with permission from Elsevier Science)

phenylethanol and vanillin. Results illustrate the influence of the sampling method on the final GC profile. As mentioned above, SDE approaches quantitative recoveries, except for compounds having a very low volatility (such as vanillin). In spite of low yields, SPME recoveries are often sufficient as these amounts are totally transferred into the GC column, and P&T-HS yields may be improved by greater stripping gas volumes. Under the conditions given in the table's footnote, sampling with SPME in the gas (1 ml headspace) or the liquid phase does not modify recoveries. Even using an abnormal headspace volume of 100 ml does not significantly affect recoveries, because the second term of equation 3 denominator is often negligible due to low air-to-water partition coefficients.

However, the ultimate goal is that the isolate exhibits the same sensorial properties as the original plant or food since the composition of an aroma is still unknown at this early stage, hence thermodynamic constants of its constituents are unknown as well. Sometimes authors informally compare the odour of an extract to that of the starting material, without any rational approach. As the sensorial representativeness can only be

Table 1. Simulated recoveries from water using different sampling techniques

	SDE solvent: CH_2Cl_2			SPME-liquid (PDMS)	
	k_{aw} (100°C)	k_{sw}	yield (%)	k_{fw}	yield (%)
vanillin	2.9e-2	1.8	0.5	1.90e-3	2.51e-5
2-propanone	59	2.8	96.6	8.35e-2	1.10e-3
2-phenylethanol	4	1.7	51.6	3.27	4.32e-2
pyridine	20.5	3.3	98.2	59.7	7.82e-1

	SPME & P&T-HS	SPME-HS (PDMS) for V_g=		P&T-HS
	k_{aw} (25°C)	1ml yield (%)	100ml yield (%)	yield (%)
vanillin	6.49e-9	2.51e-5	2.51e-5	2.59e-5
2-propanone	1.47e-3	1.10e-3	9.61e-4	5.71
2-phenylethanol	1.57e-4	4.32e-2	4.25e-2	6.26e-1
pyridine	5.70e-4	7.82e-1	7.40e-1	2.25

Variables: t: duration of SDE isolation; F_s and F_w: SDE flows of solvent and water distillate; S and W: volume of solvent and water (SDE). V_l, V_g, V_f: volumes of the liquid, of the gas and of the fiber adsorbent, respectively (for P&T-HS, V_g is the total volume of stripping gas). Numerical values are those of typical isolations for each method:

SDE: t = 60 min; F_s = 1.3 ml/min; F_w = 0.85 ml/min; S = 10ml; W = 200 ml; k_{aw} and k_{sw} come from [70].

SPME: V_l = 1 ml; V_g = 1 or 100 ml; V_f = 1.32e-4 ml for a film thickness of 7μm. Equation 5.4 of reference [58] was used to estimate k_{fw}, and k_{aw} comes from the literature (propanone and pyridine) or are calculated (vanillin and 2-phenylethanol) using "Properties Plus™" software, version 9.3-1 (Aspen Technology, Cambridge, Massachusetts, USA).

P&T-HS: V_l = 2 ml; V_g = 200 ml.

checked by sensory analysis, this was recently undertaken to evaluate the quality of alcoholic beverage extracts [75, 76, 77]. The number of literature examples is still too limited to draw any general conclusions as to which isolation method gives the most representative isolates.

3. Qualitative Analysis

3.1. SEPARATION OF VOLATILE CONSTITUENTS

Capillary GC is now the widest used technique in all laboratories analysing volatiles. It will not be discussed in this chapter, as excellent reviews have already been dedicated to its applications in this area (e.g. [5, 78, 79]).

3.1.1. Multidimensional Gas Chromatography (MDGC)

Multidimensional gas chromatography (MDGC) has greatly increased the separation capabilities of capillary GC. A preseparation is performed into the first column, then a small and complex portion of elute ("heart-cutting") is transferred to the second column whose polarity differs from the first one. The commercialisation of fully automated MDGC's has made heart-cutting easier.

3.1.2.Chiral GC columns

Chiral GC columns also became "adult" within this last decade allowing the reproducible separation of most enantiomers. Phases are generally based on cyclodextrin derivatives. Compilations of published separations have been reviewed [80]. Used as a second column in a MDGC system, it allows to transfer only optically active compounds from the first column to the chiral one [81, 82], (Fig. 3).

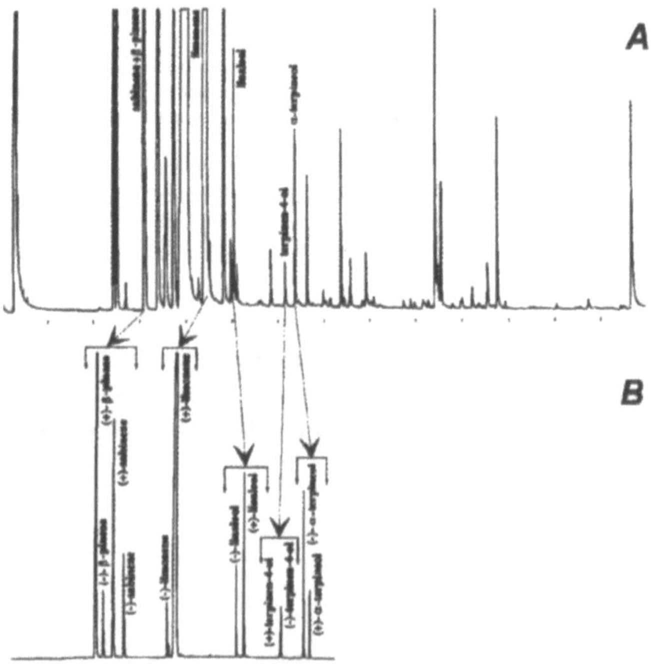

Figure 3. A. MDGC of a mandarin oil. **B.** Chromatogram of the components transferred in the chiral column. (Reprinted with permission from [81], copyright 1998, American Chemical Society)

Determination of the enantiomeric distribution of optically active compounds has now become an help in authentication of natural aroma [47, 83]. However the optical purity alone is not a conclusive criterion, and it must be associated with another technique (see 3.2.2.)

3.1.3. Preparative Capillary GC

Its recent development is linked to that of MDGC as these instruments can be equipped with a fraction collector. Much finer resolution may be achieved compared to old packed column systems. Lower injection volumes are compensated by automated repetitive injections. This is a unique tool to isolate an unknown compound from a complex mixture for further spectral analysis [84].

3.1.4. Profile Recognition

For specific applications such as quality control, it may be useful to automatically compare GC profiles of different products. For this purpose, GC peaks are characterised by their retention indices and their areas. Libraries containing matrices of indices and areas of all constituents may be generated and compared to another profile of the same type with an appropriate algorithm [85, 86, 87]. It is noteworthy that quantitative comparisons of natural products such as essential oils greatly vary according to many parameters (geographical origin, climatic conditions, plant processing). This makes building a reliable data-bank for automated recognition of a natural composition in a mixture tedious [85]. To enhance recognition reliability, the algorithm can also associate the similarity of mass spectra of individual GC peaks with those of the memorised components of, e.g. an essential oil. As little as 1% of an essential oil may be retrieved from a complex mixture [87].

3.1.5. Derivatisation

Derivatisation of compounds with low volatility has not received much attention from flavour analysts. It remains a useful technique to investigate flavour precursors such as glucosides in wine [88], and Amadori compounds in Maillard reactions [89].

3.2. IDENTIFICATION OF SENSORY-RELEVANT CONSTITUENTS

The simple detectors used in gas chromatography (FID, FPD, NPD) do not allow identification of the flavour constituents. Retention indices on capillary columns [90] have been shown to be reproducible with relative standard deviations of less than 1% being observed between measurements from different laboratories for polar (ethylene glycol type) or non-polar (methyl siloxane type) phases [91]. However, identification by retention indices on a single column is not acceptable. The use of two different polarities improves recognition of the components [92], but confusion is still possible [91]. Therefore the International Organisation of Flavor Industry (IOFI), as well as serious scientific journals strongly recommend the conjunction of 2 different techniques (e.g. retention indices and mass spectra) to identify a compound with a sufficient degree of confidence.

3.2.1. Mass Spectrometry

Classical GC-MS with electronic impact ionisation (EI) is a basic tool in any flavour or fragrance laboratory and does not need detailed description here. Its use in SIM (simultaneous ion monitoring) mode increases the signal-to-noise ratio and allows the detection of low abundant components and their quantification (method 24 of [93]). More specific results may be obtained using positive or negative chemical ionization (PCI and NCI, respectively). For instance, PCI using hydroxyl ions is able to substitute carboxylate groups from esters in the plasma to generate carboxylate ions [94]. Consequently acetates, propanoates, etc., may be easily visualised by extracting ion traces at m/z 59, 73, etc., respectively. Isomers exhibiting identical EI spectra in scan mode (e.g. alkyl-thiocyanates and isothiocyanates) may sometimes be differentiated using chemical ionisation [95].

Two uncommon ionisation techniques are of special interest: proton transfer-mass spectrometry involves $H3O^+$ as a reactant in a specially designed CI source. This mild ionisation mode leads to very few fragments and to almost only $[MH]^+$ quasi molecular ions. Consequently its ability to identify is limited, but the concentration of several known target compounds of a complex mixture may be followed on-line without any chromatographic separation. This requires that no isobaric components interfere with ions of interest. Volatiles emitted by plants, or released by a food in the breath during eating have been monitored [96].

Resonance-enhanced multiphoton ionisation uses a UV-laser excitation which offers a great ionisation selectivity as only those components that are excitable by the selected laser wavelength become charged and transferred into a time-of-flight mass spectrometer. This again permits on-line analysis of a headspace surrounding, e.g. a beverage [97] without chromatographic separation because detected ions are restricted to only the class of compounds that are excited by the laser.

Liquid chromatography-mass spectrometry (LC-MS) has a restricted application for volatiles because of the limited separation performance of HPLC and because volatiles could be lost with the solvent in the interface. However it is a useful technique for investigating non-volatile aroma precursors [44, 98, 99] or to analyse taste impact components [100]. In the above mentioned references, various types of interface were used: atmospheric pressure chemical ionisation (APCI), thermospray or electrospray ionisation.

Tandem mass spectrometry (MS-MS) is still little used in spite of its great possibilities [101]. Ions corresponding to a given m/z value may be filtered by the first mass analyser, then fragmented in the collision chamber, daughter ions being scanned by the last analyser for identification [40]. The reverse procedure (search for parent ions), as well as searches for neutral losses are also possible. Alternatively the second analyser may also operate in SIM mode. In such ways the signal-to-noise ratio is greatly increased, leading to the detection of trace components that are not detectable by classical MS in SIM mode [36]. MS-MS also helps to understand the fragmentation patterns of compounds [102] or the formation mechanism of aroma compounds using labeled precursors [40, 99, 103, 104].

3.2.2. Isotopic Ratio-Mass Spectrometry (IRMS)

This technique is mainly dedicated to authentication of natural products. For instance determination of the $^{13}C/^{12}C$ ratio involves combustion of the organic compound, followed by comparison of the $^{13}CO_2$ abundance's from the target compound to those of a reference gas in a specially designed mass spectrometer. The method has been hyphenated to GC, and sometimes to MDGC, by using on-line oxidation by copper oxide in a microfurnace [105]. Results are expressed in δ:

$$\delta^{13}C_{PDB}\,(‰) = 1000 \times [\,(^{13}C/^{12}C)_{sample} - (^{13}C/^{12}C)_{PDB}]/(^{13}C/^{12}C)_{PDB} \qquad (5)$$

The reference value for CO_2 is that given by the carbonate degradation of fossil *Belemnitilla americana* from the Peedee formation (PDB).

IRMS was first applied to vanillin [106]. These days associating IRMS results with natural enantiomeric distribution enhances the reliability of authentication's [107].

3.2.3. Fourier Transformed Infrared Spectrometry (FTIR)

GC-FTIR is very complementary to GC-MS as structural information can also be drawn from IR spectra. Substituted pyrazines sometimes exhibit similar MS spectra and retention indices. FTIR not only differentiates them, but also allows to assign the position of substituents [108].

GC-FTIR hyphenation uses either a "light pipe" or a "matrix isolation" interface to record spectra in the gas phase or in a liquid argon matrix, respectively [109, 110]. Therefore differences exist between both types of spectra. FTIR has also been hyphenated to supercritical fluid chromatography (SFC-FTIR) and applied to the analysis of sesquiterpenes [111].

3.2.4. Atomic Emission Detector (AED)

The use of an atomic emission detector not only allows the specific detection of nearly all elements founds in GC effluents, but it also enables the determination of empirical formulas for unknown peaks [112].

3.2.5. Nuclear Magnetic Resonance (NMR)

Usual ^{1}H or ^{13}C NMR will not be mentioned here as they are normally applied to pure compounds that have been previously isolated using, e.g. preparative MDGC [84]. This use does not differ from that of classical NMR in organic chemistry. However HPLC/NMR hyphenation now exists and has been applied to taste impact compounds in hops [113], but its sensitivity and resolution remain generally too low for on-line structural analysis of odorants.

A special application of ^{2}H-NMR to flavour authentication has been called SNIF-NMR (site-specific natural isotope fractionation), [114]. Integration of the deuterium signals gives the distribution of this isotope between the different sites of the compound. To ascertain the natural origin of a compound, this distribution is compared with that of an authentic natural flavour [115, 116, 117]. SNIF-NMR has also been used to study the biosynthetic pathway of raspberry ketone [118].

3.2.6. GC-Olfactometry (GC-O, also called GC-Sniffing)

GC-O does not provide the analyst with an unambiguous identification. However, combining retention indices with odour descriptors of a given compound may lead to a restricted list of possible candidates, from which the right one must be confirmed by a spectral technique (example in [36]). Nevertheless, GC-O possesses a unique sensitivity which is much superior to any electronic detector for some odour active compounds which exhibit extremely low odour thresholds. Commercially available sniffing ports normally use a make-up of humid air to avoid nose dryness and to facilitate odour recognition (odours in dry air are perceived differently).

GC-O has been used for a long time without any special procedure, leading to results having a low reliability. This situation has been improved during the last decade by 3 data-treatment procedures. CHARM [119] and AEDA [120] were proposed in 1984 and 1987, respectively. Both techniques are based on a series of dilution's of the sample extract, until no odour is perceived at the sniffing port. In other terms, the sample is diluted and sniffed as long as the threshold concentration of all components is not reached. For AEDA, odorants of an extract are characterised by their FD values (factor of dilution at which an odour is perceived for the last time). CHARM uses a computerised data treatment to reconstruct an aromagram corresponding to odors detected at the different dilution levels, as a continuous function of time. Most papers reporting the determination of aroma impact components in various foods are based on this dilution approach [121, 122].

McDaniel (OSME method, 1989) [123] proposed to record a signal by moving a cursor as a function of the perceived odour intensity. This requires "calibration" of panelists by training prior to the GC-O run. OSME has mainly been applied to wines [123, 124].

Dilution methods have been criticised [125] due to their insufficient reproducibility and their "gaps in coincident responses". Therefore, 2 groups introduced simultaneously a new data treatment procedure [126,127]. Instead of 1 or 2 panelists, like other methods, the GC-O is repeated 6 to 10 times at the same concentration with different panelists. An odourant is then characterised by its detection frequency ("NIF") by the panel. Peak areas ("SNIF") may also be used. Using such a concept, aromagrams appear to be repeatable with the same panel, as well as reproducible by 2 different panels without training [128]. Determination of impact odorants of Champagne wines using the 3 above mentioned methods has shown similar results for OSME and detection frequencies [124].

3.3. SENSOR ARRAYS

Sensor arrays are not a real analytical technique as they neither allow a separation of odour constituents, nor their identification. Sensors are generally based on semiconductor metal oxides, conductive polymers or quartz resonators. The presence of volatile organic compounds in the sample's headspace changes the sensor's conductivity or resonant frequency differently according to the nature of the sensor's material and the nature of the organic compounds [129]. Collection of signals given by an array of sensors gives a response profile more or less characteristic of the detected odour. As this approach somewhat mimics odour detection by the array of receptors in the nose, the technique is often improperly called "electronic nose". However its sensitivity and selectivity can not be compared to those of mammalian perception which is able to detect a component having an extremely low threshold in a complex mixture of less intense odorants (e.g. 1-100 ppb of 2,4,6-trichloroanisole in coffee (Rio off-flavour, [130]). Therefore sensor arrays are used in quality assurance to classify global odour profiles without giving information on their composition [131]. Neural networks may be employed for the same purpose.

Two recent attempts have been made to use sensors as GC detectors [132, 133]. Constituents of a butter aroma model were detected down to 0.3-5 ng. However, no application to GC-detection of a real food or plant extract has ever been reported.

4. Quantitative Analyses

There is no general quantification procedure that may be applied to any volatile in any matrix. Each case requires the method's validity to be checked in the real conditions in which the target compound(s) occur(s). Most important guidelines that may be drawn from the literature are given hereafter.

4.1. SAMPLE PREPARATION

Whereas sample preparation for qualitative analysis aims to recover the highest possible percentage of all sensory relevant compounds, only target ingredients need to be isolated either quantitatively, or in a known proportion. When none of these objectives can be achieved, the difficulty can be overcome using an isotopically labeled standard. This possibility, applicable to all sample preparation techniques mentioned hereafter, is the most universal and accurate method as losses of target and labeled compounds are identical and their ratio remains unchanged during the enrichment step. This ratio must be measured with a mass spectrometer (see 4.2.3.).

Only the most important sample treatments for quantitative analysis are listed hereafter.

4.1.1. Extraction

Batch extraction (liquid/liquid extraction in a separating funnel, SPME with liquid sampling) does not yield quantitative recoveries. However, using a fixed sample volume, and the same temperature and stirring conditions, reproducible extraction yields are feasible with SPME, specially when using an automated autosampler. This has even been proven to be valuable for a reproducible sampling duration, when equilibrium between the phases was not reached [134]. Consequently model standard solutions in water can be used to establish calibration curves. However, this is only applicable to the analysis of water itself. As foods and plants generally also imply the occurrence of sugars, biopolymers, etc., which could pollute the SPME fiber, headspace sampling is normally preferred (see above).

As already mentioned (2.2.1.), when other methods fail (e.g. isolation of very polar components), a solvent extraction is sometimes unavoidable. Because of questionable recoveries, a labeled internal standard must be added to the sample prior to its extraction [135].

Continuous liquid/liquid and solid/liquid extraction's (Soxhlet) arc known to reach quantitative recoveries for a variety of organic compounds. SPE on a column or a clean-up cartridge reflects a similar situation, when some precautions are respected. If the percolated volume does not exceed breakthrough, and if reproducible volumes of sample and rinsing solvent are used to evaluate recoveries of the target components and the IS, the extraction efficiencies may be calculated. Concentrations in the original sample can thus be obtained [50].

4.1.2. Headspace

Quantitation in S-HS has been recently reviewed [14]. Internal and external standardisation may be used, but standards must be prepared in the same matrix as the sample to ensure that, if both headspace concentrations are equal, concentrations in condensed phases are also equal. This difficulty does not happen with a standard addition procedure as it uses the sample itself as the solvent. For this reason, it is a widely applicable technique in HS sampling.

Multiple headspace extraction (MHE) has been developed to eliminate matrix effects. It is specially suitable for complex matrices (heterogeneous, solid, viscous, etc.). Therefore it has mainly been applied to environmental samples (e.g. soil analysis), but also to food (e.g. butter). MHE consists of successive withdrawals and quantitation of the headspace in equilibrium with a given sample in a closed vessel. Volatiles are thus extracted from the matrix when a new pure gas volume replaces the previous one. Extracted amounts decrease according to:

$$\ln(A_i) = -k(n-1) + \ln(A_1) \quad [136] \qquad (6)$$

(A_i: amount of analyte in the i-th headspace withdrawal; n: number of extractions; A_1: amount of analyte in the first withdrawal).

Consequently, the total amount A_T originally present in the sample is the sum of all extracted amounts:

$$A_T = \Sigma(A_i) = A_1/(1-e^{-k}) \quad [136] \qquad (7)$$

k can be obtained as the slope of the straight line (eq. 6) representing $\ln(A_i)$ as a function of (n-1). Using MHE, halocarbons were quantified in butter with a coefficient of variation of 1.4 to 5.5% , and recoveries were higher than 82% [136].

Quantitation is not an easy task with P&T-HS because recoveries of volatiles are highly dependent on the sample temperature, stripping duration, and different air-to-matrix partition coefficients of components. This can be partially overcome by fully automated P&T autosamplers.

An IS may be added to the sample prior to gas stripping, then a calibration curve is established by submitting a model solution containing standards and the IS to exactly the same P&T conditions. Evaluating P&T recoveries from crushed tomatoes using an aqueous model solution [137] does not take into account possible interactions of volatiles with tomato non-volatiles. Therefore, preparing the standard solution in the investigated matrix (free of target compounds) should be preferred, as proposed by Baldwin for S-HS analysis of tomatoes. He removed volatiles by distillation and used the deodorised juice as the solvent to dilute the standards [138].

The S&T-HS technique can determine the concentration of organic compounds present in the gas phase after calculation of individual partition coefficients and calibration with an aqueous solution. Concentration in the matrix can be obtained by MHE [35]. A quicker way is addition of standards to the sample. Yogurt volatiles have been quantified with a standard deviation of less than 5.5% [139].

SPME-HS is commonly used for quantitation, keeping the temperature, the stirring rate, and the sampling time constant, even under non-equilibrium conditions [134]. External calibration [29], standard addition [32], or labeled standards [29] may be used.

4.1.3. Combined Methods

For SDE isolations, an IS is usually added to the solvent flask before the distillation-extraction process [68]. When the sample matrix is very retentive (e.g. lipids), a second IS may be added to the sample flask to check recovery efficiency [140]. Both IS structures must be similar. The best IS choice remains an isotopomer of the target compound, added to the sample flask prior to isolation [141].

Steam distillation-SPE has also been used in quantitation, but its reliability is unknown as the IS was added to the extract after isolation [74].

4.2. SEPARATION AND QUANTIFICATION

4.2.1. GC

The applicability of GC to quantitation has been widely established and reviewed [79, 142]. Internal normalisation is little used in food and fragrance analysis as it requires the reconstitution of a model mixture containing all constituents of the sample. Unfortunately many reference compounds of natural flavours and scents remain unknown or unavailable. Therefore, internal and external standardisation and standard additions are used. As external standardisation requires a high reproducibility of the injected volume, internal standardisation should be preferred. Various detectors are used (mainly FID, NPD, FPD), but their linear dynamic ranges differ (10^6, 10^4, 10^3, respectively) [142]. General recommendations in the food area can be found in [93]. MDGC also exhibits quantitative performances [143].

4.2.2. HPLC

In spite of its restricted use in qualitative flavour and fragrance analysis, HPLC sensitivity is often exploited to quantitate target constituents of essential oils and foods, whose consumption is restricted or forbidden. Coumarin ([93], method 27), safrole [144], etc., belong to this family for which specific HPLC methods are published. As the methodology does not differ from normal practice of HPLC analysis, readers can refer to classical literature [141].

4.2.3. MS

Methods mentioned above for GC are valid with MS detection. It should be pointed out that the scan mode is inappropriate for quantitation, and that SIM mode gives accurate results ([93], method 24). The possibility of using labeled standards, called isotope dilution, is unique to MS. In addition to advantages already mentioned in sample preparation, the target compound and its isotopomer elute simultaneously, if their difference in molecular weight is low. Care must be taken to avoid labeling in exchangeable positions (e.g. deuterium vicinal to a carbonyl). Isotope dilution is applicable to GC/MS with electronic [141] or chemical ionization [135], FAB-MS/MS [104], LC-MS/MS [100], etc.

4.2.4. GC-FTIR

Although FTIR detection obeys Beer's law, and is suitable for quantitation, it has not significantly held the attention of flavour and fragrance analysts.

4.2.5. GC-O

Although sensory detection may not seem very quantitative, recent improvements are promising. With the GC-"SNIF" technique, peak areas follow a sigmoidal relationship as a function of the logarithm of concentration for components having a unimodal detection threshold. This sigmoid can be linearised by transforming its points into probits. Thus a straight calibration line of area probits versus the logarithm of concentration can be drawn [128]. A linear relationship has also been found between the detection frequency itself and the logarithm of concentration [127]. This is presumably because the middle part of a sigmoid can be assimilated to a straight line. These observations have allowed quantitative comparisons of aromagrams of rehydrated French beans [127], milk and yogurt [36] and brewed and instant coffee [37].

5. Data-banks

The development of computers has allowed construction of data-banks which greatly help in flavour and fragrance analysis. Many analysts have developed their own libraries, although commercial data-banks also exist in the flavour and fragrance area. Unfortunately not all compilations are commercially available in a computerised form.

5.1 GC

Retention indices have been compiled [145], also for flavour compounds [146], but their identification ability remains limited if they are not associated with a spectral library, as previously mentioned (see 3.2.).

5.2. MS

Huge MS libraries cover a wide range of areas: environment, perfumes, flavours, etc. [147]. More aroma-specific ones are also available [148, 149]. Combining a MS library and a retention index search in a single step with the same program saves time and improves identification reliability [150, 151].

5.3. FTIR

Hyphenation of GC to FTIR with to a light-pipe interface requires the use of specific vapor phase IR libraries [152,153].

5.4. NMR

The authentication of natural products has led to compilation of ^{2}H-NMR spectra of reference components isolated from plants or fermentations [154]. The list is restricted to most commercially important compounds (vanillin, benzaldehyde, etc.).

5.5 GC-O

GC-O performances may be significantly enhanced when it is associated with a data-bank of odour descriptors and retention indices [151, 155]. The limitation of such an approach obviously lies in the semantic interpretation of terms used to describe an odour as the vocabulary can not be standardised independently from the culture of the individual assessors.

5.6. MISCELLANEOUS

It is worthwhile mentioning collections of volatile components found in plants, flavours and fragrances. They include names, CAS registry numbers, natural occurrences [156,157], suppliers [158], and sometimes analytical and chromatographic data [151]. Even if they do not contain spectral information, they greatly help the analyst.

6. Conclusion

The systematic identification of all constituents of flavours and fragrances is now declining in importance. The new trend of the '90s focuses on major sensory relevant compounds which are identified and quantified with higher reliability. Technical improvements are characterised by miniaturisation (e.g. SPME), and hyphenation (e.g. sample treatment-GC, GC-new detectors).

From this review it can be concluded that no ideal nor universal technique exists for flavour analysis. It is the analyst's responsibility to choose a suitable approach for each step of the procedure: extraction, component separation, identification, quantification. As flavours and fragrances are mixtures of many chemicals with very different behaviors, they often lead to contradictory analytical requirements. This is even further complicated by the various matrices in which they occur. The right analytical strategy is often a compromise between these different requirements.

Acknowledgments: we gratefully acknowledge Dr. D. Prètre for calculating missing air-to-water partition coefficients, and colleagues who spent time reviewing this article: Profs. C. Bicchi, and A. Voilley, Drs. L. Fay, E. Prior, P. Visani, D. Welti. and Mr. P. Pollien.

7. Bibliography

[1].Bett, K. L. and Grimm, C. C. (1994) Flavor and aroma. Its measurement. *Advances in Meat Research*, **9**, 202-221.
[2].MacLeod, A. J. (1973) *Instrumental methods of meat flavour analysis,* London.
[3].Maarse, H. (1991) Introduction. In: *Volatile compounds in foods and beverages.*, Maarse, New York.
[4].Buttery, R. and Ling, L. C. (1996) Methods for isolating food and plant volatiles. In: *Biotechnology for improved foods and flavours*, Takeoka, Teranishi, Williams & Kobayashi, Washington, DC.
[5].Bicchi, C. *et al.* (1987) *Capillary gas chromatography in essential oils analysis*, Sandra, P. and Bicchi, C., Heidelberg, (D).
[6].Bicchi, C., D'Amato, A., Frattini, C. (1983) Improved method for the analysis of small amounts of essential oils by microdistillation followed by capillary gas chromatography. *Journal of Chromatography.*, **279**, 409-416.
[7].Bicchi, C., Frattini, C., Sacco, T. (1985) Essential oil of three Asiatic Artemisia species. *Phytochemistry*, **24**, 2440-2442.

[8].Garnero, J. (1977) Problèmes rencontrés au cours de l'étude de la composition chimique des huiles essentielles. *Parfums Cosmétiques Arômes*, **14**, 31-40.
[9].Dumont, J. P., Adda, J. (1972) Isolement des constituants de l'arôme des fromages: comparaison de méthodes. Le Lait, 515-516, 311-323.
[10].Joulain, D. (1986) Study of the fragrance given by certain springtime flowers. In: *Progress in essential oil research*, Brunke, Berlin.
[11].Jaulmes, P. (1963) De la distillation a la chromatographie. Mises au point de *chimie analytique*, **11**, 153-187.
[12].Joulain, D. (1993) Cryogenic vacuum trapping of scents from temperate and tropical flowers. In: *Bioactive volatiles from plants.*, Teranishi, Buttery & Sugisawa, Columbus (O, USA).
[13].Rao, B. R. (1980) Determination of the maximum emission from storage tanks for heavy fuel oil. In: *Applied headspace gas chromatography*, Kolb, London.
[14].Kolb, B. and Ettre, L. (1997) *Static headspace-gas chromatography*, Kolb, B. and Ettre, L., New York.
[15].Abeel, S. M., Vickers, A. K., Decker, D. (1994) Trends in purge and trap. *Journal of Chromatographic Science*, **32**, 328-338.
[16].Rapp, A., Knippser, W. (1980) Eine neue Methode zur Enreicherung von Dampfkomponenten. Dargestellt am Beispiel des Weines. *Chromatographia*, **13**, 698-702.
[17].Pankow, J. F., Rosen, M. E. (1988) Determination of volatile compounds in water by purging directly to a capillary column with whole column cryotrapping. *Environmental Science and Technology*, **22**, 398-405.
[18].Buttery, R. G., Teranishi, R., Ling, L. C. (1987) Fresh tomato volatiles: a quantitative study. *Journal of Agricultural and Food Chemistry*, **35**, 540-544.
[19].Nerin, C., Rubio, C., Cacho, J., Salafranca, J. (1998) Parts-per-trillion determination of styrene in yogurt by purge-and-trap gas chromatography with mass spectrometry detection. *Food Additives and Contaminants*, **15**, 346-354.
[20].Bayer, C. W. (1994) Advances in trapping procedures for organic indoor pollutants. *Journal of Chromatographic Science*, **32**, 312-316.
[21].Maier, I., Fieber, M. (1988) Retention characteristics of volatile compounds on Tenax TA. *Journal of High Resolution Chromatography*, **11**, 566-576.
[22].Rojas, J. A., Voilley, A., Thomas, C. (1991) Study of adsorption of 1-octen-3-ol on porous hydrophobic polymers. *Gas Separation and Purification*, **5**, 187-194.
[23].Toulemonde, B. and Beauverd, D. (1984) Headspace analysis: trap desorption by microwave energy. Application to the volatile components of some tropical fruits. In: *Progress in flavour research 1984*, Adda, Amsterdam.
[24].Driss, M. R., Bouguerra, M. L. (1994) Analysis of volatile organic compounds in water by purge-and-trap and gas chromatography techniques. *International Journal of Environmental and Analytical Chemistry*, **45**, 193-204.
[25].Heath, R. R., Manukian, A. (1994) An automated system for use in collecting volatiles released from plants. *Journal of Chemical Ecology*, **20**, 593-608.
[26].Kaiser, R. (1991) Trapping, investigation and reconstitution of flower scents. In: Perfumes, Müller & Lamparsky, London.
[27].Bicchi, C., D'Amato, A., David, F., Sandra, P. (1989) Capturing of volatiles emitted by living plants by means of thick film open tubular traps. *Journal of High Resolution Chromatography* , **12**, 316-321.
[28].Bicchi, C., Joulain, D. (1990) Headspace-gas chromatographic analysis of medicinal and aromatic plants and flowers. *Flavour and Fragrance Journal*, **5**, 131-145.
[29].Zhang, Z., Pawliszyn, J. (1993) Headspace solid-phase microextraction. *Analytical Chemistry*, **65**, 1843-1852.
[30].Yang, X., Peppard, T. (1994) Solid phase microextraction for flavour analysis. *Journal of Agricultural and Food Chemistry*, **42**, 1925-1930.
[31].Bicchi, C., Panero, O., Pellegrino, G. M., Vanni, A. C. (1997) Characterisation of roasted coffee and coffee beverages by solid phase microextraction-gas chromatography and principal component analysis. *Journal of Agricultural and Food Chemistry*, **45**, 4680-4686.
[32].Steffen, A., Pawliszyn, J. (1996) Analysis of flavour volatiles using headspace solid-phase microextraction. *Journal of Agricultural and Food Chemistry*, **44**, 2187-2193.
[33].Pelusio, F., Nilsson, T., Montanarella, L., Tilio, R., Laesen, B., Facchietti, S. *et al.* (1995) Headspace solid phase microextraction analysis of volatile organic sulfur compounds in black and white truffle aroma. *Journal of Agriculture and Food Chemistry*, **43**, 2138-2143.
[34].Chin, H. W., Bernhard, R. A., Rosenberg, M. (1996) Solid phase microextraction for cheese volatile compound analysis. *Journal of Food Science*, **61**, 1118-1128.

[35].Chaintreau, A., Grade, A., Munoz-Box, R. (1995) Determination of partition coefficients and quantitation of headspace volatile compounds. *Analytical Chemistry*, **34**, 3300-3304.
[36].Ott, A., Fay, L. B., Chaintreau, A. (1997) Determination and origin of the aroma impact compounds of yogurt flavour. *Journal of Agricultural and Food Chemistry*, **45**, 850-858.
[37].Pollien, P., Krebs, Y., Chaintreau, A. (1998) Comparison of a brew and an instant coffee using a new GC-olfactometric method. *Association Scientifique Internationale du Café*, 191-196, Paris.
[38].Shimoda, M., Shigematsu, H., Shiratsuchi, H., Osajima, Y. (1995) Comparison of the odour concentrates by SDE and adsorptive column method from green tea infusion. *Journal of Agricultural and Food Chemistry*, **43**, 1616-1620.
[39].Pickenhagen, W., Velluz, A., Passerat, J. P., Ohloff, G. (1981) Estimation of 2,5-dimethyl-4-hydroxy-3(2H)-furanone (furaneol) in cultivated and wild strawberries, pineapples and mangoes. *Journal of the Science of Food and Agriculture*, **32**, 1132-1134.
[40].Blank, I., Fay, L. B. (1996) Formation of 4-hydroxy-2,5-dimethyl-3(2H)-furanone and 4-hydroxy-2(or 5)-ethyl-5(or 2)-methyl-3(2H) furanone through Maillard reaction based on pentose sugars. *Journal of Agricultural and Food Chemistry*, **44**, 531-536.
[41].Mussinan, C. J., Walradt, J. P. (1975) Organic acid from fresh California strawberries. *Journal of Agricultural and Food Chemistry*, **23**, 482-484.
[42].Stevens, K. L., Flath, R. A., Lee, A., Stern, D. J. (1969) Volatiles from grapes. Comparison of Grenache juice and grenache rosé wine. *Journal of Agricultural and Food Chemistry*, **17**, 1102-1106.
[43].Sen, A., Laskawy, G., Schieberle, P., Grosch, W. (1991) Quantitative determination of damascenone in foods using a stable isotope dilution assay. *Journal of Agricultural and Food Chemistry*, **39**, 757-759.
[44].Calvey, E. M., Matusik, J. E., White, K. D., Deorazio, R., Sha, D. Y., Block, E. (1997) Allium chemistry - supercritical fluid extraction and LC-APCI-MS of thiosulfinates and related compounds from homogenates of garlic, onion, and ramp - identification in garlic and ramp and synthesis of 1-propanesulfinothioic acid S-allyl ester. *Journal of Agricultural & Food Chemistry*, **45**, 4406-4413.
[45].Um, K. W., Bailey, M. E., Clarke, A. D., Chao, R. R. (1992) Concentration and identification of volatile compounds from heated beef fat using supercritical CO2 extraction gas-liquid chromatography/mass spectrometry. *Journal of Agricultural and Food Chemistry*, **40**, 1641-1646.
[46].Ericson, A. P., Matthews, R. F., Teixeira, A. A., Moye, H. A. (1992) Recovery of grapefruit oil from processing waste water using styrene-divinylbenzene resins. *Journal of Food Science*, **57**, 186-189.
[47].Bitteur, S., Rosset, R. (1988) Use of octadecyl-bonded silica and styrene-divinylbenzene copolymer for the recovery of blackcurrent aroma compounds from a food plant waste water. *Journal of Food Science*, **53**, 141-147.
[48].Bitteur, S., Rosset, R. (1987) Comparison of octadecyl-bonded silica and divinylbenzene copolymer sorbents for trace enrichment purposes. Fundamental aspects II. Linear chromatographic behavior of the extracting support. *Chromatographia*, **23**, 163-168.
[49].Parliment, T. H. (1985) Volatile sample preparation using bonded phase adsorbents. In: *Semiochemistry: flavours and pheromones*, Acree & Soderlund, Berlin (D).
[50].Mangas, J. J., Gonzalez, M. P., Rodriguez, R., Blanco, D. (1996) Solid-phase extraction and determination of trace aroma and flavour components in cider by GC-MS. *Chromatographia*, **42**, 101-105.
[51].Krings, U., Kelch, M., Berger, R. G. (1993) Adsorbents for the recovery of aroma compounds in fermentation processes. *Journal of Chemical Technology and Biotechnology*, **58**, 293-299.
[52].Aishima, T., Ozawa, Y. (1989) A simple method for concentrating volatile flavour compounds in aqueous systems using RP-HPLC. *Chromatographia*, **28**, 405-411.
[53].Sun, J. J., Fritz, J. S. (1992) Chemically modified resins for solid-phase extraction. *Journal of Chromatography*, **590**, 197-202.
[54].Parliment, T. H. (1981) Concentration and fractionation of aromas on reverse-phase adsorbents. *Journal of Agricultural and Food Chemistry*, **29**, 836-841.
[55].Coulibaly, K., Jeon, I. J. (1992) Solid-phase extraction of less volatile flavour compounds from ultrahigh-temperature processed milk. *Journal of Agricultural and Food Chemistry*, **40**, 612-616.
[56].Ibañez, E., Bernhard, R. A. (1996) Solid-pase microextraction (SPME) of pyrazines in model reaction systems. *Journal of the Science of Food and Agriculture*, **72**, 91-96.
[57].Vergnais, L., Masson, F., Montel, M. C., Berdague, J. L., Talon, R. (1998) Evaluation of solid-phase microextraction for analysis of volatile metabolites produced by staphylococci. *Journal of Agricultural & Food Chemistry*, **46**, 228-234.
[58].Pawliszyn, J. (1997) *Solid phase microextraction*, New York.
[59].Goosens, E. C., de Jong, D., de Jong, G. J., Brinkman, U. A. T. (1998) On-line sample treatment-capillary gas chromatography. *Chromatographia*, **47**, 313-345.

[60].Voilley, A., Lamer, T. (1990) *La pervaporation et ses applications potentielles dans l'industrie agro-alimentaire.* IAA, octobre, 881-888.

[61].Bengtsson, E., Trägardh, G., Hallström, B. (1992) Concentration of apple juice aroma from evaporator condensate using pervaporation. *Lebensmittel - Wissenschaft und Technologie*, **25**, 29-34.

[62].Spinnler, H. E., Dufosse, L., Souchon, I., Latrasse, A., Piffaud-Juffard, C., Voilley, A. *et al.* (1998) Production de gamma-decalactone par bioconversion. Patent, FR 2 705 971-A1.

[63].Burger, B. V., Burger, W. J. G., Burger, I. (1996) Trace determination of volatile organic compounds in water using permeation through a hollow fiber membrane and carrier gas stripping. *Journal of High Resolution Chromatography*, **19**, 571-576.

[64].Yang, M. J., Pawliszyn, J. (1996) Membrane extraction with a sorbent interface. *LC-GC International*, May, 283-296.

[65].Mihara, S., Bando, S., Harada, K., Ishizuka, N. (1990) Stirring continuous extraction of aqueous organic compounds with fluororesin as water/organic solvent separator. *Journal of Agricultural and Food Chemistry*, **38**, 999-1003.

[66].Sommerer, N., Garem, A., Mollé, D., Septier, C., Le Quéré, J. L., Salles, C. (1998) Isolation of a peptidic fraction from the goat cheese water-soluble extract by nanofiltration for sensory evaluation studies. In: *Food Flavors; formation and packaging influences.* In press.

[67].Likens, S. T., Nickerson, G. B. (1964) *ASBC Proceedings*, 5.

[68].Godefroot, M., Sandra, P., Verzele, M. (1981) New method for quantitative essential oil analysis. *Journal of Chromatography*, **203**, 325-335.

[69].Rijks, J., Curver, J., Noy, T., Cramers, C. (1983) Possibilities and limitations of steam distillation-extraction as a pre-concentration technique for trace analysis of organics by capillary gas chromatography. *Journal of Chromatography*, **279**, 395-407.

[70].Pollien, P., Chaintreau, A. (1997) Simultaneous distillation-extraction: theoretical model and development of a preparative unit. *Analytical Chemistry*, **69**, 3285-3292.

[71].Reineccius, G. (1993) Biases in analytical flavour profiles introduced by isolation method. In: *Flavor measurement*, Ho & Manley, New-York.

[72].Maignial, L., Pibarot, P., Bonetti, G., Chaintreau, A., Marion, J. P. (1992) Simultaneous distillation-extraction under static vacuum: isolation of volatiles at room temperature. *Journal of Chromatography*, **606**, 87-94.

[73].Pollien, P., Ott, A., Fay, L. B., Maignial, L., Chaintreau, A. (1998) Simultaneous Distillation-Extraction: Preparative recovery of volatiles under mild conditions in batch or continuous operations. *Flavour and Fragrance Journal*, In press.

[74].Dix, K. D., Fritz, J. S. (1992) Steam distillation with resin extraction for isolation and concentration of organic compounds from aqueous samples. *Analytica Chimica Acta*, **236**, 43-49.

[75].Abbott, N., Bayonove, C., Etievant, P., Issanchou, S. (1993) Production of wine extracts for chemical and sensory analysis with an odour representative of that of the original wine. In: *Actes du Symposium International "Connaissance Aromatique des Cépages et Qualité des Vins"*, Bayonove, Crouzet & Flanzy. Montpellier, (F).

[76].Moio, L., Chambellant, E., Lesschaeve, I., Issanchou, S., Schlich, P., Etievant, P. (1995) Production of representative wine extracts for chemical and olfactory analysis. *Italian Journal of Food Science*, **3**, 1995, 265.

[77].Abbott, N., Etievant, P., Langlois, D., Lesschaeve, I., Issanchou, S. (1993) Evaluation of the representativeness of the odour of beer extracts prior to analysis by GC eluate sniffing. *Journal of Agricultural and Food Chemistry*, **41**, 777-780.

[78].Bertsch, W. *et al.* (1981) *Recent advances in capillary gas chromatography*, Bertsch, Jennings & Kaiser. Heidelberg (D).

[79].Grant, D. W. (1995) *Capillary gas chromatography*, Grant, D. W., Chichester.

[80].(1993) Collection of enantiomer separation factors obtained by capillary gas chromatography on chiral stationary phases. *Journal of High Resolution Chromatography*, **16**, 312-323.

[81].Mondello, L., Catalfamo, M., Proteggente, A. R., Bonaccorsi, I., Dugo, G. (1998) Multidimensional capillary GC-GC for the analysis of real complex samples - 3 - Enantiomeric distribution of monoterpene hydrocarbons and monoterpene alcohols of mandarin oils. *Journal of Agricultural & Food Chemistry*, **46**, 54-61.

[82].Bartschat, D., Beck, T., Mosandl, A. (1997) Stereoisomeric flavour compounds - 79 - Simultaneous enantioselective analysis of 3-butylphtalide and 3-butylhexahydrophtalide stereoisomers in celery, celeriac, and funnel. *Journal of Agricultural & Food Chemistry*, **45**, 4554-4557.

[83].Casabianca, H., Graff, J. B., Fleig, F., Grenier, C. (1998) Enantiomeric distribution studies of linalool and linalyl acetate. *Journal of High Resolution Chromatography*, **21**, 107-112.

[84].Werkhoff, P., Güntert, M., Krammer, G., Sommer, H., Kaulen, J. (1998) Vacuum headspace method in aroma research: flavour chemistry of yellow passion fruits. *Journal of Agricultural and Food Chemistry*, **46**, 1076-1093.

[85].Chien, M-J. (1988) A computer database of essential oils. In: *Flavors and fragrances: A world perspective.*, Lawrence, Mookherjee & Willis, Amsterdam.

[86].Dimov, N., Tsoutsoulova, A., Stojanov, E. (1988) Pattern recognition methods for discrimination of essential oils (rose oils) by their gas chromatograms. *Perfumer & Flavorist*, **12**, 45-50.

[87].Gavini, P. (1994) Reconnaissance globale de profils chromatographiques de matrices complexes: application aux aromes et aux parfums, Thesis, Nice (F).

[88].Voirin, S. (1990) Connaissance de l'arome du raisin: analyse et synthese de precurseurs heterosidiques, Thesis, Montpellier (F).

[89].Wittmann, R., Eichner, K. (1989) Nachweis von Maillard-Produkten in Malzen, Bieren und Braucouleuren. *Zeitschrift für Lebensmittel -Untersuchung und -Forschung*, **188**, 212-220.

[90].Van den Dool, H., Kratz, P. D. (1963) A generalisation of the retention index system including linear temperature programmed gas-liquid partition chromatography. *Journal of Chromatography*, **11**, 463-471.

[91].IOFI (1991) The identification of individual components in flavourings and flavoured foods. *Zeitschrift für Lebensmittel -Untersuchung und -Forschung*, **192**, 530-534.

[92].Angelini, E. (1992) Traitement de données chromatographiques pour l'amélioration de la reconnaissance de substances:application aux arômes et aux parfums., Thesis, Nice (F).

[93].Working Group on Methods of Analysis. (1973) *Recommended methods*, International Organisation of the Flavour Industry (IOFI), Geneva (CH).

[94].Bruins, A. P. (1979) Negative ion chemical ionization mass spectrometry in the determination of components in essential oils. *Analytical Chemistry*, **51**, 967-972.

[95].Brevard, H., Brambilla, M., Chaintreau, A., Marion, J. P. (1992) Occurence of elemental sulphur in capers (Capparis spinosa L.) and first investigation of the flavour profile. *Flavour and Fragrance Journal*, **7**, 313-321.

[96].Lindinger, W., Hansel, A. (1997) Analysis of trace gases at ppb levels by proton transfer reaction mass spectrometry (PTR-MS). *Plasma Sources Sci.Technol.*, **6**, 111-117.

[97].Zimmermann, R., Heger, H. J., Yeretzian, C., Nagel, H., Boesl, U. (1996) Application of laser ionisation mass spectrometry for on-line monitoring of volatiles in the headspace of food products: roasting and brewing of coffee. *Rapid Communications in Mass Spectrometry*, **10**, 1975-1979.

[98].Humpf, H. U., Schreier, P. (1992) 3-Hydroxy-5,6-epoxy-beta-ionol beta-D-glucopyranoside and 3-hydroxy-7,8-dihydro-beta-ionol beta-D-glucopyranoside: new C13 norisoprenoid glucoconjugates from sloe tree (Prunus spinasa L.) leaves. *Journal of Agricultural and Food Chemistry*, **40**, 1898-1901.

[99].Herderich, M., Roscher, R., and Schreier, P. (1996) Application of atmospheric pressure ionization liquid chromatography-tamdem mass spectrometry for the analysis of flavour precursors. In: *Biotechnology for improved foods and flavours*, Takeoka, Teranishi, Williams & Kobayashi, Washington, DC.

[100].Hau, J., Cazes, D., Fay, L. B. (1997) Comprehensive study of the "beefy meaty peptide". *Journal of Agricultural and Food Chemistry*, **45**, 1351-1355.

[101].Fay, L. B. (1998) Application du couplage chromatographie en phase gazeuse/spectrometrie de masse en tandem (CPG/SM/SM) à l'analyse de produits alimentaires. *Analusis*, **26**, M28-M32.

[102].Fay, L. B., Huynh-Ba, T., Blank, I. (1997) Study of the fragmentation of 3(2H)-furanones by mass-spectrometry. *Journal of Agricultural and Food Chemistry*, **45**, 4057-4064.

[103].Amrani-Hemaimi, M., Cerny, C., Fay, L. B. (1995) Mechanism of formation of alkylpyrazines in the Maillard reaction. *Journal of Agricultural and Food Chemistry*, **43**, 2818-2822.

[104].Stämpfli,A., Blank, I., Fumeaux, R., Fay, L. B. (1994) Study of the decomposition of the Amadori compound N-(1-deoxy-d-fructose-1-yl)-glycine in model systems: quantification by fast atom bombardment tandem mass spectrometry. *Biological Mass Spectrometry*, **23**, 642-646.

[105].Nitz, S., Weinreich, B., Drawert, F. (1992) Multidimensional gas chromatography-isotope ratio mass spectrometry: system description and technical requirements. *Journal of High Resolution Chromatography*, **15**, 387-391.

[106].Kaunzinger, A., Juchelka, D., Mosandl, A. (1997) Progress in the authenticity assessment of vanilla. 1. Initiation of authenticity profiles. *Journal of Agricultural and Food Chemistry*, **45**, 1752-1757.

[107].Mosandl, A. (1995) Enantioselective capillary gas chromatography and stable isotope ratio mass spectrometry in the authenticity of flavours and essential oils. *Food Review International*, **11**, 597-664.

[108].Papavoine, F., Maignial, L., Brevard, H., Chaintreau, A., Marion, J. P. (1992) Contribution de la spectrometrie infra-rouge à transformée de Fourier dans la détermination structurale de composés d'arôme. Application à la détermination structurale des pyrazines disubstituées. *Rivista Italiana EPPOS*, Numero speciale Febbraio 1992.

[109].Le Quéré, J. L. (1992) Le couplage CPG/IRFT dans l'analyse des arômes. *Analusis*, **20**, M46-M49.
[110].Schneider, J. F., Demirgian, J. C., Stickler, J. C. (1986) A comparison of GC/IR interfaces: the light pipe versus matrix isolation. *Journal of Chromatographic Science*, **24**, 330-335.
[111].Morin, P., Caude, M., Richard, H., Rosset, R. (1987) Couplage de la chromatographie en phase dioxide de carbone supercritique avec la spectrometrie infrarouge à transformée de Fourier. *Analysis*, **15**, 117-127.
[112].Wilie, P. L., Sullivan, J. J., Quimby, B. D. (1990) An investigation of gas chromatography with atomic emission detection for the determination of empirical formulas. *Journal of High Resolution Chromatography*, **13**, 499-506.
[113].Hoetzel, A., Schlotterbeck, G., Albert, K., Bayer, E. (1996) Separation and characterisation of hop bitter acids by HPLC-1H NMR coupling. *Chromatographia*, **42**, 499-505.
[114].Martin, G. (1992) L'analyse isotopique par résonnance magnetique nucléaire. *Annales des Falsifications et de l'Expertise Chimique et Toxicologique*, **85**, 249-270.
[115].Danho, D., Naulet, N., Martin, G. J. (1992) Deuterium, carbon and nitrogen isotopic analysis of natural and synthetic caffeines. Authentication of coffee extracts. *Analysis*, **20**, 179-184.
[116].Remaud, G. S., Martin, Y. L., Martin, G. G., Martin, G. J. (1997) Detection of sophisticated adulterations of natural vanilla flavours and extracts - Application of the SNIF-NMR method to vanillin and p-hydroxybenzaldehyde. *Journal of Agricultural and Food Chemistry*, **45**, 859-866.
[117].Remaud, G., Debon, A. A., Martin, Y. L., Martin, G. G., Martin, G. J. (1998) Authentication of bitter almond oil and cinnamon oil: application of the SNIF-NMR method to benzaldehyde. *Journal of Agricultural and Food Chemistry*, **45**, 4042-4048.
[118].Fronza, G., Fuganto, C., Guillou, C., Reniero, F. (1998) Natural abundance 2H nuclear magnetic resonance study of the origin of raspberry ketone. *Journal of Agriculture and Food Chemistry*, **46**, 248-254.
[119].Acree, T. E., Barnard, J., and Cunningham, D. G. (1984) The analysis of odour-active volatiles in gas chromatographic effluents. In: *Analysis of volatiles*, New York.
[120].Ullrich, F., Grosch, W. (1987) Identification of most intense volatile flavour compounds formed during auto oxidation of linoleic acid. *Zeitschrift für Lebensmittel -Untersuchung und -Forschung*, **184**, 277-282.
[121].Grosch, W. (1994) Determination of potent odorants in foods by aroma extract dilution analysis (AEDA) and calculation of odour activity values (OAVs). *Flavour and Fragrance Journal*, **9**, 147-158.
[122].Blank, I. (1996) Gas chromatography-olfactometry in food aroma analysis. In: *Techniques for analysing food aroma*, Marsili, New York.
[123].McDaniel, M. R., Miranda-Lopez, R., Watson, B. T., Michaels, N. J., and Libbey, L. M. (1989) Pinot noir: a sensory/gas chromatographic approach. In: *Flavors and off-flavours*, Charalambous, Amsterdam.
[124].Priser, C. (1997) Contribution à la caractérisation objective de l'arôme de vin de Champagne par des méthodes sensorielles et physico-chimiques, Thesis, Dijon (F).
[125].Abbott, N., Etievant, P., Issanchou, S., Langlois, D. (1993) Critical evaluation of two commonly used techniques for the treatment of data from extract dilution sniffing analysis. *Journal of Agricultural and Food Chemistry*, **41**, 1698-1703.
[126].Ott, A., Pollien, P., Montigon, F., and Chaintreau, A. (1996) Improved headspace-GC-Sniffing technique: screening of aroma impact flavourings. 4th International Symposium on Hyphenated Techniques in Chromatograpy. Bruge (B).
[127].Van Ruth, S. M., Roozen, J. P., Cozijnsen, J. L. (1996) Gas chromatography/sniffing port analysis evaluated for aroma release from rehydrated french beans (Phaesolus vulgaris). *Food Chemistry*, **56**, 343-346.
[128].Pollien, P., Ott, A., Montigon, F., Baumgartner, M., Munoz-Box, R., Chaintreau, A. (1997) Hyphenated headspace-gas chromatography-sniffing: screening of impact odorants and quantitative aromagram comparisons. *Journal of Agricultural and Food Chemistry*, **45**, 2630-2637.
[129].Bartlett, P. N., Eliott, J. M., Gardner, J. W. (1997) Electronic noses and their application in the food industry. *Food Technology*, **51**, 44-48.
[130].Spadone, J. C., Takeoka, G., Liardon, R. (1990) Analytical investigation of Rio off-flavour in green coffee. *Journal of Agricultural and Food Chemistry*, **38**, 226-233.
[131].Talou, T., Sanchez, J. M., and Bourrounet, B. (1996) New trends in electronic noses for flavour chemistry. In: *Flavour science*, Taylor & Mottram, Cambridge (UK).
[132].Mielle, P., Hivert, B., and Mauvais, G. (1995) Are gas sensors suitable for on-line monitoring and quantification of volatile compounds? Poster, 1st Symposium on Biocatalysts for Flavour Production Bioflavour 95, Dijon (F).
[133].Hofmann, T., Schieberle, P., Krummel, C., Freiling, A., Bock, J., Heinert, L. et al. (1997) High resolution gas chromatography/selective odourant measurement by multisensor array (HRGC/SOMSA): a useful approach to standardise multisensor arrays for use in the detection of key food odorants. *Sensors and Actuators*, **B 41**, 81-87.

[134].Ai, J. (1997) Solid phase microextraction for quantitative analysis in nonequilibrium situations. *Analytical Chemistry*, **69**, 1230-1238.
[135].Grosch, W., Semmelroch, P., Masanetz, C. (1993) Quantification of potent odorants in coffee. *Association Scientifique Internationale du Café*, 545-549.
[136].Uhler, A. D., Miller, L. J. (1994) Multiple headspace extraction gas chromatography for the determination of volatile halocarbon compounds in butter. *Journal of Agricultural and Food Chemistry*, **36**, 772-775.
[137].Burger, B. V., Teranishi, R., Ling, L. C. (1987) Fresh tomato aroma volatiles: a quantitative study. *Journal of Agricultural and Food Chemistry*, **35**, 540-544.
[138].Baldwin, E. A., Nisperos-Carriedo, M. O., Baker, R., Scott, J. W. (1991) Quantitative analysis of flavour parameters in six Florida tomato cultivars (Lycopersico *esculentum Mill). Journal of Agricultural and Food Chemistry*, **39**, 1135-1140.
[139].Ott, A., Baumgartner, M., Chaintreau, A. (1999) Aroma comparisons of traditional and mild yogurt: headspace-GC quantitation of volatiles and origin of α-diketones. *Journal of Agricultural and Food Chemistry*, in press.
[140].Personal results.
[141.Bade-Wegner, H., Holscher, W., Vitzhum, O. G. (1993) Quantification of 2-methylisoborneol in roasted coffee by GC-MS. *Association Scientifique Internationale du Café*, 537-544.
[142].Katz, E. (1987) *Quantitative analysis using chromatographic techniques*, Chichester.
[143].Beens, J., Boelens, H., Tijssen, R. (1998) Quantitative aspects of comprehensive two-dimensional gas chromatography. *Journal of High Resolution Chromatography*, **21**, 47-54.
[144].Carlson, M., Thompson, R. D. (1998) Liquid chromatographic determination of safrole in sassafras-derived herbal products. *Journal of the Association of Official Analytical Chemists*, **80**, 1023-1028.
[145].The Sadtler capillary GC standard retention index library and data base, Sadtler Research Laboratories, Philadelphia, PA.
[146].Kondjoyan, N. and Berdagué, J-L. (1996) *A compilation of retention indices for the analysis of aromatic compounds*, INRA, Laboratoire flaveur.
[147].(1995) *The Wiley registry of mass spectral data.*
[148].Adams, R. P. (1995) *Identification of essential oil components by gas chromatography/mass spectrometry*, Adams, Carol Stream (I, USA).
[149].Almanza, R. (1996) *INRAMASS, LRSA-INRA.*
[150]Vernin, G., Petitjean, M., Metzger, J. (1983) Identification of volatile compounds of flavours and fragrances by VPC-MS-SPECMA bank. Conception and realisation of the data bank. *Parfums Cosmétiques Arômes*, **51**, 43-52.
[151].Brevard, H., Fay, L. B., Krebs, Y., Simian, H. (1996) Produits naturels: isolement et identification de composés à l'état de trace. Rivista Italiana EPPOS, Acte des 14ièmes journées internationales de huiles essentielles. Num. Speciale Gennaio '96, 73-79.
[152].Joulain, D. *Database of FTIR spectra in vapor phase*, Hewlett-Packard Co.
[153].Pouchert, C. J. (1998) *The Aldrich library of FT-IR spectra: vapor phase*, Aldrich Chem., Milwaukee (WI, USA).
[154].*Database of SNIF-NMR data from authentic samples*, EUROFIN, Nantes, (F)
[155].Acree, T., Arn, H. (1997) Flavornet. *http://WWW.nysaes.cornell.edu/flavournet/index.html.*
[156].(1996) *VCF96: database of volatile compounds in food.*, TNO nutrition and food research.
[157].(1997) *ESO97: database of essential oils.*, Boelens, H.
[158].(1998) *FRM98: database of flavour (raw) materials.*, Boelens, H.

FLAVOUR MATRIX INTERACTIONS

A. J. TAYLOR
Samworth Flavour Laboratory, Division of Food Sciences
University of Nottingham, Sutton Bonington Campus
Loughborough LE12 5RD

Abstract

The subject of flavour-matrix interactions in foods is tremendously broad and involves chemical and physical interactions of many different types. Flavour molecules can be volatile or non-volatile compounds and the food matrices range from water to complex biological tissues like meat. While the physical properties of the flavour compounds are important in determining their interaction with the matrix, the state of the matrix is equally important. An example is starch, which can be present in foods as a viscous solution (e.g. a sauce), a rubbery solid (e.g. bread dough) or a glass (e.g. a low moisture snack product). Trying to generalise the interaction of a volatile with starch in all its forms is not practicable and consideration of the physical state, the mechanism of the interaction and other parameters like water content and temperature is essential. Some food ingredients have been extensively studied, while others are considered difficult and have been largely ignored. One of the factors limiting research is the lack of a complete set of physical data (solubility, vapour pressure etc.) for the many compounds associated with flavour. Because of the complexity of the interactions and the lack of physical data, some workers have approached the problem by investigating complex real food systems, followed by analysis of the data to determine the key physicochemical factors involved in the flavour-matrix interactions. Other researchers have attempted to study complex systems using a fundamental approach based on defined mechanisms. Both approaches have yielded useful data.

1. Introduction

Food is complex in terms of its composition and in terms of its structure. For example, a beverage like beer (which might be considered a simple system) contains water, ethanol, a range of fusel alcohols, hop acids, hop breakdown products, tannins, malt flavour compounds, oligosaccharides from starch, proteins and protein fragments, as well as inorganic compounds. Beer is not just an aqueous solution containing these components, it also contains structures that interact with the flavour components and change the way they are sensed by people during consumption. The foam on beer, for

K.A.D. Swift (ed.), Current Topics in Flavours and Fragrances, 123–138.

instance, is produced by release of gas from the solution after dispensing and involves surface-active proteins as well as polysaccharide stabilisers. Since it forms a layer on top of the beer, the release of volatile flavour compounds is modified by the foam and some of the flavour components which have surface active properties may also be incorporated into the foam structure. At the other extreme, there are biological tissues like meat where the structure of the muscle tissue is complex [1] and thus this provides many opportunities for flavour compounds to interact with the structural components. A manufactured product like bread relies on its structure for its characteristic textural properties and the matrix of starch and protein, containing gas bubbles, provides a complex medium within which the flavour compounds can interact. Foods therefore have a diverse composition and structure, thus making generalisations about their interactions with flavours difficult. Flavours are also complex as they can be volatile or non-volatile and represent many different chemical classes (see Table 1 for some examples).

Table 1. Examples of common flavour compounds

Compound Class	Example	Flavour
Sugar	Sucrose Glucose Fructose	Sweet
Acid	Malic Citric Lactic	Sour
Pyrazines	Dimethyl pyrazine	Roasted Toasted Nutty
Aldehydes	Hexenal 1-Octen-3-ol 3-Methyl butanol	Grassy, Green Mushroom Malty
Terpenoids	Limonene Menthol	Lemony Minty
Esters	Hexyl acetate Methyl cinnamate Ethyl deca-2,4-dienoate	Cox's apple Sweet, Honey Pears

Flavour molecules can be ionic or non-ionic and may contain functional groups, which interact with specific chemical groups on other food components like proteins. Their relative hydrophilicity or hydrophobicity will also determine their interaction with the fat and water components of food matrices. Those compounds, which are amphoteric, or have surface-active properties, are especially prone to interaction with food components (or with other flavour molecules) and readily form structures within the food matrix. This further complicates the issues, and it is necessary to consider the fundamental mechanisms by which flavours can interact with food components and food structures to gain an understanding of flavour-matrix interactions. Most interactions have been studied using volatile compounds as the interactions are easier to measure. However, around 2600 chemicals have been reported as volatile constituents of foods [2]. They represent not only the wide range of chemical classes indicated in Table 1, but also a range of physical properties like solubility, vapour pressure, and partition coefficients. These determine how a given material will interact with the other components of the food matrix.

Volatiles have also been widely studied as they represent an important part of overall flavour and take part in interactions to a greater extent than the non-volatiles. Typically, the whole art of the flavourist, whether in perfumery or in food, is to make a mixture of flavour chemicals, which, when added to a matrix, will deliver the required volatile profile such that the material is perceived as possessing the desired characteristic flavour desired. Flavourists have traditionally used their skill and experience to reformulate flavours. Understanding the mechanisms of flavour-matrix interactions provides an additional tool to help formulate good quality flavours, which can perform in any matrix.

This review will examine the interaction(s) of volatile flavours with various food matrices to indicate the differences found in foods and the various approaches taken to study the interactions. It should be remembered that ultimately, the purpose of this work is linked to sensory perception of the food and we should briefly consider how volatile compounds released from foods are sensed by humans.

The perception of flavour consists of several, related steps [3]. Initially, the packaging, external appearance or arrangement of food can influence us to believe that what we are about to eat will be flavoursome. Restaurants go to great lengths to deliver a pleasant ambience prior to, and during, a meal and the placement and decoration of food on the plate is of huge importance. Our second experience is the odour of food, prior to consumption, which is sensed orthonasally; i.e. the volatiles are sniffed through the nostrils and give us our first flavour signal. Next, the food is consumed and volatile flavour signals, released from the food in the mouth, travel via the orthonasal route to sensors in the nose. Lastly, in many foods, some components of the flavour persist and give a feeling of well being or satisfaction. Although all parts of this process are important, and contribute to the overall flavour perception, this review limits itself to the orthonasal situation from food prior to consumption.

The volatile profile sensed orthonasally is essentially the profile found in the headspace above a food. For convenience, many workers have measured the headspace volatile profile when it is in equilibrium with the liquid or food phase. In real life, equilibrium is rarely found and consumers encounter a volatile profile that changes with time. Measuring such changes is not easy although there are now some reports of such dynamic investigations [4,5].

The release of flavours from foods "in-mouth" during eating involves many different mechanisms and is outside the scope of this review. Information on the mechanisms can be found in reviews by Overbosch *et. al.* [6], Plug and Haring [7], Taylor [8,9] and Bakker [10]. More information on interactions between flavours and matrices can be found in earlier reviews by Solms [11], Solms & Guggenbuehl, [12] and Land [13].

2. Interactions of Volatile Flavours with Aqueous Solutions

2.1 SIMPLEST CASE – EQUILIBRIUM ABOVE AQUEOUS SOLUTIONS

Theory can predict the partition of volatiles between an aqueous solution and the gas phase above it (the headspace) providing the systems contain low levels of volatiles and providing the interactions between the volatiles and the aqueous phase can be defined. The subject has been reviewed previously [6,12,14] with emphasis on i) the measurement of partition between the liquid and gas phases and ii) the relevance of the commonly measured air-water partition to more complex flavour-release systems. The partition behavior of a single volatile, or simple mixtures of volatiles, has been studied using a combination of experimental data and data derived from fundamental thermodynamic parameters. The partition coefficient between the gas and liquid phases (K^i_{gl}) for a volatile compound (i), for example, can be measured directly in a system by determining the concentrations in the gas and liquid phase (equation 1) or calculated from fundamental physicochemical parameters according to equation (2).

$$K^i_{gl} = \frac{C^g_i}{C^l_i} \qquad \text{Equation (1)}$$

$$K^i_{gl} = \left(\frac{\gamma_i P^0_i(T)}{P_T} \right) \cdot \frac{\bar{V}_l}{\bar{V}_g} \qquad \text{Equation (2)}$$

where $P^0_i(T)$ is the vapour pressure for the pure component i (Pa) at temperature T, P_T, the total pressure in the gas phase (Pa) and $\bar{V}_l$, $\bar{V}_g$ are the molar volumes of the liquid and the gas phases respectively ($m^3.mol^{-1}$). If the volatile is highly diluted in the liquid phase, the activity coefficient γ_i can be assumed to be independent of the concentration of the volatile in the liquid, and is equal to a constant value γ_l^∞.

In this case, the product $\gamma_l^\infty * P_i^0(T)$ is a constant (Henry's constant), so that the value of K_{gl}^i depends only on the temperature. In all but the simplest systems, γ_i depends on the other components present in the aqueous phase and its value varies from the ideal value of 1.0. Although partition data can be calculated, experimental data is also required to check that the value chosen for γ_i is appropriate and to ensure that other, unknown interactions are not occurring. As mentioned previously, most of the published work involves model systems containing just a few volatiles where the effect of one interacting factor (e.g. sucrose) is studied. There are some papers that broaden these parameters but, the application of partition to a real food product like beer, involves so many interactions that the reliability of any mathematical model, based solely on theory, would be questionable.

Partition of volatile compounds is affected by soluble non-volatile flavour components like sugars, acids and salts. Buttery and his group [15] reported changes in headspace concentrations due to interaction with food components in 1971 and there have been many other papers on the subject since then. Voilley and her group at ENSBANA, Dijon, France have investigated this phenomenon extensively. One of the first papers studied the change in concentrations of acetone and octanol in the gas phase above aqueous solutions containing sucrose, calcium chloride and citric acid [16]. Other workers have examined similar systems such as the change in the partition of pentyl acetate above water and sugar solutions [17] and the effect of salt, glucose and malic acid on the partitioning of diacetyl between water and air [18]. In this latter study, salt increased the amount of diacetyl in the headspace significantly, while glucose and malic acid were not so effective. Some authors have described the general increase of volatile concentration in the headspace when a solute is added to the solution as "salting out" and a decrease as "salting in". However, "salting in" and "salting out" are specific terms that originate from work on protein solubility where addition of low levels of salt can aid solubilisation, while, at higher concentrations, the protein is precipitated. This change in solubility may also occur with some volatiles on addition of solute. However, there are other mechanisms by which the addition of solute can change the headspace concentration and it is incorrect to use the terms "salting in" and "salting out" to describe the general change in headspace concentration on addition of a solute. Voilley *et. al.* [16] showed, that in the system they studied, the change in headspace concentration when sugar was added was due to the increase in the mole fraction of acetone and octanol in the liquid phase. In brief, if a volatile is present in water and in a sucrose solution at the same concentration, the amount of water in the sucrose solution is less and, with a constant concentration of volatile, the mole fraction (moles of volatile divided by moles of solvent) actually increases. Since partition depends on the mole fraction in the aqueous phase (see equation 2) the amount in the headspace increases. Simultaneous measurement of the activity coefficient for the Voilley *et.al.,.* [16] system, showed no change with sucrose concentration, confirming that the change in mole fraction was responsible for the change in headspace concentrations. In contrast, calcium chloride did not significantly increase the concentration of acetone in the headspace even though the mole fraction changed as the salt was added up to 32%. This was because the activity coefficient also changed and the two processes effectively cancelled each other out. At high concentrations of acetone, however, some increase in headspace concentration was seen presumably due to true "salting out" i.e. some acetone was no longer in solution and contributed directly to vapour pressure in the headspace, rather than through the partition coefficient.

There are many other examples of similar effects but the Voilley paper [16] provides in-depth analysis which demonstrates the different mechanisms and shows, quite clearly, that flavour-matrix interactions, even in these simple systems are complex. The paper also illustrates another issue, namely that it is useful to study volatiles and non-volatiles with different physical properties, otherwise it is not certain that the mechanism found with one compound holds good for other types of molecule.

2.2 NON-EQUILIBRIUM CONDITIONS

Although it is convenient to measure interactions in model systems where the volatiles have reached equilibrium between the liquid and gas phases, equilibrium rarely occurs in real foods. In the model systems used to study equilibrium, the volatile concentration in both the liquid and gas phases is the same throughout the phases (i.e. the volatiles are homogeneously distributed). In real foods, mixing of the phases either does not occur or may not be adequate to ensure that the volatile is evenly distributed throughout the phases. In this case, mass transfer effects must be taken into account. There are various standard texts on these types of systems and, for a full description, readers are referred to Reid *et. al.* [19].

Overbosch and coworkers [6] presented an overview of the processes that occur when food is eaten and, the static diffusion model presented by them can be applied to the headspace situation in foods. The model concentrates on diffusion in the food phase. Inspection of the Overbosch mathematical equation for static diffusion reveals that the partition coefficient plays a key role in the mass transfer effect and a plot of relative release rate against partition coefficient showed a sigmoidal relationship. Overbosch *et. al.* [6] also presented a convective diffusion model to represent the release of volatiles in the mouth where the gas phase is flowing over the food product. The model uses convective diffusion as well as penetration theory to predict the release of volatiles. With convective diffusion, the rate of release decreased in a fairly regular manner as the partition coefficients decreased from 10^{-1} to 10^{-4}.

The penetration theory model, however, showed marked differences between the release rates for volatiles with partition coefficients of 10^{-1} and 10^{-2} compared to those with values of 10^{-3} and 10^{-4}. Some validation of the models has been carried out by following the release of volatiles using Flame Ionisation Detection [6,20].

Recently, Marin *et. al.* [5] have developed a model to study the release of volatiles from food packages when they are opened and to predict how the headspace concentration and profile changes over relatively short time intervals. In this situation, the volatiles are in equilibrium with the food phase initially and the headspace is then diluted as air mixes with the headspace of the packet. In reality, the airflow through the headspace is both slow and random, which causes great problems in trying to obtain reproducible direct experimental data with which the process can be followed. The approach adopted by Marin *et. al.* [5] was to build a model system where the conditions could be controlled, although they did not fully represent the real-life situation. Volatile release from the model was expressed in mathematical terms and then validated with experimental data to confirm that the mathematical model was correct. They were then able to consider what would happen in conditions where the Reynolds number was low and mass transfer in the gas phase became an issue. This model again demonstrated that the air-water partition coefficient was a governing factor although at K_{aw} values around 10^{-3} the overall release was also affected by conditions in the gas phase. This model takes into account mass transfer in both the gas and food phases and can be applied not only to aqueous systems but also to viscous solutions or gels, which are common matrices in food products.

Release of volatiles from other systems under non-equilibrium systems has not been studied comprehensively from a fundamental point of view. An exception is the dynamic measurement of flavour volatile release from liquid foods in a model system [4]. The technique followed the release of volatiles of differing polarity from an aqueous ethanolic solution with time and derived time-release curves from which the relative contributions of the systems components could be studied.

Studies have also been undertaken to look at systems that are closer to real foods and then develop empirical relationships that explain release characteristics. An example is the work of Roberts and Acree [21,22] where the release of a range of volatiles (with different hydrophilicity/hydrophobicity values) from a model food system consisting of oil and aqueous phases was studied. The system was designed to simulate some of the processes that occur during eating but incorporated some elements that are applicable to the headspace situation, which forms the basis of this current review. Cayot *et. al.* [23] studied the release of isoamyl acetate from a system composed of starch, sugar and, skimmed milk which has been processed to simulate a dessert cream. The volatile affected the rheology of the system and the volatile also interacted to different extents with starch from different origins (potato, waxy maize). The authors concluded that, even in this well-defined system, isoamyl acetate interactions were complex and it was not easy to elucidate all the sources of interaction. Guichard [24,25] described the interactions of flavour volatiles in jam systems with an emphasis on the effect of different types of pectin. These studies are indicative of a desire to apply fundamental studies to systems that are very close to real food systems, so that the volatile interactions with the food components can be understood.

3. Interactions with Starch

As mentioned in the summary to this chapter, interactions of volatiles with starch are dependent on the state of the starch. Solms group has contributed much to our understanding of volatile binding to biopolymers and the paper from Rutschmann *et. al.* have [26] studied the interaction of volatiles with dilute aqueous solutions of starch in which the volatile was present in high concentrations (200mmol volatile per mol glucose equivalent). This is way above normal flavour concentrations and care should be taken in extrapolating results to the low levels of volatiles normally found in starch-containing foods. Starch is known to form complexes with lipids and monoglycerides through the helices of amylose, which create a hydrophobic interior and where the number of glucose molecules per turn of helix depends on the molecule complexed [27]. Cyclic compounds like limonene and menthol can also be complexed [23] although the binding is weaker than compounds with linear alkyl chains. Many of the studies have used potato starch as it is essentially lipid-free and there is no competition for the helices from lipid.

Competition was observed when two volatiles were mixed with starch [28]. Volatile compounds that possess a hydrophobic and a hydrophilic region seem to be bound most strongly which suggests that both the hydrophobic interior of the starch helix and the hydrophilic exterior are involved in binding [29]. The loss of flavour during the ageing of bread has been attributed to complexation of the flavour molecules by inclusion into the starch [30].

Generally, to form a volatile-lipid helical complex, some form of heat processing in the presence of water is required and the starch needs to be gelatinised. Complexation into the helices of amylose occurs when the starch molecules are sufficiently mobile. This may occur in aqueous solutions or at lower water contents when heat or mechanical energy contribute to the starch mobility. During extrusion, starch can form inclusion complexes of flavour compounds even though the water content is in the intermediate region of 20-40% [31].

Nuessli *et. al.* [32] prepared potato starch-flavour complexes in water and then freeze dried the material. Analysis of the starch-flavour interactions was carried out by Differential Scanning Calorimetry (DSC) of the rehydrated, freeze dried material and by X ray diffraction of the dry samples. A range of flavour molecules representing linear (decanal), branched chain (geraniol), cyclic (carvone) and various bicyclic compounds (e.g. napthol, fenchone and camphor) were studied. DSC analysis gave information on the melting characteristics of the starch-flavour interaction with different melting temperatures for the different flavour compounds. X-ray diffraction showed the typical V amylose pattern but the patterns were different for each flavour compound and not the same as those observed for lipid-amylose interactions. The current suggestion is that lipid is present inside the amylose helices whereas the flavour compounds are present between the helices.

Binding of this nature reduces the effective concentration of volatile in the food phase and thus reduces the headspace concentration. The interaction affects not just the flavour quality but also the texture of the starch product [32]. There are also questions about the conditions under which the complexed molecules might be released and the sensory effects of such interactions [33].

3.1 STARCH AT LOW MOISTURE CONTENTS

The interactions of volatiles with materials at low water contents are often referred to as sorption. In the case of starch, it is important to clarify its state when considering its interactions with flavour volatiles. Native starch is rarely found in foods as it is both unpalatable and indigestible. Food processing is usually designed to convert it into gelatinised starch in the presence of water, which may lead to inclusion of flavour volatiles in helices as described above. After gelatinisation, many starch containing products are further processed to reduce moisture content and create attractive textures such as those found in biscuits, snack foods or breakfast cereals. In many of these low moisture cereal products, starch can be found in the glassy state and there has been much interest in the mobility of large and small molecules in glassy/rubbery food systems [34]. Mobility in the starch systems undergoes a marked decrease as the material goes through the glass transition. This has been used to achieve physical and chemical stability of foods, particularly during storage and distribution. More recent work has shown that, although the mobility of the biopolymers is affected by glass transition, smaller molecules are less affected and do not undergo a step change in mobility [35].

Examples of volatile sorption to starch can be found in the work by Maier (1975; cited in Overbosch *et. al.* [6]) where the uptake of high concentrations of volatile onto starch was studied gravimetrically and then by physical techniques. High concentrations of volatile were used to speed sorption and to make detection of volatiles easier. However, at high concentrations, the behaviour of the system may be different; for example, a layer of volatile can be sorbed (or condense on the sample) and attract more volatile leading to a liquid layer. Interpretation of the results from these systems needs care. In an effort to circumvent these problems, Maier [36] described the use of Inverse Gas Chromatography (IGC) to study sorption. In IGC, the biopolymer is used as the stationary phase, a sample of volatile is injected and the retention of volatile is measured. The technique was also used by Gilbert [37] and by Gilbert and Roshdy [38]. The technique is attractive, in that relatively low concentrations of volatile can be used, but there are difficulties in maintaining the humidity of the sample and in ensuring that the material in the stationary phase is not affected by the temperatures used in IGC. Improvements to IGC have been made and a new device with humidity control has been developed by Surface Measurement Systems Limited (smsuk@dircon.co.uk).

Other examples of volatile sorption to starch include the work by Menting *et. al.* [39] and LeThanh *et. al.* [40] where the sorption of a range of volatiles onto starch and maltodextrin were studied. Again, high concentrations of volatile were used and in the case of the latter paper it was not clear whether the starch used in the experiments was gelatinised or native. Sorption of volatiles at lower concentrations was studied by Hau *et. al.* [41-43]. The dynamics of sorption to starch in the glassy and rubbery states were monitored and, from the data, mechanistic models of sorption were proposed.

3.2 OTHER FOOD BIOPOLYMERS

Most work has focused on starch because it is an ingredient in many foods and because it exhibits specific binding through amylose helices. Other polysaccharides show smaller interactions with volatiles [44], although they can affect the dynamic headspace concentration through their viscous or gel properties. This slows the mass transfer of some volatile from the food phase into the headspace. The interactions of aroma compounds with some polysaccharides has been investigated by dynamic exponential dilution [45].

As mentioned previously, there are limited reports on the dynamic release of volatiles from such systems. Carr *et. al.* [46] used a combination of sensory analysis and dynamic headspace analysis to study the effects of different biopolymers on volatile release. Increasing gel strength correlated with a decrease in sensory intensity and with results from the dynamic headspace analysis. Similar data was obtained by Guinard and Marty [47]. Wilson and Brown [48] reported a similar sensory decrease in gelatine gels with increasing gelatine concentration. Morris [49] studied the release of volatiles from viscous solutions with differing rheological properties by sensory analysis. They proposed that volatile release was dependent on a critical gel concentration, which they termed C*.

Since volatile release depends on so many, inter-related factors, it is often difficult to separate the factors and determine the nature of the mechanisms that change the headspace concentration. It is essential to realise that the behaviour of volatile compounds depends on their physical properties as well as their interactions with the matrix. Since volatile flavour compounds include so many different chemical classes (alcohols, pyrazines etc.) it is difficult to generalise the behaviour of volatiles in a particular food matrix. For some compounds, the air-water partition coefficient may be the major factor determining release. In others, binding may occur to the matrix to a substantial extent, decreasing the amount of free volatile and therefore decreasing the headspace concentration. For others, mass transport through a viscous food product may be the rate-limiting step. Undoubtedly, for some compounds a combination of these effects will determine their release into the headspace.

4. Oil-Water Partition

As described elsewhere [14] oils and fats are not usually found as bulk phases in foods but more often as some sort of emulsion. Various reviews of volatile interactions with fats and fat-containing foods have been published recently and the reader is referred to the review by Hatchwell [50] for an overview of the situation. The basic nature of volatile-fat interactions are understood and this has assisted in the development of low-fat foods where fat is often replaced by other macromolecules to provide bulk, mouthfeel and texture to the product. Measurement of headspace profiles from full-fat and reduced-fat products show marked differences unless the flavour volatiles are reformulated. The interactions between fat (at various inclusion levels) and volatile flavours in yogurts and in biscuits have been described [51,52].

Emulsion systems present challenges to experimenters because of their complexity, which is due to interactions between all the components. Many flavours are surface active themselves and adding them to emulsions can change the emulsion structure and properties. This is especially noted in products with high flavour loads e.g. toothpaste. The effect of emulsion structure on volatile release was reported by Bakker & Mela [53] who used both water-in-oil and oil-in water emulsions. Interactions were measured using volatile release and sensory properties but only one volatile, diacetyl, was studied.

5. Interactions of Volatile Flavours with Proteins

Interactions of volatiles with proteins tend to be of two types. Volatiles with an aldehydic or keto group can bind strongly to the amino side groups of proteins in an irreversible manner. Alternatively, proteins can contain regions which are strongly hydrophobic or hydrophilic and volatiles may be bound at these sites although less strongly. O'Neill [54] gave an overview of the subject and a short description of some of the key points is presented here. Most of the work has studied dilute aqueous solutions of native proteins.

Since aldehydes and ketones bind to the amino groups of proteins, the state of the protein (native or denatured) has little impact on the extent of binding unless more amino groups are made available during protein denaturation. On the other hand, interactions which involve hydrophobic regions are greatly affected, as denaturation causes protein unfolding and disturbs the hydrophobic pockets.

Beta-lactoglobulin has been widely studied as it is an important milk protein and binds ketones and aldehydes (commonly found in milk products like cheese). Binding behaviour of lactoglobulin has been determined using the well-known Hill and Scatchard plots which show the degree of binding and the number of binding sites per molecule [55]. The review by O'Neill [54] describes the physical chemistry of binding as well as the equations and procedures used to determine these binding parameters.

Boudaud and Dumont [56] reviewed the interaction of lactoglobulin with flavour components and raised interesting questions about the relative roles of binding and interaction in these systems. They pointed out that (-lactoglobulin shows strong homology with retinol binding protein, which suggests that "hydrophobic molecules could be trapped in the (–barrel common to both molecules". Jouenne and Crouzet [57] have also presented work on (-lactoglobulin-flavour interactions. The review by Batt *et. al.* [58] raises other interesting issues and concludes that lactoglobulin is the most studied food protein. In terms of its interactions with flavour volatiles, it is also well-characterised [59].

Methods for measuring interactions range from headspace analysis to affinity chromatography. Headspace gives the overall binding behaviour and data from protein solutions of different concentrations can provide some information on the type of binding and the degree of binding. Affinity chromatography involves immobilising the protein onto a HPLC column and then injecting volatiles to determine how much binds, a technique that is analogous to IGC. The technique has been used in pharmacology but has also been applied to (-lactoglobulin [60]. Other chromatographic methods for measuring protein-volatile interactions have been described (see for example, [56]).

Interactions of volatile with other proteins have also been reported such as casein and whey [61], soy [62] and bovine serum albumin [63]. Harvey *et. al.* [64] studied the effect of protein on aroma transfer at a lipid water interface using a custom designed apparatus. Mottram *et. al.* [65] reported a different type of volatile-protein interaction between sulfur-containing volatiles and proteins with thiol groups. The interaction was noted initially because of the differences between headspace and SDE (simultaneous distillation extraction) analyses of disulfide volatiles from meat. When authentic disulfide compounds were added to cooked meats, only ~20% could be extracted. The binding in this case has been attributed to sulfhydryl-disulfide interchanges, which although observed in flour when glutathione and flour proteins interact, have not been previously noted in flavour-protein interactions.

6. Encapsulation

Flavour-matrix interactions have been used to encapsulate flavours either to protect them against adverse conditions prior to use or to control their release from foods during processing or during eating. The subject has been reviewed by Risch and Reineccius [66]. Encapsulation makes use either of the molecular binding mechanisms or the macroscopic effects of matrices (e.g. high viscosity or the glassy state). Just two examples are presented here to demonstrate the types of interaction that occur.

6.1 CYCLODEXTRIN BINDING

Cyclodextrins are circular molecules containing six (alpha), seven (beta) or eight (gamma) glucose molecules [67]. They possess a cavity, which is hydrophobic in nature and into which many volatile molecules can fit [68]. Preparation can be as easy as making a solution of cyclodextrin and adding volatile until the complex precipitates out of solution. The material can then be dried. When rehydrated in foods, release is due to the equilibrium between the bound and free volatiles. The use of some cyclodextrins in foods has recently been allowed.

6.2 'GLASSY MATRICES'

Flavours can be encapsulated in various glassy matrices. Historically, citrus flavours, which are prone to oxidation, were mixed with sugar and extruded to prevent deleterious flavour changes. Since then a variety of processes has been developed (see for example [69]). Such glassy materials are used commercially under various trade names (e.g.Durarome, Firmenich) and confer exceptional stability on the material with no oxidation detectable for periods of years. Since they dissolve in water, the encapsulated flavours can be released readily.

7. Conclusion

Even in aqueous solutions, flavour volatiles interact with solutes, the solvent, other volatiles and their release from the solution is affected by the presence of various hydrocolloids. In real food systems, there is a wide variety of interactions which determine the profile and concentrations of volatile molecules found in the headspace and, thereby, the aroma perceived when humans sense these molecules.

8. Acknowledgements

The author is grateful to colleagues in the EU COST96 group on *Interactions of food matrix with small ligands influencing flavour and texture* who have provided much of the background to this review.

9. Bibliography

[1].Lawrie, R.A. (1991) *Meat Science*. 5th Edition, Cambridge, Pergamon Press.
[2].Fisher, C. & Scott, T.R. (1997). *Food Flavours: Biology and Chemistry*. Cambridge, Royal Society of Chemistry.
[3].Laing, D.G. & Jinks, A. (1996). Flavour perception mechanisms. *Trends in Food Science & Technology*. 7, 387-389 and 422-423.
[4].Elmore, J.S. & Langley, K.R. (1996). Novel vessel for the measurement of dynamic flavour release in real time from liquid foods. *Journal of Agricultural and Food Chemistry*. **44**, 3560-3563.

[5].Marin, M, Baek, I. & Taylor, A.J. (1998). Flavour release from aqueous solutions under dynamic headspace dilution conditions. In: *Modelling Chemical Change in Food.* (edited by B. Wedzicha) Royal Society of Chemistry in press.
[6].Overbosch, P., Afterof, W.G.M. & Haring, P.G.M. (1991). Flavor release in the mouth. *Food Reviews International* **7**, 137-184.
[7].Plug, H. & Haring P.G.M. (1993). The role of ingredient-flavour interactions in the development of fat-free foods. *Trends in Food Science Technology.* **4**, 150-152.
[8].Taylor, A.J. (1996). Volatile flavor release from foods during eating. *Critical Reviews in Food Science and Nutrition* **36**, 765-784.
[9]Taylor, A.J. & Linforth, R.S.T., (1996) Flavor release in the mouth. *Trends in Food Science and Technology* 7, 444-448.
[10].Bakker, J., Brown, W.E., Hills, B.P., Boudaud, N., Wilson, C & Harrison, M. (1996). Effect of the food matrix on flavour release and perception. In: *Flavour Science: Recent Developments* (edited by A.J. Taylor & D.S. Mottram). Pp. 369-374. Cambridge: Royal Society of Chemistry.
[11].Solms, J. (1986). Interactions of non-volatile and volatile substances in foods. In: *Interactions of food components*, (edited by Birch, G.G.; Lindley, M.G). Pp189-210, Elsevier, London.
[12].Solms, J. & Guggenbuehl, B. (1990). *Physical aspects of flavor.* New York :John Wiley.
[13].Land, D.G., (1996). Perspectives on the effects of interactions on flavour perception: an overview. In: *Flavor-Food Interactions* (edited by R.J. McGorrin & J.V. Leland). Pp. 2-11. Washington DC: American Chemical Society.
[14].Taylor, A.J. (1998). Physical chemistry of flavour. *International Journal of Food Science and Technology* in press.
[15].Buttery, R.G., Bomben, J.L., Guadagni, D.G. & Ling, L.C. (1971). Some considerations of the volatiles of organic flavour compounds in foods. *Journal of Agricultural and Food Chemistry,* **19**, 1045-1048.
[16].Voilley, A., Simatos, D. & Loncin, M. (1977). Gas phase concentration of volatiles in equilibrium with a liquid aqueous phase. *Lebensmittel Wissenschaft Technologie,* **10**, 45-49.
[17].Kieckbusch, T.G. & King, C.J. (1979). Partition coefficients for acetates in food systems. *Journal of Agricultural and Food Chemistry.* **27**, 504-507.
[18].Land, D.G. & Reynolds, J. (1981). The influence of food components on the volatility of diacetyl. In: *Flavour '81* (edited by P.Schreier). Pp. 701-705. Berlin: Walter de Gruyter.
[19].Reid, R.C., Prausnitz, J.M. & Poling, B.E. (1987). *The properties of gases and liquids.* New York: McGraw Hill, 4th Edition
[20].Darling, D.F., Williams, D & Yendle, P. (1986). Physico-chemical interactions involved in aroma transport processes from solutions. In: *Interactions of Food Components* (edited by G.G. Birch & M.G. Lindley), Elsevier, London .
[21].Roberts, D.D. & Acree, T.E. (1996). Model development for flavour release from homogenous phases. In: *Flavour Science: Recent Developments* (edited by A.J. Taylor and D.S. Mottram). Pp. 399-404. Cambridge, Royal Society of Chemistry.
[22].Roberts, D.D. & Acree, T.E. (1996). Retronasal flavor release in oil and water model systems with an evaluation of volatility predictors. In: *Food-Flavor Interactions.* (edited by R.J. McGorrin and J.V. Leland) American Chemical Society Symposium Series 633, Washington D.C., Pp179-187.
[23].Cayot, N., Taisant, C. & Voilley, A. (1998). Release and perception of isoamylacetate from a starch-based food matrix. *Journal of Agricultural and Food Chemistry,* **46**, 3201-3206.
[24].Guichard, E. (1996). Interactions between pectins and flavor compounds in strawberry jam. In: *Flavor-Food Interactions* (edited by R.J. McGorrin & J.V. Leland). Pp. 118-129. Washington DC: American Chemical Society.
[25].Guichard, E. (1994). Interactions between pectins and flavor compounds in strawberry jam. *Abstracts of Papers of the American Chemical Society,* Vol.208, No.Pt1, pp.60-AGFD
[26].Rutschmann, M.A., Heiniger, J., Pliska, V. & Solms, J. (1989). Formation of inclusion complexes of starch with different organic compounds. I. Method of evaluation of binding profiles with menthone as an example. *Food Science and Technology* **22**, 240-244.
[27].Rutschmann, M.A. & Solms, J. (1990). Formation of inclusion complexes of starch with different organic compounds. IV. Ligand binding and variability in helical conformations of V-amylose complexes. *Food Science and Technology* **23**, 84-87.
[28].Godshall, M.A. & Solms, J. (1992). Flavour and sweetener interactions with starch. *Food Technology.* June 140-145.
[29].Yamamoto, M., Harada, S., Sano, T., Yasunaga, T. & Tatsumoto, N. (1984). Kinetic studies of the complex formation in the ternary system of amylose, SDS and iodine. *Biopolymers* **23**, 2083-2096.

[30].Schoch, T.J. (1965). Starch in bakery products. *Bakers Digest* **39**, 48-50.
[31].Guzman, L.B., Lee, T.C., & Chichester, C.O. (1992). Lipid binding during extrusion cooking. In : *Food Extrusion Science and Technology*. (edited by J.L. Kokini, C-T. Ho, M.V. Karwe.) Pp 427-436, New York, Marcel Dekker Inc.
[32].Nuessli, J., Sigg, B., CondePetit, B. & Escher, F. (1997). Characterization of amylose-flavour complexes by DSC and X-ray diffraction. *Food Hydrocolloids*, **11**, 27-34.
[33].Bakker, J. (1995). Flavor interactions with the food matrix and their effects on perception. In: *Ingredient interactions*. (edited by A.G. Gaonkar) Pp. 411-440 New York, Marcel Dekker Inc.
[34].Slade, L. & Levine, H. (1993). The glassy state phenomenon in food molecules. In: *The glassy state in foods*. (edited by J.M.V. Blanshard & P.J. Lillford). Pp. 35-101. Nottingham: Nottingham University Press.
[35].Hemminga, M.A., Roozen, M.J.G.W. & Walstra, P. (1993). Molecular motions and the glassy state. In: *The glassy state in foods*, (edited by J.M.V. Blanshard & P.J. Lillford) Pp. 157-189, Nottingham, Nottingham University Press.
[36].Maier, H.G. (1969). Inverse gas chromatography. *Journal of Chromatography*, **45**, 57-62.
[37].Gilbert, S.G. (1984). Inverse gas chromatography. *Advances in Chromatography*. **23**, 199-228.
[38].Gilbert, S.G. & Roshdy, T.H. (1989). The use of inverse gas chromatography in Food Science research. In: *Flavors and off-flavors*. Proceedings of 6th International Flavor Conference. (edited by G. Charalambous). Pp.439-453. The Netherlands, Elsevier Science BV.
[39].Menting, L.C., Hoogstad, B. & Thijssen, H.A.C. (1970). Diffusion coefficients of water and organic volatiles in carbohydrate-water systems. *Journal of Food Technology*, **5**, 111-126.
[40].Lethanh, M., Thibeaudeau, P., Thibaut, M.A. & Voilley, A. (1992). Interactions between volatile and non-volatile compounds in the presence of water. *Food Chemistry*, **43**, 129-135.
[41].Hau, M.Y.M., Gray, D.A. & Taylor, A.J. (1996). Binding of volatiles to starch. In: *Flavor-Food Interactions* (edited by R.J. McGorrin & J.V. Leland). Pp. 109-117, Washington DC: American Chemical Society.
[42].Hau, M.Y.M., Hibberd, S. & Taylor, A.J. (1996). Modelling the uptake of diacetyl in starch. In: *Flavour Science: Recent developments*. (edited by A.J. Taylor & D.S. Mottram). Pp. 437-441. Cambridge, Royal Society of Chemistry.
[43].Hau, M.Y.M., Gray, D.A. & Taylor, A.J. (1998). Binding of volatiles to starch at low water contents. *Flavour and Fragrance Journal* **13**, 77-84.
[44].Israelides, C., Vlyssides, A. Linforth, R.S.T. & Taylor, A.J. (1998). Binding aroma volatiles to macromolecules. In: *EU COST ACTION Interaction of food matrix with small ligands influencing flavour and texture*. Vol. 2 (editor A.B. Zacharoff). EU Directorate General, Brussels.
[45].Langorieux, S. & Crouzet, J. (1994). Study of aroma compounds-polysaccharide interactions by dynamic exponential dilution. *Lebensmittel Wissenschaft Technologie*, **27**, 544-549.
[46].Carr, J., Baloga, D., Guinard, J-X., Lawter, L., Marty, C. & Squire, C. (1996). The effect of gelling agent type and concentration on flavor release in model systems. In: *Flavor-Food Interactions* (edited by R.J. McGorrin & J.V. Leland). Pp. 98-108. Washington DC: American Chemical Society.
[47].Guinard, J.X., Marty, C. (1995) Time-intensity measurement of flavor release from a model gel system: Effect of gelling agent type and concentration. *Journal Food Science.*, **60**, 727-730.
[48].Wilson, C.E., and Brown, W.E. (1997). Influence of food matrix structure and oral break down during mastication on temporal perception of flavour. *Journal Sensory Studies*, **12**,(1), 69-86.
[49].Morris, E.R. (1994). Rheological and organoleptic properties of food hydrocolloids. In: *Food Hydrocolloids: Structure, Properties and Functions*. (edited by K. Nishinari & E.Doi), Pp 201-208. New York, Plenum Press.
[50].Hatchwell, L.C. (1996). Implications of fat on flavor. In: *Flavor-Food Interactions* (edited by R.J. McGorrin & J.V. Leland). Pp. 14-23. Washington DC: American Chemical Society.
[51].Brauss, M.S., Balders, B., Linforth, R.S.T. & Taylor, A.J. (1998). Fat content, baking time, hydration and temperature affect flavour release from biscuits in model and real systems. *Flavour Fragrance Journal*. Submitted.
[52].Brauss, M.S., Linforth, RST, Cayeux, I, Harvey B, Taylor, A.J. (1998). Altering the fat content in a model yogurt system; Instrumental and sensory effects. *Journal of Agricultural and Food Chemistry* Submitted.
[53].Bakker, J. & Mela, D.J. (1996). Effect of emulsion structure on flavor release and taste perception. In: *Flavor-Food Interactions* (edited by R.J. McGorrin & J.V. Leland). Pp. 36-47. Washington DC, American Chemical Society.
[54].O'Neill, T.E. (1996). Flavor binding by proteins: an overview. In: *Flavor-Food Interactions* (edited by R.J. McGorrin & J.V. Leland). Pp. 59-74. Washington DC: American Chemical Society.

[55].O'Neill, T.E. & Kinsella, J.E. (1987). Binding of alkanone flavour to (-lactoglobulin – effects of conformational and chemical modification. *Journal of Agricultural and Food Chemistry* **35**, 770-774.
[56].Boudaud, N & Dumont, J-P. (1996). Interaction between flavor components and B-lactoglobulin. In: *Flavor-Food Interactions* (edited by R.J. McGorrin & J.V. Leland). Pp. 90-97. Washington DC: American Chemical Society.
[57].Jouenne, E. & Crouzet, J. (1996). Interaction of aroma compounds with beta lactoglobulin. In: *Flavour Science: Recent Developments* (edited by A.J. Taylor & D.S. Mottram). Pp. 425-432. Cambridge: Royal Society of Chemistry.
[58].Batt, C.A., Brady, J. & Sawyer, L. (1994). Design improvements of (-lactoglobulin. *Trends in Food Science and Technology*, **5**, 261-265.
[59].Charles, M., Bernal, B. & Guichard, E. (1996). Interactions of (-lactoglobulin with flavour compounds. In: *Flavour Science: Recent developments*, (edited by A.J. Taylor and D.S. Mottram) Pp. 1433-436. Cambridge: Royal Society of Chemistry.
[60].Guichard, E & Sostman, K. (1998). Measurement of interactions between (-lactoglobulin and different flavour compounds: determination of binding constants by affinity chromatography. In: *Interaction of food matrix with small ligands influencing flavour and texture* Vol 2. Pp31-36. EU COST96 publication EUR 18162 Brussels.
[61].Hansen, A.P. & Booker, D.C. (1996). Flavor interactions with casein and whey protein. In: *Flavor-Food Interactions* (edited by R.J. McGorrin & J.V. Leland). Pp. 75-89. Washington DC: American Chemical Society.
[62].O'Keefe, S.F., Wilson, L.A. Adoraccion, A.P. & Murphy, P.A. (1991). Determination of the binding of hexanal to soy glycinin and (-conglycinin in an aqueous model system using a headspace technique. *Journal Agricultural and Food Chemistry*, **39**, 1022-1028.
[63].Damadoran, S. & Kinsella, J.E. (1981). Interactions of carbonyls with soy protein: Thermodynamic effects. *Journal of Agricultural and Food Chemistry*, **29**, 1249-1253.
[64].Harvey, B.A., Druaux, C. & Voilley A. (1995). Effect of protein on the retention and transfer of aroma compounds at the lipid-water interface. *Food Macromolecules and Colloids*, Royal Society of Chemistry,**156**, 154-163.
[65].Mottram, D.S., Nobrega, I.C., Dodson, A.T. & Elmore, J.S. (1996). Changes in thiol and disulfide flavour compounds resulting from the interaction with proteins. In: *Flavour Science: Recent Developments* (edited by A.J. Taylor & D.S. Mottram). Pp. 413-418. Cambridge: Royal Society of Chemistry.
[66].Risch, S.J. & Reineccius, G.A. (1995). *Encapsulation and controlled release of food ingredients.* ACS Symposium Series #590, Washington DC, American Chemical Society.
[67].Qi, Z.H & Hedges, A.R. (1995). Use of cyclodextrins for flavor. In : *Flavor Technology.* (edited by C-T. Ho, C-T. Tan & C-H. Tong). Pp.231-243, Washington DC, American Chemical Society.
[68].Hedges, A.R., Shieh, W.J. & Sikorski, C.T. (1995). Use of cyclodextrins for encapsulation and treatment food products. In: *Encapsulation and controlled release of food ingredients.* (edited by S.J. Risch & G.A. Reineccius). Pp.60-73, Washington DC, American Chemical Society.
[69].Levine, H., Salde, L., VanLengerich, B. & Pickup, J.G. (1991). Glassy matrices containing volatile and/or labile components and processes for preparation and use thereof. US Patent 5,009,900, April 23 1991.

BIOTRANSFORMATIONS IN THE FLAVOUR INDUSTRY

RALF G. BERGER, JAN A. M. DE BONT, GERRIT EGGINK, M. MANUELA DA FONSECA, MAIK GEHRKE, JEAN-BERNARD GROS, FREDERIK VAN KEULEN, ULRICH KRINGS, CHRISTIAN LARROCHE, DAVID J. LEAK, MARIET J. VAN DER WERF. ‡

1. Summary

Bacteria and fungi synthesise or degrade a vast number of natural and xenobiotic substrates. Biotransformations and bioconversions occur, if a single substrate structure is altered by an identifiable redox, hydrolysis, or addition type reaction, or by a sequence of these reactions. Generally competing with chemosynthesis, biocatalysts possess inherent advantages: They functionalise chemically inert carbons, modify one functionality in a multifunctional molecule selectively or specifically, introduce chirality, resolve racemates, and operate under ambient conditions. The production of natural flavour and aroma ingredients could benefit from these characteristics. The biotransformation of volatile terpenes particularly well demonstrates perspectives, but

‡ Ralf G. Berger, Institut für Lebensmittelchemie, Zentrum Angewandte Chemie der Universität, Wunstorferstr. 14, D 30453 Hannover, Germany.
Jan A. M. de Bont, Division of Industrial Microbiology, Department of Food Science, Wageningen Agricultural University, P.O. Box 8129, NL-6700 EV Wageningen, The Netherlands.
Gerrit Eggink, ATO-DLO, Postbus 17, NL-6700 AA Wageningen, The Netherlands.
M. Manuela da Fonseca, Instituto Superior Técnico, Secção de Biotecnologia, Av. Rovisco Pais, P-1096 Lisboa Codex, Portugal.
Maik Gehrke, Institut für Lebensmittelchemie, Zentrum Angewandte Chemie der Universität, Wunstorferstr. 14, D 30453 Hannover, Germany.
Jean-Bernard Gros, Laboratoire de Génie Chimique Biologique, Université Blaise Pascal, 24 Avenue des Landais, F-63177 Aubière cedex, France.
Frederik van Keulen, Instituto Superior Técnico, Secção de Biotecnologia, Av. Rovisco Pais, P-1096 Lisboa Codex, Portugal.
Ulrich Krings, Institut für Lebensmittelchemie, Zentrum Angewandte Chemie der Universität, Wunstorferstr. 14, D 30453 Hannover, Germany.
Christian Larroche, Laboratoire de Génie Chimique Biologique, Université Blaise Pascal, 24 Avenue des Landais, F-63177 Aubière cedex, France.
David J. Leak, Dept of Biochemistry, Imperial College of Science, Technology and Medicine, Exhibition Road, London, SW7 2AY, UK.
Mariet J. van der Werf, Division of Industrial Microbiology, Department of Food Science, Wageningen Agricultural University, P.O. Box 8129, NL-6700 EV Wageningen, The Netherlands.

K.A.D. Swift (ed.), Current Topics in Flavours and Fragrances, 139–170.

also current problems. Solvent tolerant bacteria hold much promise for a lipophilic biotechnology. Genetic engineering will help to create tailor-made biocatalysts. Downstreaming of products, based on *in situ* solvent extraction or adsorption, is required for an efficient bioprocess. Cost calculations show that high-yielding biotransformations have commercial potential.

2. Introduction

For many centuries, plant essential oils were the major source of flavour compounds. As the structures of the compounds present in these essential oils were elucidated, more and more synthetic flavours became available. However, at present, many flavour compounds are still extracted from botanical sources because of the consumer's preference for natural products [1]. This increasing demand for natural flavours has resulted in a shortage of several of these natural flavours, such as peppermint and strawberry aroma. Biotechnological methods offer an alternative for the production of natural flavours.

In food, over 6,300 volatile chemicals have been identified. The complex chemical mixture responsible for the aroma of a particular food can number as high as 1,000 compounds for an individual food. Many of these compounds contribute to the overall flavour of the particular foodstuff, although generally only a few of them, the character impact compounds, furnish a major contribution towards its characteristic aroma profile.

Biotechnological methods can be used for the production of complex mixtures as well as for the production of single compounds. So far, not much information is available in literature on the biotechnological production of complex mixtures. The production of complex mixtures requires a continuous organoleptic evaluation of the fermentation broth. The flavour industry produces complex flavour mixtures by biotechnological methods, but this occurs under a wave of secrecy. The most promising use of biotechnological production methods in the flavour industry seems in the production of character impact compounds. A few recent publications have presented an excellent summary of flavours produced by enzymes and microorganisms ([2,3]; Table 1).

Biotechnological methods for the production of pure flavours can be divided in two groups: fermentation and biotransformation. Fermentation involves "de novo"-synthesis, using microorganisms growing on cheap substrates like sugars. Examples of flavours produced on a commercial scale by fermentation are butanoate and citrate (Table 1). Biotransformation and bioconversion involve the alteration (single or multistep) of a precursor or intermediate into a more valuable product. This paper deals only with biotransformation processes for the production of flavours.

Table 1. Flavour compounds produced by biotechnology

Class	Examples
Alcohols	Ethanol, butanol, 1-octen-3-ol
Aldehydes	Acetaldehyde, hexanal
Ketones/methylketones	Diacetyl, 2-heptanone, acetophenone, 2-undecanone
Acids	Acetate, propanoate, butanoate, citrate, succinate
Esters	Ethylbutanoate, isopentyl acetate, ethyl propanoate
Lactones	δ-Dodecalactone, γ-decalactone
Aromatics	Vanillin, benzaldehyde, phenylacetaldehyde
Pyrazines	2-Methoxy-3-isopropylpyrazine, tetramethylpyrazine
Terpenes	Nootkatone, carvone, (-)-menthol
Ionones	Derivatives of α- and β-Ionone
Sulfur compounds	3-Mercaptohexanol, *p*-menth-1-en-8-thiol, furfurylthiol
Carbohydrates	Isomerised glucose, ascorbate
Amino acids	Glutamate, phenylalanine
Peptides	Aspartame, thaumatin
Mononucleotides	5'-IMP, 5'-GMP

Three principle biotransformation techniques can be distinguished; (1) use of enzymes, (2) use of microorganisms, and (3) plant cell and tissue culture. The most straightforward biotransformation technique is the use of enzymes. Many enzymes are commercially available in large quantities (e.g. lipases, esterase). An example of the commercial application of an enzymatic biotransformation process is in the production of natural benzaldehyde from apricot and peach pits [4]. However, many enzymes which might be useful in the production of flavours are not commercially available and/or are very unstable. Therefore, in many instances whole cells, either from microorganisms or from plants, are used as biocatalysts for the bioformation of flavours. Examples of microbial biotransformation processes which are run commercially in the flavour industry are the formation of γ-decalactone from ricinoleic acid [5], and the production of furfurylthiol from furfural and cysteine [6]. Also, plant cells or tissues can be used as the biocatalyst. In general, the use of plant cell cultures for the production of bioflavours has more disadvantages than the use of enzymes or microorganisms, but they can be advantageous for the production of complex secondary metabolites [7].

3. Problems Encountered in Flavour Biotransformation Studies

At the present time, the production of flavours by biotransformation is in many instances, more of an academic exercise than a commercial reality. Although many interesting biotransformation processes have been described producing interesting natural flavours, the number of industrial biotransformation processes is limited. This is due to a variety of technical difficulties associated with biotransformation (Table 2). Despite these problems, a number of biotransformation processes for the production of flavours has been commercialised in the last decade.

Table 2. Problems prohibiting flavour biotransformation processes from being commercialised

- Substrate and product toxicity.
- Low product concentrations.
- Low yields.
- Long reaction times.
- Transient product accumulation; need to terminate the process at the correct point of time.
- Volatility of the products; loss of the product with the outgoing airstream.
- Multiple bioconversion pathways resulting in the formation of a mixture of products including off-flavours.
- Irreproducibility; unknown induction of enzymatic bioconversion activity.
- Instability of the biocatalyst.
- Organism morphology; difficulties in growing fungi and plant cells/tissues.

4. Advantages of Biotransformation

In view of several specific advantages of biotransformation processes, the commercial importance of biotransformation for the production of flavours will certainly grow further in the near future:

4.1. PRODUCTION OF NATURAL FLAVOURS

Recent market surveys have demonstrated that consumers prefer foodstuffs that can be labelled natural [1]. This has resulted in a premium price for flavours isolated from natural sources. Flavours produced by biotransformation are classified “natural“, if the substrate was natural.

4.2. PRODUCTION OF UNIFORM PRODUCTS AT A CONSTANT PRODUCTIVITY

So far, most natural flavours are obtained from plant materials. This production method has several disadvantages. These include a variability in the composition and yield of the final products from different geographical sources, but also of one source depending on the weather, the risk of plant diseases, a seasonal variation in supply, the fact that plant materials imported from tropical or subtropical regions can be the object of trade restrictions, the socio-political instability of the regions where several of these plants grow, and ecological drawbacks as several of the interesting plants are on the verge of extinction. This has resulted in large fluctuations of quality, price and supply of the natural flavours extracted from plants. Thus, the increased demand for certain flavours cannot be met by traditional production methods, and a shortage of supply of certain flavours has been encountered. By applying biotransformation the flavour industry becomes independent from external influences, and is able to produce natural flavours at a constant quality and productivity all year round.

4.3. REGIO- AND STEREOSELECTIVITY

The absolute configuration of a flavour, especially with respect to its stereochemical composition, strongly influences the sensory properties of a compound [8]. The methods used for the production of flavours must, therefore, result in optically pure compounds. While chemical methods usually lead to the formation of a mixture of isomers and by-products, biotechnological methods are suited to achieve this type of transformation, as enzymes generally show a pronounced regio- and stereoselectivity.

4.4. ENVIRONMENTALLY FRIENDLY PRODUCTION METHODS

The recent campaign for safer and environmentally friendly technology demands the use of milder reaction conditions, which are more (energy) efficient and cause fewer environmental problems. In this respect, biotransformation is especially suited as they proceed at ambient conditions of temperature and pressure, even in the case of multistep conversions, and show fewer side-reactions.

4.5. PRODUCTION OF NATURAL FLAVOURS WHICH ARE NOT FOUND IN APPRECIABLE AMOUNTS IN NATURE

Several character impact compounds, such as nootkatone (grapefruit-like), ß-damascenone (honey-sweet), or raspberry ketone are only present in trace quantities in plants, thus rendering technical-scale isolation expensive or even impossible. With biotransformation, a specific bioconversion pathway can be selectively enhanced, ultimately resulting in the production of high concentrations of a single compound.

5. Toxicity/Solvent Tolerance

The partial removal of lipids from a food to decrease the energy content has regularly shown disappointing results. The reason is that most flavour compounds preferentially accumulate in the lipophilic fraction of food; as a result, the sensory attributes of lipid reduced foods are often not up to standard. This common physical trait has a profound effect on microbial production methods for these compounds. During a production process employing whole cells, these compounds also will preferentially partition to the cell membranes of the microbes. As a result, the compounds are toxic for the producing organisms, even at low concentration of the product. Very often, the substrates are also toxic.

5.1. COMPOUNDS IN MICROBIAL MEMBRANES

A relationship has been established between the toxicity of a compound for an organism and the partitioning of a compound to octanol from the water phase [9]. The decimal logarithm of the partition coefficient of a solute in a standard octanol-water system is termed log $P_{O/W}$, and this parameter has been taken as an indicator for the partitioning of compounds from the aqueous medium to the membrane of organisms [10]. Generally, it is accepted [11] that compounds have their destroying effects on organisms at the level of the cytoplasmic membrane [9], where they preferentially accumulate. As a consequence of the high concentration of compounds in this compartment, the cell is no longer able to perform essential biochemical reactions and eventually loses its integrity. Various analytical techniques [10,12] can be employed to determine the dose-response relationship of a compound in a membrane. The different techniques reveal similar patterns, supporting the view that the actual membrane concentration of a compounds is an important parameter.

Actual membrane concentrations depend on (1) the concentration of compounds in the water phase and (2) the partitioning of the compounds from the water phase to the membrane. Based on work with ^{14}C-labelled solvents a correlation was found between the log $P_{O/W}$ value of a solvent and its partitioning ($P_{M/W}$) between membrane and water ([10]; see section 7.2.2.2):

$$\log P_{M/W} = 0.97 \times \log P_{O/W} - 0.64$$

Using this equation it is possible to calculate the actual membrane concentration of a compound if its concentration in the water phase is known. Limonene is taken as an example here. This compound potentially is a substrate for the production of a variety of odorous monoterpenoids. However, the compound is quite toxic to bacteria, and most of the potential products are even more toxic. Limonene has a log $P_{O/W}$ = 4.5, and, based on the formula above, this results in a log $P_{M/W}$ of 3.7. In other words, limonene partitions 5,300 times better to a bacterial membrane than to water. The maximal solubility of limonene in water is in the order of 0.1 mmol L^{-1}. Consequently, in a two-phase water-limonene system the concentration of limonene in the membrane will be 530 mmol L^{-1}. No common microorganism will survive this high actual membrane concentration. The situation is even more problematic, if a terpenoid is produced from limonene. The resulting product will have a lower log $P_{O/W}$ and, hence, a lower log $P_{M/W}$. But product concentrations will have to remain in the mM-range in order not to inactivate the whole cell. Consequently, high product concentrations will not be achievable.

5.2. SOLVENT-TOLERANT BACTERIA

Special organisms have been isolated recently that have the ability to survive by far higher concentrations of lipophilic compounds than all other organisms. These solvent-tolerant organisms allow a new degree of freedom in coping with toxic products. They can tolerate solvents in a two-phase system even for lipophilic compounds, such as heptanol (log $P_{O/W} = 2.4$). These organisms may find application in the production of flavour compounds in two ways:

(1) They may be used as such in an aqueous environment because they will withstand much higher product concentrations than normal bacteria.

(2) They may be applied in two-phase water-solvent systems. A rather hydrophilic solvent such as hexane or octanol may be used as a second phase to which the desired flavour compound partitions selectively.

5.3. MECHANISMS OF SOLVENT TOLERANCE

The mechanisms underlying solvent tolerance have been subject of considerable debate, and a number of systems have been discussed. The effects of solvents on both lipid ordering and bilayer stability of biological membranes have been investigated in most detail [9]. At best, such mechanisms can modify the physical properties of membranes to such an extent that the effects of solvents are temporarily and partially counterbalanced. The active removal of solvents from the cytoplasmic membrane would be a dynamic and longer-lasting mechanism for solvent tolerance. However, solvents such as limonene, are uncharged lipophilic molecules, and export systems for such compounds would seem difficult.

Recently, however, in the case of toluene an active efflux system has been identified [13]. Three proteins were involved that strongly resembled proteins involved in proton-dependent multidrug-efflux systems of the resistance-nodulation-division (RND) family [14]. These efflux pumps are composed of three proteins that are thought to span the inner and outer membranes of Gram-negative bacteria. Members of this family are involved in such diverse activities as exporting heavy metals, oligosaccharides, and antibiotics [12]. The toluene pump also exported other solvents and it seems its substrate specificity is very broad. Most likely, a number of flavour compounds, including limonene and some terpenoids, will be exported as well by such pumps. Organisms containing such a pump (Fig. 1) may be very useful hosts for heterologous genes encoding enzymes involved in the biotransformations of flavours.

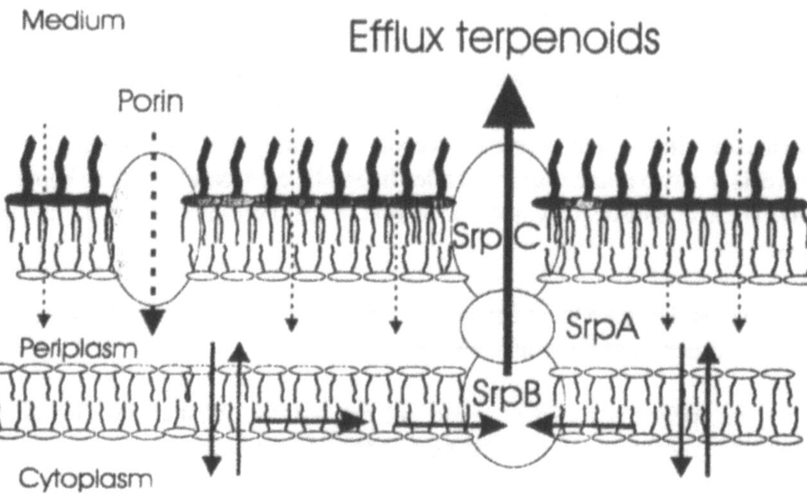

Figure 1. Schematic drawing of the efflux of a toxic terpenoid from the bacterial membrane. A terpenoid, or any other lipophilic flavour compound produced in the cell, will tend to accumulate in the membrane. But due to the so called SrpABC pumping system, the concentration of the flavour compound can remain relatively low in the membrane. An impermeable outer membrane should prevent a rapid re-entry of the product in the cell

6. Genetic Engineering: Successes and Perspectives

The food industry is no stranger to the benefits (and associated controversies) of genetic engineering. Genetically altered food crops with modified properties such as herbicide tolerance or enhanced post-harvest processing characteristics are now a commercial reality. In principle, the same techniques can be applied to enhance or modify the flavour quality of food crops [e.g. see 15], although the multifactorial nature of organoleptic properties make this a significantly more complex task. Thus, although this might be considered as an application of genetic engineering to "*in situ* flavour biotransformations", it is outside the scope of this article to consider the potential of this approach. Nonetheless, it should be noted that with the cloning, expression and recent availability of crystal structures for terpene cyclases [16,17], the authors of a recent review have included the alteration of terpenoid based pathways in foodstuffs to impart desirable flavour properties, as a biotechnological possibility [18].

6.1. PRESENT STATUS

Although the genetic modification of flavour properties of food crops may be some way off, the potential to modify the flavour properties of biologically processed/fermented foods, such as cheese, fermented meats and beverages through genetic manipulation of the natural organisms involved is already being realised. Genetic engineering of lactic acid bacteria, in particular *Lactococcus lactis*, has advanced to such a stage that the cloning, re-introduction and expression of individual genes or operons is now routine [19,20]. To allow these systems to be used in the food industry, so called "food grade" cloning and expression vectors based on natural lactococcal plasmids and promoters, and avoiding the use of antibiotic based selection are being developed [21]. Additionally, techniques of "reverse genetics" or gene replacement have been developed [22,23]. Where there is sufficient homology between an introduced gene and a wild type chromosomal homologue then, in recombination proficient strains, the introduced gene will be incorporated into the host chromosome via a Campbell-type single cross-over event. A second cross-over within the homologous region will then eliminate the vector entirely. Usually in this type of experiment the vector is unable to replicate independently in the host strain to which it has been introduced and is, therefore, lost. In this way it is possible to replace a wild type gene either with a deleted version, to knock out a specific function, or with a modified version (e.g. with a modified promoter to allow inducible expression under the control of an exogenous inducer) of the gene. Thus, all of the requisite techniques for manipulation of lactococcal metabolism and physiology are available, and reports are starting to appear on the use of these techniques in process improvement including flavour development [24-26]. However, before the impact of this technology can be fully appreciated and applied to the production of specific flavours (in a bioreactor) it will be necessary to understand the process of flavour development during maturation and fermentation in more detail. Molecular biological techniques are invaluable for this as well. For instance, promoter probe vectors can be used to detect which promoters are switched on during the process, and reporter gene construct can be used to assess the level of expression. Thus the complexities of flavour development in biologically processed foods are starting to be unravelled [27,28] and manipulated towards commercial goals. Although work on *Lactococcus lactis* is the most advanced, there is no practical reason why other GRAS microorganisms involved in flavour development should not be amenable to the same approach, and some progress is evidently being made. However, it should be noted that polyploid eukaryotic microbes pose some additional problems, particularly with techniques such as gene replacement.

While enhancement of flavour development by genetic manipulation of organisms involved in traditional food biotechnology might reasonably be classified as "*in vivo* biotransformations", the term biotransformation is more commonly associated with processes yielding single products using precultivated cells or enzymes derived from them as single or multiple step biocatalysts. On one level, genetic engineering has already had a major impact on the economic feasibility of using biocatalytic processes, as the production costs of enzymes have been significantly reduced through cloning and over-expression. Additionally, the use of engineered N- or C-terminal affinity tags can simplify the process of enzyme purification. Thus, many of the commonly used enzymes such as lipases, proteases and glycohydrolases are available as products from recombinant cells. Indeed, their availability is a major determinant in deciding a biotransformation strategy. Notwithstanding, because the technology of gene cloning and gene expression is now relatively mature, where previously a potentially useful enzyme activity might not have found application because of low abundance, genetic engineering provides a logical route to solve this problem. Good examples of this are already available, e.g. cloning of the gene encoding acetolactate decarboxylase (ALDC) and expression in brewing yeast [29]. Recombinant yeasts expressing the ALDC gene have been shown to produce much less diacetyl during the fermentation, thus reducing the length of the subsequent maturation process. (An alternative strategy is to add the free or immobilised enzyme during, or at the end of fermentation). Conversely, in dairy products where the "buttery" flavour of diacetyl is a benefit, elimination of the ALDC gene by reverse genetics has been shown to enhance diacetyl production [25].

6.2. PERSPECTIVE APPLICATIONS OF GENETIC ENGINEERING

In the future genetic engineering is likely to have a major impact on the intrinsic efficiency of biotransformation processes *per se*, through our increasing ability to rationally redesign and, hence, modify enzyme specificity through protein engineering [30]. Additionally, the same genetic engineering techniques cited above for lactococci are likely to have a significant impact on our ability to exploit novel multistep biotransformations in which readily available natural products (e.g. monoterpenes) are converted either to known, or novel flavour compounds. Exploiting the natural pathways for *de novo* synthesis of flavour compounds from carbon substrates such as carbohydrates by converting microbial or plant cells into "cell factories" is a more difficult task (see below), but is nonetheless a realistic long term target.

6.2.1. Process Improvement by Genetic Modification of Biocatalysts

One of the primary reasons for adopting a biocatalytic process as opposed to a chemical transformation is because of the inherent specificity of enzyme reactions. In flavour production a completely biological route from source to product will also enable the added value "natural" classification to be used [31]. Specificity implies reaction specificity (e.g. production of an aldehyde from a primary alcohol without further oxidation to the carboxylic acid), regioselectivity (e.g. selective attack at only one out of a number of equivalent sites) and stereoselectivity. Depending on the nature of the biotransformation, the inherent substrate specificity of the biocatalyst may be an advantage or a hindrance. From a commercial enzyme producer's point of view the ideal catalyst is one that has a high regio- and stereoselectivity but a wide substrate range, making it applicable to a number of processes. However, for a flavour biotransformation it may be useful to have a catalyst which can discriminate between a number of similar substrates e.g. deaminases, aminotransferases and decarboxylases which can produce specific types of volatile aroma compounds from certain groups of amino acids (allowing them to be used with crude proteolytic digests). While it may be possible to find enzymes with the required specificity by screening, an alternative approach is to modify the specificity of an existing enzyme by protein engineering (i.e. making defined alterations in the gene sequence which lead to the substitution of specific amino acids). Equally it may be desirable to have enzymes with specific combinations of properties. Again, as an alternative to extensive screening programmes it is feasible to engineer properties, such as the pH optimum, thermostability and resistance to substrate or end product inhibition [32-35]. An example where this might be useful is in the application of glycosidases for the release of volatile flavours from glycoside conjugates [36] present in plant wastes such as peelings or skins.

It should be noted that although classical protein engineering implies a rational alteration based on knowledge of the contribution of different amino acids to catalysis [30], less knowledge based approaches using "molecular breeding" or enzyme evolution [37,38], and hybrid formation [39], assisted by powerful selection strategies may provide a more rapid route to success.

6.2.2. Genetic Engineering of Natural Product Catabolic Pathways

Most plant flavour compounds are not intermediates in primary metabolism, but are effectively secondary metabolites. As such, their biosynthesis from the carbon substrates used for growth and energy generation is lengthy and, as outlined below, would be difficult to increase significantly. An alternative strategy, and one in which biotransformations are likely to play a major role, is to exploit enzymes involved in natural product catabolism to make the flavour compounds from available, natural precursors e.g. the production of vanillin from an intermediate in eugenol catabolism by *Arthrobacter globiformis* [31], or from ferulic acid using *Pseudomonas fluorescens* [40]. The major advantage of this strategy is that isolating an organism which degrades a specific compound is relatively facile, and catabolic enzymes are usually expressed at much higher levels than those associated with the biosynthesis of secondary metabolites. Hence, the initial screening for potentially useful enzymes is straightforward. Furthermore, given that many volatile flavour compounds are alcohols and ketones/aldehydes, the intermediates produced by catabolism of natural oils, fats and abundant terpenoids are potential targets. Genetic engineering can have a significant role to play here in both gene deletion (resulting in the accumulation of intermediates) and in heterologous gene expression. With respect to the latter, it should be noted that in bacteria the initial catabolic steps are often encoded in an operon (i.e. closely together and under a common control mechanism), so it may be possible to clone the genes encoding a number of steps in a relatively small DNA fragment. As an example, Chang *et al* [41] reported the cloning of the genes encoding the entire catabolic pathway (to intermediary metabolites) of limonene metabolism in *Bacillus stearothermophilus* and their expression (from a 9.6kb fragment) in *E. coli* enabling the recombinant strain to grow on limonene and produce a variety of intermediates including perillyl alcohol and α-terpineol. Heterologous expression of parts of pathways or use of specific gene deletion techniques (reverse genetics) should also make it possible to circumvent the problem of multiple catabolic, or dead end pathways acting on a substrate, which appears to be a particular feature of terpene catabolic pathways [42,43].

As well as targeting known compounds as products, such an approach can occasionally give rise to previously uncharacterised flavours/fragrances, e.g. a study of the α-pinene catabolic pathway in a strain of *Pseudomonas fluorescens* led to the discovery of the aldehyde intermediate "isonovalal" [44], the production of which, at one stage, was being developed as a commercial biotransformation process by a company of the same name.

6.2.3. Pathway Manipulation

Various forms of pathway manipulation are desirable, ranging from the addition of functionality at the beginning or end of a pathway to the increase of metabolic flux towards a specific product. Examples of the former include extending the substrate range of yeast (*Saccharomyces cerevisiae*) to use substrates such as lactose [45] or starch [46], and the restriction of diacetyl production by expressing the ALDC gene in yeast [29]. As described earlier, addition of functionality at the beginning or end of a pathway can also be achieved by using free or immobilised enzymes in the fermentation process, but in practice the process will be more economic if a recombinant strain expressing all of the required activities, is used. Increased production of specific flavour compounds has also been achieved by expressing or deleting specific functions which lead to a diversion of metabolism towards specific intermediates. Thus, increased production of methylbutyl acetate in sake yeast was achieved by over-expression of the LEU4 gene which increased production of 3-methylbutanol [47], a precursor of 3-methylbutyl acetate. Increased production of inosine from *Bacillus subtilis* has also been achieved by deleting the gene encoding IMP-dehydrogenase [48].

Despite these successes in pathway manipulation, we should make a sharp distinction between making a step increase in the low level production of a flavour compound and developing a biocatalyst for the high efficiency multistep conversion of a cheap starting material to a valuable single product. This is because it is clear that there are very few pathways in which there is a single rate determining step which would be the obvious target for cloning and over-expression. Instead, each step will usually exert a certain level of control over the whole pathway and increasing the expression of an enzyme catalysing just one of these steps may have little, and sometimes no effect on overall flux rates. Thus, while there have been many reports of microorganisms making small amounts of potentially useful flavour compounds [49], increasing these or the productivities of plant cell cultures (or the plants themselves) to become the major product, would require a considerable amount of work in metabolic control analysis [50,51] to establish the most important genes to target.

7. Process Development

7.1. DOWNSTREAMING OPTIONS

Traditionally, flavours are recovered from their plant or animal sources by solvent extraction, vacuum distillation or steam distillation [52]. With the advent of microbial and enzymatic production systems, the flavour industry has seeked to adapt the proven techniques and existing equipment to these novel flavour sources. Microbiological volatiles are either secreted into the extracellular environment, where they unfold their bioactivity, or retained inside the cell. Biotechnologically generated flavour compounds, in a more or less complex mixture, are usually found in the cultivation broth, and, hence, relatively easy to recover: The peak concentration during the transformation is monitored using a rapid analytical technique, and the entire batch is then submitted to an *off line* downstreaming sequence. (Cultured plant cells tend to store flavours in the cytosol, another complication of these systems).

However, several reasons suggest that an *in situ* removal of lipophilic flavours would be superior to the present industrial standard (see also Table 2 and section 5). These reasons are:

1) Biochemical instability of the target compound due to the presence of degrading enzymes.
2) Low water-solubilities cause substrate and product losses via the waste air stream.
3) Substrate and/or product inhibition phenomena.
4) Non-stationary concentrations in conventional batch processes.

Classical aroma technology is not suitable for *in situ* recovery and shows further disadvantages, such as thermal stress on sensitive chemicals, inactivation of the biocatalyst, use of large amounts of organic solvents, and high energy consumption. Some promising approaches face the problem and offer *in situ* recovery by different ways.

7.1.1. Pervaporation

Well known applications of pervaporation in the food and flavour industry are wine and beer dealcoholisation, juice concentration, and extraction of aroma compounds from various dilute media [53]. Pervaporation means a partial vaporisation of a liquid through a special membrane driven by a pressure gradient. Volatiles dissolve in the polymer material at the upstream side of the membrane and then diffuse into the polymeric network. They are finally evaporated in the downstream vacuum phase, from which they need to be condensed. A range of applications for the (*off-* and *on line*) recovery from fermentation broths has been described [53,54].

7.1.2. Supercritical Fluid Extraction

Supercritical fluid extraction is a rapid, non-selective method for sample preparation prior to the analysis of compounds in natural product matrices. Extraction yields are affected by the solubility of the solute in the fluid, diffusion through the matrix, and adsorption onto the matrix. Applications are the extraction of carotenoids, lipids, flavour and fragrance compounds etc. [55]. The market potential and economics of supercritical extraction for the separation of essential oils and flavours from spices and herbs has been reviewed [56]. A first application, the recovery of 4-decanolide from a fungal culture, was reported [57]. The isobaric solubility of 4-decanolide in CO_2 was studied, and the complete extraction of 4-decanolide loaded XAD 2 resin was achieved by using high CO_2 throughput rates.

7.1.3. Other Options

Two-phase *in situ* extraction of flavours could make use of the experience gathered from the extractive production of ethanol and other alcohols. The step from absorption into a solvent phase to adsorption is made using an inorganic or organic polymer solid as the second phase. Adsorption is the accepted standard to enrich analytes from gaseous samples and to accumulate trace organics from, e.g. drinking water. Both absorptive and adsorptive recovery deserve particular attention.

7.2. EQUILIBRIUM PROPERTIES: VAPOUR PRESSURE, WATER SOLUBILITY, ACTIVITY COEFFICIENT

Biotransformation processes leading to aromas and fragrances typically involve compounds exhibiting high vapour pressures (volatile molecules) and low water solubility. As outlined above, these features may give rise to a significant substrate and/or product loss during the course of a reaction, especially if carried out in an aerated bioreactor filled with a pure aqueous phase. Under these circumstances knowledge of the above-mentioned properties is of crucial importance.

Measurement of water solubility of hydrophobic compounds is difficult due to the low concentrations that have to be quantified. It is also well known that prediction of this property remains a major problem [58]. These considerations are especially true in the case of terpenes and terpenoids where experimental data remains scarce and is often poorly consistent. For example, Schmid *et al* [59] reported a water solubility of 0.011 mmol L^{-1} for limonene at 25 °C, while Weidenhamer *et al.* [60] reports 0.095 mmol L^{-1} for the same compound. These results highlight the need for providing both, accurate experimental procedure, and reliable estimation methods for the determination of vapour-liquid equilibria (VLE) data (water solubility, vapour pressure) for terpenes and terpenoids.

7.2.1. Experimental Determination of VLE Data

7.2.1.1. Basis for Methods. Turner *et al* [61] have recently published a review on the measuring of vapour-liquid equilibria for aqueous-organic systems. All techniques have in common the ability to ascertain the temperature, pressure, and concentration of one or both (liquid and gas) phases at thermodynamic equilibrium. They always consider the following relationship:

$$y_i = \gamma_i \; x_i \; \frac{P_i^0}{P} \qquad (1)$$

with y_i and x_i the molar fraction of the solute i in the gas and liquid phase, respectively, g_i the activity coefficient, P_i^0 the vapour pressure of the solute, and P the total pressure in the system. The ratio y_i/x_i is a partition coefficient equivalent to the Henry's law coefficient for low solubility mixtures.

The activity coefficient varies from unity, for a pure compound, to a limiting value γ_i^∞, theoretically obtained at infinite dilution of the solute, i.e. when the solvent activity coefficient is 1. Practically, it is generally considered that this situation is met when x_i is less than 0.01 [58].

In the case of organics that are sparingly water soluble, the limiting solubility may be attained in the infinite dilution domain. At thermodynamic equilibrium, γ_i^∞ can be estimated as the reciprocal of the solubility limit in the molar fraction:

$$\gamma_i^\infty = \frac{1}{x_i} \qquad (2)$$

Equation (2) is valid only if the solvent is insoluble in the solute [58].

7.2.1.2. Principle of Measurement Techniques. Existing experimental techniques may be classified into five distinct groups, and among them, only three are suitable for low solubility systems [61]. These are static methods, mechanical recirculation systems, and separate measurement of solubility and pure species vapour pressure.

Static and mechanical recirculation methods both involve systems driven to equilibrium in a suitable container. One or both (liquid or gas) phases are then analysed to determine equilibrium concentrations. In static procedures, the solute/solvent mixture is sealed in a flask placed in a thermostat held at the desired temperature, while the second approach provides mechanical mixing of the vapour and liquid phases with each other to increase the rate at which the sample equilibrates. This last technique can be performed in a continuous way, by bubbling air through a liquid solution. Assuming equilibrium between the liquid phase and the bubbling vapour (eq. 1), a material balance on this system gives the following relationship [61]:

$$\ln \frac{C_{i0}}{C_i} = \frac{G \gamma_i^\infty P_i^0}{V P C_{solv}} t \qquad (3)$$

where G is the gas molar flow rate, V the liquid volume, and C_{solv} the molar solvent concentration in the liquid phase. C_{i0} and C_i are the solute concentrations in the liquid phase at time 0 and at time t, respectively.

The last group involves solubility measurements, i.e. building of a saturated solution. This can be done according to three main techniques which are:

1) Two liquid-phase contact (TLPC) methods.
2) Turbidity or cloud point measurement.
3) Vapour-liquid contact (VLC) [61].

7.2.1.3. Application to Terpenes and Terpenoids in Water. VLE data for a series of title compounds have been obtained using an original combination of above-mentioned methods. First, solubility has been determined by means of a TLPC method, from which γ_i^∞ has been deduced using equation (2). Continuous mechanical recirculation (stripping) experiments, carried out in an aerated (0.3 vvm), stirred (1,000 rpm) bioreactor (3 L solution kept at 25 °C), have then allowed determination of P_i^0 (eq. 3). Results obtained show that, although solubility among monoterpenes is extremely variable (ranging from 0.037 mmol L^{-1} to 19 mmol L^{-1}), they result in molar fractions always less than 0.01, i.e. in the infinite dilution region (Table 3).

Table 3 Terpenes and terpenoids solubility measured in water at 25 °C, activity coefficient at infinite dilution γ_i^∞ and vapour pressures P° obtained. The confidence interval is given at 95% level. ND: Not Determined

Compound	Solubility (mmol L^{-1})	γ_i^∞	$\gamma^\infty P°$ (mm Hg)	P° (mm Hg)
(R)-(+)-limonene	0.150 ± 0.015	$3.7x10^5 \pm 3.7x10^4$	$5.93x10^5 \pm 5.93x10^4$	1.60 ± 0.22
(-)-α-pinene	0.037 ± 0.004	$1.5x10^6 \pm 1.5x10^5$	$5.95x10^6 \pm 5.95x10^5$	3.97 ± 0.55
(-)-β-pinene	0.081 ± 0.006	$6.86x10^5 \pm 5.14x10^4$	ND	ND
myrcene	0.22 ± 0.02	$2.52x10^5 \pm 2.52x10^4$	$4.75x10^5 \pm 4.75x10^4$	1.88 ± 0.26
1-(S)-endo-(-)-borneol	3.00 ± 0.24	$1.85x10^4 \pm 1.48x10^3$	$8.55x10^2 \pm 2.00x10^2$	0.05 ± 0.01
endo-(+)-fenchyl alcohol	5.37 ± 0.32	$1.03x10^4 \pm 0.62x10^3$	$1.17x10^3 \pm 1.64x10^2$	0.11 ± 0.02
(-)-α-pinene-oxide	2.55 ± 0.48	$2.18x10^4 \pm 4.14x10^3$	$1.78x10^4 \pm 3.38x10^3$	0.82 ± 0.15
(+)-limonene-oxide	4.61 ± 0.16	$1.20x10^4 \pm 0.42x10^3$	$7.45x10^3 \pm 3.72x10^2$	0.62 ± 0.04
(-)-carveol	19 ± 2	$2.92x10^3 \pm 2.92x10^2$	ND	ND
(±)-linalool	10.11 ± 0.61	$5.49x10^3 \pm 0.33x10^2$	ND	ND
α-terpineol	12.25 ± 1.10	$4.53x10^3 \pm 0.41x10^2$	ND	ND
(+)-carvone	8.80 ± 0.22	$6.31x10^3 \pm 0.16x10^2$	ND	ND

7.2.2 Data Correlation

7.2.2.1. Vapour Pressure. Many equations and correlations for estimating vapour pressure P^0 of pure compounds were presented in literature. Those that relate P^0 to temperature are commonly derived by integration of the Clausius-Clapeyron equation [62]; they contain at least two coefficients that have to be estimated for each compound. Several methods, generally based on group contribution methodology, exist for this purpose [62]. Some of them are available as computer software; the method developed by ACD (Advanced Chemistry Development Inc., Toronto, Canada) appears suitable, at least for terpenes and terpenoids, since it gives P^0 values in good agreement with experimental data. It can thus be concluded that useful methods for vapour pressure estimation are now available.

7.2.2.2. Solubility, Activity Coefficient. Up to now, the most successful approach for estimating water solubility is the use of log $P_{O/W}$-based regression equations ([10], see section 5.1). However, one of the most recent correlations published in this area [63] has been shown to need adaptations to fit results obtained with terpenes and terpenoids. These modifications involve the addition of new factors and groups as shown in equation (4):

$$\log S = 3.796 - 0.854 \log P_{O/W} - 0.00728\, MW + \Sigma f_i + \Sigma n_i g_i \qquad (4)$$

where S is the solubility in mmol L^{-1} at 25 °C, MW the solute molecular weight (Da), Σf_i the summation of all correction factors applicable to a compound. The term g_i deals with corrections corresponding to groups present in the molecule, and n_i is the number of occurrences of each group i. The term log $P_{O/W}$ may be calculated either from data in Meylan and Howard [64], or using the SRC's computer program LOGKOW (http//esc.syrres.com). Additional factors with respect to the original ones [63] include diols (geminated or not, value 1.27), and single aliphatic ring (one ring per molecule, value 0.57) contributions. New groups are 1) olefinic (double bond outside of a ring, value 0.367), 2) ketone (C=O attached to a non-cyclic carbon, value 0.381), and 3) epoxide (value 0.323).

Although a good fit of experimental data can be obtained with this equation (Fig. 2), this approach presents two limits. First, it is unable to predict solubility at

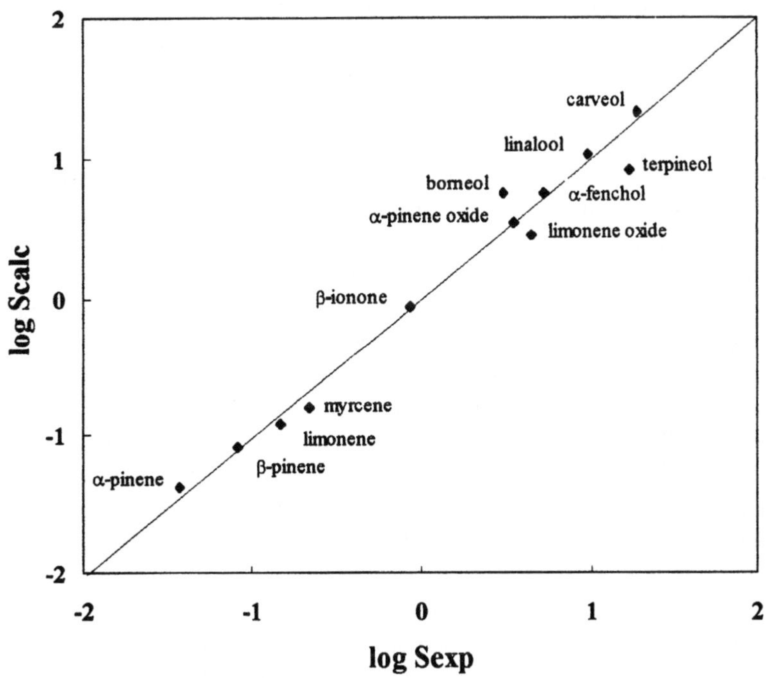

Figure 2. Comparison of solubility estimated using equation (4) with experimental data for the monoterpenes series

temperatures differing from 25 °C; secondly, it is unsuitable for calculating the behaviour of the solutes in real biotransformation media that contain, among other constituents, various ionic compounds. These aspects can, up to now, only be taken into account by UNIFAC (UNIquac Functional group Activity Coefficient)-type models, but these are known to perform poorly in the infinite dilution domain [65].

Estimation of VLE data in the highly diluted region, especially estimation of water solubility, remains a challenge. The validity of estimations given by existing equations must be checked by comparison with experimental data when applied to a specific class of compounds which are not included in the data bank used for parameter identification. This situation is met for the terpenes and terpenoids family. Results presented in this paragraph show a methodology to improve existing equations.

7.3. THE USE OF SOLVENTS IN PROCESS DEVELOPMENT

Biotransformation reactions for the production of flavour and fragrances need often to be carried out in aqueous-organic systems or even in pure organic solvents. The reasons for this have been cited above (see section 7). As in the case of carboxylic acid esters, the enhancement of reaction rates in media with low water activity is significant and has led to a series of industrially produced flavours [66-68]. Thus, organic solvents may play an important role in promoting stability, overcoming inhibition and low solubility problems, and acting as an extraction phase for *in situ* product removal.

Biotransformations to produce flavours with potential industrial interest rarely involve a single enzymatic step. In addition, most interesting biotransformations generally include at least one reaction catalysed by an oxido-reductase [43], i.e., they are coenzyme requiring enzymatic reactions. For instance, dehydrogenases, a sub-class of oxido-reductases, require an electron acceptor to carry out their oxidative activity. These facts suggest that the use of whole microbial cells might be advantageous to catalyse the various enzymatic steps of the biotransformation. The reasons are obvious, the multi-enzymatic system is naturally present in the selected microbial cell (or has been engineered in a host organism) and, as such, 1) enzyme isolation costs are avoided, and 2) co-factor regeneration is achieved by the processes of cell respiration. Also, the use of whole cells in less conventional media can lead to specificities comparable with those of isolated enzymes. To achieve goal 2) above, it is necessary to maintain a high percentage of cell viability within the reaction system so that cells, either in free or immobilised form, can be re-used in consecutive batches or catalyse a continuous process with an economically acceptable half-life activity.

In process terms, the approach of maintaining cell viability is usually easier and less expensive than the addition of a second enzyme and substrate to regenerate the co-factor. In aqueous-organic systems, the selection of a biocompatible solvent is thus crucial in order to maintain cell viability and to minimise harmful effects of the solvent towards the biocatalyst (see section 5). In the next sections some representative examples of the benefits presented by aqueous-organic reaction systems for the biotransformation of terpenes will be discussed.

7.3.1. Improved Stability

Stability of α-pinene, pinene-oxide and borneol was investigated in aqueous (phosphate buffer pH 6.8)/organic (2 ml iso-octane containing 50 mM of the terpene) systems at various phase ratios. At a 5(aq.): 1(org.) ratio, both α-pinene and borneol were stable. No disappearance of either of the two terpenes was observed during 4 days of incubation. Pinene oxide however, was very unstable. Only 50% or less of the initial amount could be detected after 24 hours. By decreasing the amount of aqueous phase present, the stability of pinene oxide was improved significantly (Fig. 3).

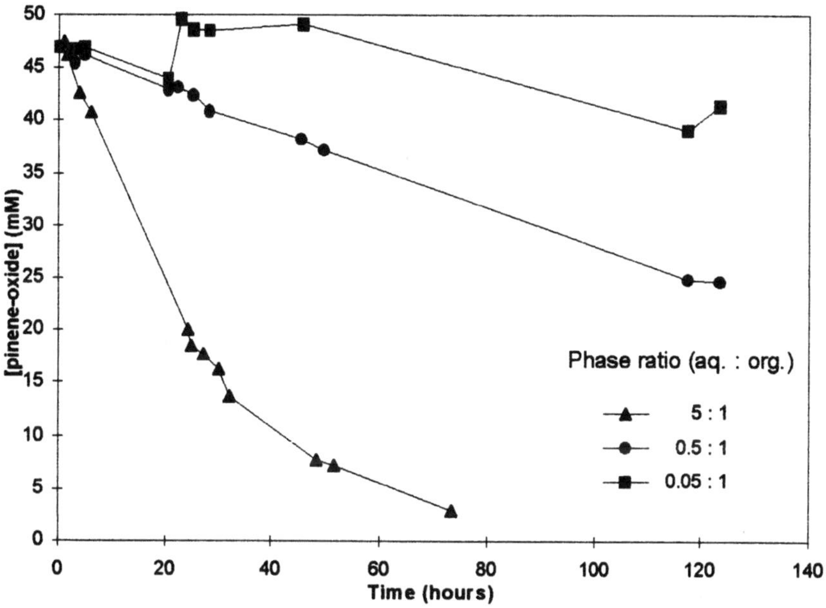

Figure 3. Time-course of pinene oxide concentration (followed by gas chromatography of the organic phase) in biphasic systems at 30 °C, 200 rpm and various phase ratios

7.3.2. Microbial Growth in the Presence of an Inhibitory Epoxide

Under certain circumstances it might be convenient to carry out the biotransformation with actively growing cells. In order to assess the potential toxic effect of the presence of terpene epoxides during microbial growth, we carried out growth experiments of *Pseudomonas putida* NCIMB 8248 in the presence of α-pinene oxide as such or dissolved in an organic reservoir (in the real system this would be the extraction phase). Besides hydrocarbons, two other solvents, namely α-pinene and 1-dodecanol were tested, the first on the basis of its low toxicity [69], and the second, because its polarity is an indicator of its better ability to dissolve the epoxide.

The inhibitory effects on growth caused by α-pinene oxide were overcome by the presence of any of the organic solvents (Table 4). This was probably a result of the absence of direct contact between the cells and a dispersed phase of the epoxide.

Table 4. Growth of *P. putida* NCIMB 8248 on 0.2% (w/v) glutamic acid with 2% (v/total volume) of organic solvent as a reservoir for 0.5% (v/total volume) α-pinene oxide

Organic solvent	log $P_{O/W}$	Presence of α-pinene oxide	Final dry weight (g L^{-1})	Duration of lag phase (h)	Growth rate (h^{-1})
-	-	No	0.64	0.0	0.91
		Yes	0.06	-	-
iso-octane	4.5	No	0.62	0.6	0.76
		Yes	0.55	0.7	0.89
α-pinene	4.9	No	0.55	1.2	0.53
		Yes	0.65	1.8	0.70
1-dodecanol	5.0	No	0.78	1.2	0.69
		Yes	0.85	0.6	0.80
n-dodecane	6.6	No	0.87	2.8	0.64
		Yes	0.61	0.8	0.76

7.3.3 Conditions for the Maintenance of Cell Viability

Since co-factor regeneration might be a prerequisite for continued biotransformation activity, the oxygen consumption rate of resting cells of P. *putida* was measured in a reaction chamber fitted with an oxygen electrode as an indication of their viability. Results in Table 5 indicate that α-pinene oxide strongly hinders cell respiration, although this toxicity effect could be overcome, at least in the short term, using 1-dodecanol as reservoir. The duration of the respiration tests ranged between 15 min and one hour.

Table 5. Relative initial oxygen consumption rate of resting cells of *P. putida* NCIMB 8248 (grown on 0.2% (w/v) glutamic acid) at 30 °C in the presence of 2% (v/total volume) of organic phase containing or not α-pinene oxide; cells were harvested at the end of the exponential phase; in blank experiments 100% activity ranged between 0.35-0.70 g O_2 consumed (g^{-1} DW·h^{-1}), depending on the physiological state of the cell suspension

Organic solvent	log $P_{O/W}$	α-pinene oxide (% v/v)	Relative respiration activity (%)
-	-	0.05	25
-	-	0.20	0-10
α-pinene	4.9	-	53
		0.20	47
1-dodecanol	5.0	-	97
		0.20	98

7.3.4. Use of the Solvent as Reservoir

Cells of *Rhodococcus erythropolis* DCL14, a strain recently isolated at the University of Wageningen, contain a high limonene-1,2-epoxide hydrolase activity when grown on monoterpenes [70]. Limonene-1,2-epoxide is unstable in an aqueous phase and has a solubility of approximately 4.6 mM [71], while the reaction product, limonene-1,2-diol, is stable and dissolves well in water.

This biotransformation was carried out in single (aqueous) phase and in biphasic (5(aq.): 1(org.)) systems using cells grown on limonene. The latter approach was aimed at obtaining higher reaction rates and productivities by overcoming the instability and solubility problems of the substrate. It was found that *R. erythropolis* cells are highly hydrophobic, since they accumulated at the aqueous-organic interface. This was confirmed by the hexadecane hydrophobicity test [72].

Results in Fig. 4 indicate that both iso-octane and hexadecane performed well as epoxide reservoirs. The specific activity of the cells increased 2.5 times at high limonene-1,2-epoxide concentrations when an organic phase was present, as compared to the saturated aqueous system. This was probably a combined effect of cell migration from the aqueous phase, resulting in biomass accumulation around the organic droplets, which, in turn, facilitated the cell access to high concentrations of the epoxide.

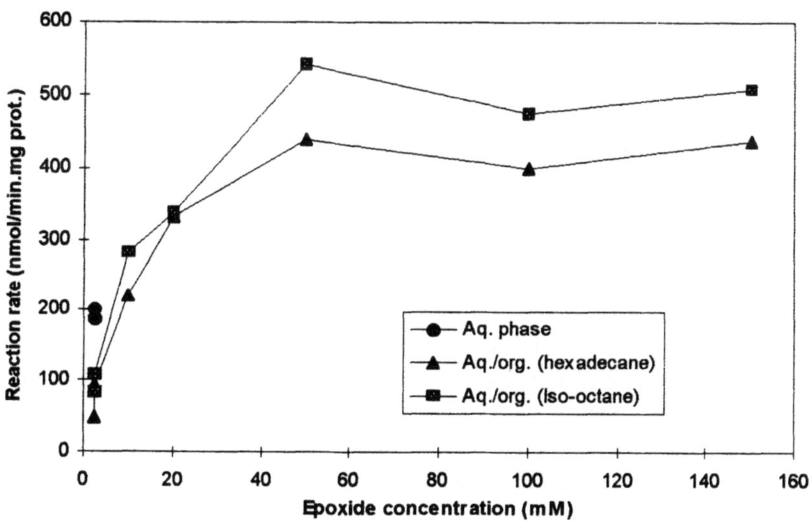

Figure 4. Formation rates of limonene-1,2-diol in single aqueous and biphasic aqueous-organic systems (30 °C and 200 rpm; initial epoxide concentrations are referred to the aqueous phase)

In this case, the use of a biphasic system, besides improving the reaction rate and the stability of the substrate, enabled *in situ* product extraction, since the diol partitioned almost completely to the aqueous phase, while the substrate remained in the organic phase.

A decrease in the solvent average drop size might well lead to a further increase of the cell specific biotransformation activity. In engineering terms it is clear that this model reaction, as carried out in aqueous-organic systems, offers considerable potential once the phase ratio and the dispersion of the organic phase have been optimised.

7.4. SOLID PHASE ADSORPTION - SELECTIVE PROPERTIES IN AQUEOUS AND ORGANIC SOLUTIONS

It is commonly known that activated carbon has a high adsorption capacity. It removes a great number of organic compounds from gaseous phases or from aqueous solutions, especially from drinking waters. The same material shows attributes which render its application in a bioprocess difficult: When common desorption solvents are used, the recovery of the adsorbed products is rarely quantitative; and further, due to the small average particle size, activated carbon cannot be used in low pressure adsorption columns [73]. These problems can be overcome with porous hydrophobic polymers based on a polystyrene skeleton. An increasing number of applications of an *in situ* recovery of flavour compounds have been described using polystyrene polymers, such as XAD-4, XAD-16, or Lewatit 1064 [74,75].

The recovery of microbially generated aroma compounds from fermentation broths can also be performed by adsorption. Cross-linked polystyrene resins are proven to be suitable adsorbents, even for *in situ* recovery, because of their enormous adsorption capacity, favourable desorption properties, chemical/physical/microbial inertness, and re-usability [75]. Sufficient flavour product yields in microbial production systems cannot be gained without feeding suitable precursor substrates and/or applying *in situ* recovery. This is true, e.g., in the case of microbial oxygenation of terpene hydrocarbons, such as limonene, pinenes, or myrcene. However, in these applications of non-specific polystyrene adsorbent material, another major obstacle occurs. The polymer has a pronounced preference for lipophilic compounds, thus leaving the valuable oxyfunctionalised target components in the cultivation broth, whilst the terpene precursors will be distributed into the polystyrene resin. In order to improve or even reverse the adsorption characteristics in favour of more polar compounds, derivatisation of commercial polystyrene material by introduction of polar functional groups appears to be promising.

7.4.1. Derivatisation of Adsorbent Material

7.4.1.1. General Methods. A required functional group can be introduced on a support in two ways. It can either be incorporated during the synthesis of the support or by chemical modification of a preformed support matrix. As far as macromolecular materials are concerned both routes have been used. The chemical modification route is particularly attractive with polystyrene-based resins, as their aromatic rings can be readily modified by an electrophilic substitution or through other simple reactions. The two most versatile routes are chloromethylation [76] and lithiation [77]. In combination, they provide a set of methods to attach a wide variety of both electrophilic and nucleophilic species (Figure 5). Hodge and Hunt have reviewed further strategies for the derivatisation of commercial adsorbent material, using simple one-step reactions [78].

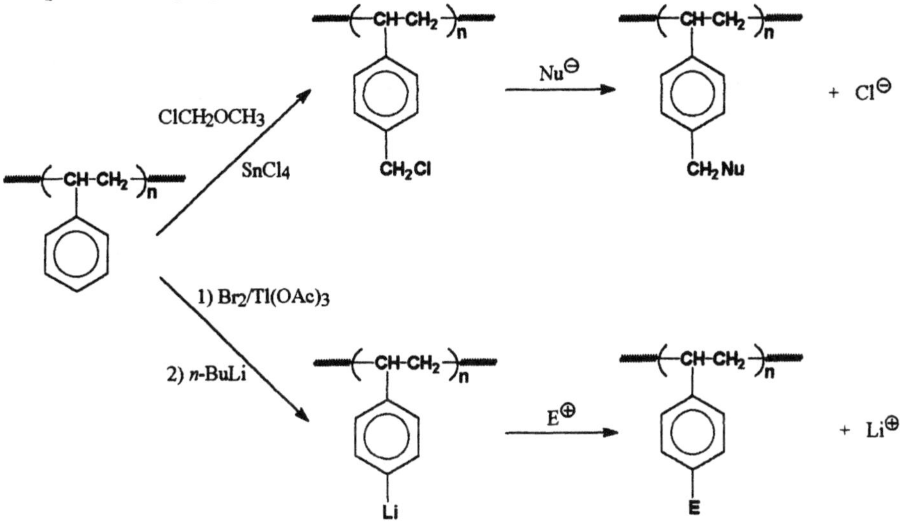

Figure 5. Routes for the chemical modification of polystyrene resins

7.4.1.2. Derivatisation Procedures. Several derivatives were synthesised from Amberlite XAD-16 by different methods covering all the mentioned strategies (Table 6). Incorporation of small functional groups like bromine was achieved with a degree of functionalisation of < 75 % of all the available aromatic rings. With increasing size of the functional group being introduced, the degree of functionalisation decreased. For example, benzoylation of the polystyrene resin led to only 13 % derivatisation of all the available aromatic rings. Sulfonated resin with a high degree of functionalisation (73 %) was chosen to determine the effect of derivatisation on the adsorption preference. It was converted to the $Na^{\oplus}$-form to prevent acid catalysed side reactions on the adsorbates.

Table 6. Functionalisation of Amberlite XAD-16 by different derivatisation procedures

Precursor			Product			
Ⓟ-X	mmol X g^{-1} polymer	DF[b]	Ⓟ-X	mmol X g^{-1} polymer	DF[b]	Ref
XAD-16	-	-	Ⓟ-CH_2Cl	4.92[c]	0.67	[78]
XAD-16	-	-	Ⓟ-Br	4.50	0.73	[77]
XAD-16	-	-	Ⓟ-SO_3Na	4.14	0.73	[80]
Ⓟ-Br[a]	4.50	0.73	Ⓟ-SCH_3	2.00	0.23	[77]
Ⓟ-Br[a]	4.50	0.73	Ⓟ-COC_6H_5	1.12	0.13	[80]
Ⓟ-CH_2Cl[a]	4.92	0.67	Ⓟ-$CH_2NH(CH_3)_2$	1.45	0.16	[81]

[a)] Ⓟ = Polymer Amberlite XAD-16,[b)] degree of functionalisation as proportion of styrene rings substituted,[c)] as by elemental analysis.

7.4.2. Experimental Determination of the Adsorption Preference

7.4.2.1. Adsorption from the Aqueous Phase. A model solution that contained both 40 mmol L^{-1} of (*R*)-limonene and of (*R*)-carvone was prepared to simulate the separation of oxyfunctionalised terpenes from their hydrocarbon precursors. Batch adsorption assays were carried out in a 2% methanolic aqueous solution to allow for the low solubility of limonene in water. The determination of isothermal points was performed by varying the adsorbent mass/solution volume relation at a constant initial liquid phase concentration [81]. An adsorption time of 16 h was chosen to ensure equilibrium conditions.

The efficiency of resins used in solid phase extraction is most commonly determined by the molar loading of adsorbate onto one gram of resin. Using a non-functionalised adsorbent material like Amberlite XAD-16 in such a binary adsorption system, it was shown that competitive adsorption took place (Figure 6). As expected, the molar loading (measured as desorbed mass) of the more polar solute, carvone, was lower than that of the non-polar precursor, limonene. A change of the adsorption preference was achieved by using sulfonated resin Amberlite XAD-16 (counterion $Na^{\oplus}$) with a high degree of functionalisation (Figure 7). The derivatisation procedure led to a more polar resin surface which now adsorbed the more polar adsorbate, carvone, preferably. This change of adsorption preference was accompanied by a significant loss of overall adsorption capacity caused by an increased competitive adsorption of water molecules.

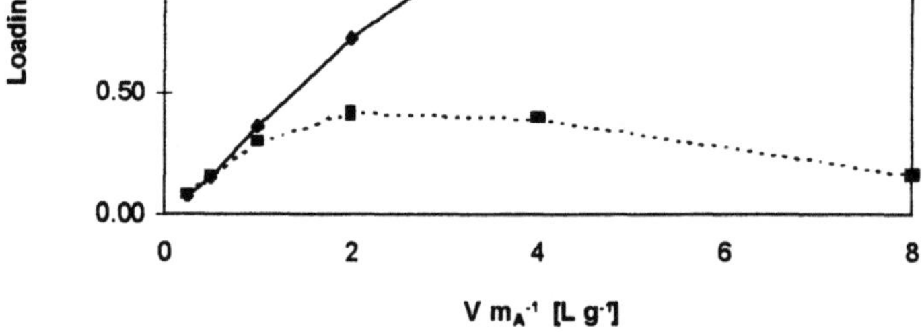

Figure 6. Co-adsorption isotherms of limonene and carvone from aqueous solution on Amberlite XAD-16

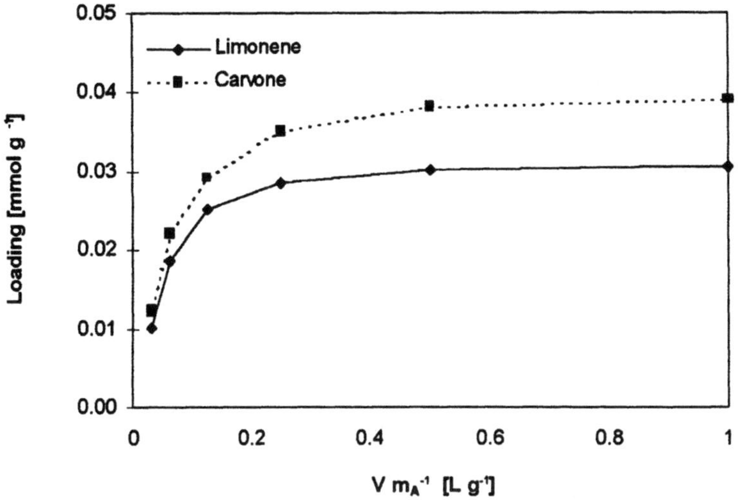

Figure 7. Co-adsorption isotherms of limonene and carvone from aqueous solution on functionalised adsorbent material XAD-16-SO_3Na (degree of functionalisation: 0.73).

7.4.2.2. Salting-out Effect. The adsorption efficiency can be increased by weakening the adsorbate-water interactions. For non-dissociated molecules, this can be achieved by increasing the ionic strength of the aqueous solution. This phenomenon, called salting-out effect, is in practice performed by adding a certain amount of an indifferent electrolyte (NaCl) to the original aqueous sample [82].

In the case of the binary component system limonene-carvone, the salting out effect led to an improved selectivity of the desired oxyfunctionalised adsorbate (Figure 8). After adding 300 g NaCl L^{-1} to the aqueous solution, water molecules were consumed for solvating the additional ions which in turn led to a lower water solubility of limonene and carvone. Polar parts of the functionalised sulfonated resin were now able to adsorb the more polar adsorbate carvone to a larger extent, because the availability of the competitive adsorbate water was reduced.

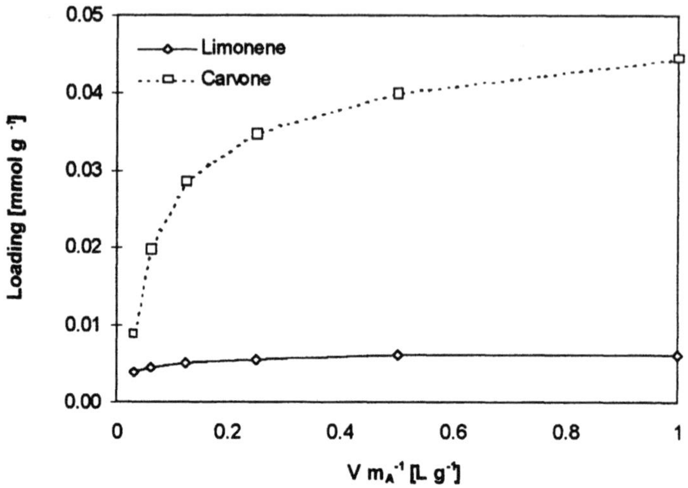

Figure 8. Co-adsorption isotherms of limonene and carvone from aqueous solution containing 300 g NaCl L^{-1} on functionalised adsorbent material XAD-16-SO_3Na (degree of functionalisation: 0.73)

7.4.2.3. Kinetic Studies. As equilibrium states may not be reached under all industrial conditions, the adsorption kinetics of sulfonated resins were studied. In the absence of salt two-third of the equilibrium amount of the terpenes was found adsorbed on sulfonated resin after five minutes. In the beginning, adsorption of limonene was faster, but the thermodynamically preferred adsorption of carvone led to a gradually higher molar loading starting after about 20 min of batch adsorption. Addition of 300 g of NaCl increased the over-all adsorption capacity almost threefold with most of the limonene being adsorbed in the first 45 minutes. After this period of kinetically controlled adsorption the thermodynamically preferred adsorption of carvone started to predominate. These results show that the change of the adsorption preference caused by sulfonation of the polystyrene resin is due to a thermodynamic effect.

7.4.2.4. Adsorption from the Organic Phase. Looking at the kinetic preference for limonene the use of highly sulfonated resins in adsorption columns will not lead to the separation being aimed at. A high concentration of salts in a bioprocess medium would facilitate the adsorptive separation of terpenoids, but, from obvious reasons, is not applicable in most practical processes. Indeed, the incorporation of polar groups provided additional hydrophilic interactions, but the increased co-adsorption of water molecules overlapped this effect, and an acceptable separation was not achieved.

An improved separation might be obtained in non-aqueous solutions, in which the solvent is not competing for active adsorption sites. Therefore, further investigations were focused on batch adsorption experiments carried out in organic solutions. Using *n*-hexane as a solvent, batch adsorption isotherms were measured by varying the concentration of oxyfunctionalised product and hydrocarbon precursor. When the binary system limonene - carvone was adsorbed on a highly sulfonated resin in *n*-hexane, a clear separation between oxyfunctionalised terpene and its hydrocarbon precursor was observed. The hydrocarbon precursor almost completely remained in the organic phase.

As a result, an adsorptive *in situ* product recovery during a bioprocess, in which terpenes are converted into oxyfunctionalised derivatives, seems to be feasible. In practice a two-step solid phase extraction process is required. In a first step the oxyfunctionalised terpene and its hydrocarbon precursor are non-specifically adsorbed from an aqueous solution by a commercial high-capacity polystyrene-divinylbenzene adsorbent material, such as Amberlite XAD-16. Both are eluted with a small amount of an non-polar organic solvent (e.g. *n*-hexane). This solution is then separated in a second solid phase extraction by using modified resins containing polar functional groups that are able to form hydrogen bonds with the target product. The oxyfunctionalised terpene will be adsorbed specifically, while the precursor will remain in the organic phase. Appropriate desorption will yield the desired biotransformation product. The precursor may be separated from the organic phase by distillation and re-used in the fermentation process.

8. Economic Considerations

In the United States more than 65% of all flavours used are labelled as natural with a value of $ 9 billion. Depending on the production volume and price these flavour compounds belong either to the category of fine chemicals (1 to 1,000 tons and price between $ 5-500 kg^{-1}) or to the category of specialities (price above $ 500 kg^{-1}). Both the chemical and pharmaceutical industry are exploring the potential of microbial conversions for the production of fine and speciality chemicals, and even production of bulk chemicals is considered [83]. Apart from the various general advantages of biotechnology as discussed above (see section 4), the aroma compounds formed are classified as natural by European and US food legislation, if the substrates used are natural [84]. For non-volatile flavours biotechnology is already one of the key technologies. Due to technical and economic reasons application of biotechnology to the production of volatile flavour chemicals is developing rather slowly. One of the economic reasons is that the production volumes of many of the flavour compounds are small allowing only very small investments in research and development, and in bioprocess facilities. For terpene biotransformations a number of additional factors adversely affect the development of bioprocesses. Microbial toxicity, low yields, low production rates, multiplicity of terpene metabolites and further breakdown of the products result in low product concentrations and high downstreaming costs (Table 2, [43]).

A solvent tolerant *Pseudomonas putida* strain (GS1), capable of converting (+)-limonene to one single, stable product, perillic acid, was recently isolated [85]. Oxidation products of limonene are used as flavourings and as antimicrobial agents in foods and pharmaceuticals. Furthermore, recent studies have shown that these terpenoids may have a preventive effect on the development of cancer. *P. putida* GS1 is able to convert limonene in the presence of a co-substrate (i.e. glycerol) to high concentrations of perillic acid. Up to 3 g L^{-1} of perillic acid was produced within 48 h which could easily be purified by selective extraction. Of the added limonene 30% was converted with a maximal rate of 60 mg L^{-1} h^{-1} (Figure 9). Currently the relevant genes are being cloned and will be overexpressed in other hosts in order to obtain higher production rates and yields of perillic acid and its intermediates. Preliminary results show that the enzyme catalysing the conversion of limonene to perilyl alcohol belongs to the category of mono-oxygenases.

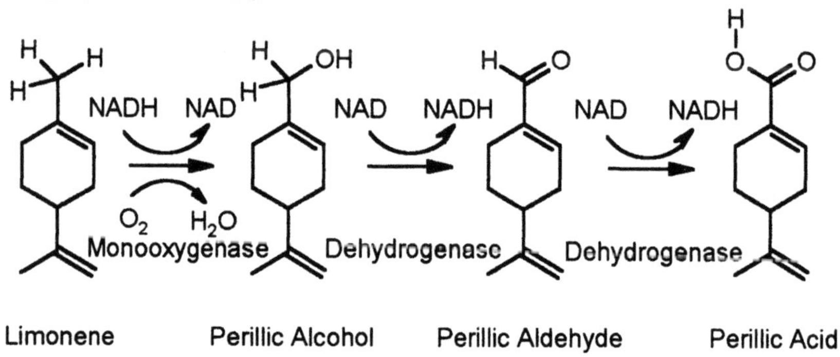

Figure 9. High-efficiency transformation of limonene by a strain of *P. putida*

The economic potential of mono-oxygenase catalysed reactions of apolar substrates has been reviewed by Witholt and co-workers [83,86] based on extensive studies and experience with the three component alkane hydroxylase system of *Pseudomonas oleovorans* and the two component xylene oxygenase of *P. putida*. A prediction of the process economics has been made based on experience with high cell density cultivations of recombinant *E. coli* strains in which growth and alkane/styrene oxidation are uncoupled and the relevant enzymes are overexpressed. With cell densities of 25 to 40 g L^{-1} dry mass a volumetric productivity for alkane oxidation of 10 mmol L^{-1} h^{-1} (1.5 g L^{-1} h^{-1} octanoic acid) and for styrene oxidation of 2.2 mmol (0.22 g L^{-1} h^{-1} styrene oxide) has been reached. These productivities are already significant and only somewhat lower than productivities reached in several commercial (bulk) fermentation processes (i.e. lactic and gluconic acid fermentations with productivities of 2.5 g L^{-1} h^{-1} and 10 g L^{-1} h^{-1} , respectively).

Further improvements are necessary. The high biocatalytic activity of the biomass should be maintained for a longer period, e.g. a 24 h production phase instead of 3 to 5 h [86]. Continuous recovery of the product may be necessary to avoid toxicity problems. If we assume that the stability of the biocatalytic activity can be significantly improved, a production of 1 to 3 tons of product per cubic metre reactor per year is feasible. The production costs of these products can be estimated to vary between $ 20 and $ 200 per kg depending on the production volume (between 10 and 500 tons per year). An increase of fermentor volume from 10 m^3 to 100 m^3 results in a fourfold decrease of the production costs. In the limonene bioconversion experiments with *P.*

putida GS1 cell densities were lower than 1 g L^{-1}. Therefore an increase of the biomass may already result in a considerable further improvement of the volumetric productivity of perillic acid. A biomass of 20 to 40 g L^{-1} dry mass should be attainable and result in a 10 fold increase of the conversion rate to a level of 0.6 g L^{-1} h^{-1}, which is similar to the levels found with the alkane and styrene mono-oxygenase systems.

According to Janssens *et al* [3] the price of a microbial flavour should range between $ 200 and $ 2,000 per kg to be competitive. Based on the data presented here it should be possible to produce natural volatile flavours at a price of $ 20 to 200 per kg. Thus, considering both the technical and economical potential one may conclude that these bioprocesses will become increasingly important in the production of natural flavour and aroma ingredients. Perhaps the biggest challenge at this moment is to bring together (in focused R & D projects) flavour companies with their specific demands and the extensive scientific expertise (genetics, microbiology, bioprocess technology) which is required to develop reliable and economic bioprocesses.

9. Acknowledgements

This work was supported by the EC under project no. BIO4-CT95-0049, in the context of which we acknowledge fruitful suggestions from the various partners in the project. We also wish to acknowledge Nathalie Hernandez, Constança Correia, Carla de Carvalho, and Immacula Fichan for carrying out some of the experimental work.

10. References

[1] McNutt, K.W., Powers, M. and A.E. Sloan (1986) Food colors, flavors, and safety: a consumer viewpoint., *Food Technology*, January, 72-78.

[2] Welsh, F.W., Murray, W.D., Williams, R.E. (1989) Microbiological and enzymatic production of flavor and fragrance chemicals. *CRC Critical Reviews in Biotechnology*, **9**,105-169.

[3] Janssens, L., De Pooter, H.L., Schamp, N.M., Vandamme, E.J. (1992) Production of flavours by microorganisms. *Process Biochemistry*, **27**, 195-215.

[4] Cheetham, P.S.J. (1993) The use of biotransformations for the production of flavours and fragrances. *Trends in Biotechnology*, **11**, 478-488.

[5] Gatfield, I.L. (1997) Biotechnological production of flavour-active lactones. *Advances in Biochemical Engineering/Biotechnology*, **55**, 221-238.

[6] Schaft, P. van der, Geel, I van, Jong, G. de. ter Burg, N. (1994) Microbial production of natural furfurylthiol. In: *Trends in Flavour Research* (Maarse, H., van der Heij, D.G., eds) Elsevier, Amsterdam. pp. 437-448.

[7] Scragg, A.H. (1997) The production of aromas by plant cell cultures. *Advances in Biochemical Engineering/Biotechnology*, **55**, 239-267.

[8] Boelens, M. H., Boelens, H., van Gemert, L.J. (1993) Sensory properties of optical isomers. *Perfumer & Flavorist*, (6), 1-16.

[9] Weber, F. J., Bont, J. A. M. de (1996) Adaptation mechanisms of microorganisms to the toxic effects of organic solvents on membranes. *Biochimica Biophysica Acta*, **1286**, 225-245.

[10] Sikkema, J., Bont, J.A.M. de, Poolman, B. (1994) Interaction of cyclic hydrocarbons with biological membranes. *Journal of Biological Chemistry*, **269**, 8022-8028.

[11] Sikkema, J., Bont, J. .A. M. de, Poolman, B. (1995) Membrane toxicity of cyclic hydrocarbons. *Microbiological Reviews*, **59**, 201-222.

[12] Heipieper, H-J., Weber, F. J., Sikkema, J., Keweloh, H., Bont, J. A. M. de (1994) Mechanisms behind resistance of whole cells to toxic organic solvents. *Trends in Biotechnology*, **12**, 409-415.
[13] Kieboom, J., Dennis, J. J., Bont, J. A. M. de, Zylstra G. J. (1997) Identification and molecular characterisation of an efflux pump involved in *Pseudomonas putida* S12 solvent tolerance. *Journal of Biological Chemistry*, **273**, 85-91.
[14] Nikaido, H. (1996) Multidrug efflux pumps of gram-negative bacteria. *Journal of Bacteriology*, **178**, 5853.
[15] Wang, C.L., Chin, C.K., Ho, C.T., Hwang, C.K., Polashock, J.J., Martin, C.E. (1996) Changes of fatty acids and fatty acid-derived flavour compounds by expressing the yeast delta-9 desaturase gene in tomato. *Journal of Agricultural and Food Chemistry*, **44**, 3399-3402.
[16] Starks, C.M., Back, K., Chappell, J., Noel, J.P. (1997) Structural basis for cyclic terpene biosynthesis by tobacco 5-epi-aristochene synthase. *Science*, **277**, 1815-1820-1824.
[17] Lesburg, C.A., Zhai, G., Cane, D.A., Christianson, D.W. (1997) Crystal structure of pentalenene synthase: mechanistic insights on terpenoid cyclization reactions in biology. *Science*, **277**, 1820.
[18] Bohlmann, J., Meyer-Gauen, G., Croteau, R. (1998) Plant terpenoid synthases: molecular biology and phylogenetic analysis. *Proceedings of the National Academy of Sciences USA*, **95**, 4126-4133.
[19] de Vos, W.M., Simons, G.F.M. (1994) Gene cloning and expression systems in lactococci. In: *Genetics and biotechnology of lactic acid bacteria*, (eds M.J. Gasson and W.M. de Vos) pp 52-105, Blackie, Glasgow, UK.
[20] de Vos, W.M. (1996) Metabolic engineering of sugar catabolism in lactic acid bacteria. *Antonie van Leeuwenhoek International Journal of General and Molecular Microbiology*, **70**, 223-242.
[21] Platteeuw, C., van Alen-Boerrigter, I., van Schalkwijk, S., de Vos, W.M. (1996) Food-grade cloning and expression system for *Lactococcus lactis*. *Applied and Environmental Microbiology*, **62**, 1008-1013.
[22] Biswas, I., Gruss, A., Ehrlich, S.D., Maguin, E. (1993) High-efficiency gene inactivation and replacement system for Gram-positive bacteria. *Journal of Bacteriology*, **175**, 3628-3635.
[23] Leloup, L., Ehrlich, S.D., Zagorec, M., Morel-Deville, F. (1997) Single-crossover integration in the *Lactobacillus sake* chromosome and insertional inactivation of the ptsI and lacL genes. *Applied and Environmental Microbiology*, **63**, 2117-2123.
[24] Platteeuw, C., Hugenholtz, J., Starrenburg, M. van Alen-Boerrigter, I., de Vos, W.M. (1995) Metabolic engineering of *Lactococcus lactis*: influence of the overproduction of alpha-acetolactate synthase in strains deficient in lactate dehydrogenase as a function of culture conditions. *Applied and Environmental Microbiology*, **61**, 3967-3971.
[25] Swindell, S.R., Benson, K.H., Griffin, H.G., Renault, P. Ehrlich, S.D., Gasson, M.J. (1996) Genetic manipulation of the pathway for diacetyl metabolism in *Lactococcus lactis*. *Applied and Environmental Microbiology*, **62**, 2641-2643.
[26] de Ruyter, P.G.G.A., Kuipers, O.P., Meijer, W.C., de Vos, W.M. (1997) Food-grade controlled lysis of *Lactococcus lactis* for accelerated cheese ripenıng. *Nature Biotechnology*, **15**, 976-979.
[27] Yvon, M., Thirouin, S, Rijnen, L., Fromentier, D., Gripon, J.C. (1997) An aminotransferase from *Lactococcus lactis* initiates conversion of amino acids to cheese flavour compounds. *Applied and Environmental Microbiology*, **63**, 414-419.
[28] Fox, P.F., Wallace, J.M. (1997) Formation of flavour compounds in cheese. *Advances in Applied Microbiology*, **45**, 17-86.
[29] Kronlof, J., Linko, M. (1992) Production of beer using immobilized yeast encoding α-acetolactate decarboxylase. *Journal of the Institute of Brewing*, **98**, 479-491.
[30] Cleland, J.L., Craik, C.S. (1996) *Protein Engineering: Principles and Practice*. Wiley-Liss Inc. N.Y.
[31] Cheetham, P.S.J. (1997) Combining the technical push and business pull for natural flavours. *Advances in Biochemical Engineering/Biotechnology*, **55**, 1-50.
[32] Masui, A., Fujiwara, N., Yamamoto, K., Takagi, M., Imanaka, T. (1998) Rational design for stabilization and optimum pH shift of serine protease aprN. *Journal of Fermentation and Bioengineering*, **85**, 30-36.
[33] Wind, R.D., Uitdehaag, J.C.M., Buitelaar, R.M., Dijkstra, B.W., Dijkhuizen, L. (1998) Engineering of cyclodextrin product specificity and pH optima of the thermostable cyclodextrin glycosyltransferase from *Thermoanaerobacterium thermosulfurigenes* EM1. *Journal of Biological Chemistry*, **273**, 5771-5779.
[34] Querol, E., Perezpons, J.A., Mozovillarias, A. (1996) Analysis of protein conformational characteristics related to thermostability. *Protein Engineering*, **9**, 265-271.
[35] Shaw, A., Bott, R. (1996) Engineering enzymes for stability. *Current Opinion in Structural Biology*, **6**, 546-550.
[36] Winterhalter, P., Skouroumounis, G.K. (1997) Glycoconjugated aroma compounds: occurrence, role and biotechnological transformation. *Advances in Biochemical Engineering/Biotechnology*, **55**, 73-106.
[37] Akanuma, S., Yamagishi, A., Tanaka, N., Oshima, T. (1998) Serial increase in the thermal stability of 3-isopropylmalate dehydrogenase from *Bacillus subtilis* by experimental evolution. *Protein Science*, **7**, 698.

[38] Moore, J.C., Jin, H.M., Kuchner, O, Arnold, F.H. (1997) Strategies for the in vitro evolution of protein function: enzyme evolution by random recombination of improved sequences. *Journal of Molecular Biology*, **272**, 336-347.

[39] Hayashi, K., Singh, A., Aoyagi, C., Nakatani, A., Tokuyasu, K., Kashiwagi, Y. (1996) Chimeric ß-glucosidases with increased heat-stability. In: *Biotechnology for improved foods and flavours* (Takeoka, G.R., Teranishi, R., Williams, P.J., Kobayashi, A. eds) ACS Symposium series **637** pp 134-145 American Chemical Society, Washington DC.

[40] Narbad, A., Gasson, M.J. (1998) Metabolism of ferulic acid via vanillin using a novel CoA dependent pathway in a newly isolated strain of *Pseudomonas fluorescens*. *Microbiology*, **144**, 1397-1405.

[41] Chang, H.C., Gage, D.A., Oriel, P.J. (1995) Cloning and expression of a limonene degradation pathway from *Bacillus stearothermophilus* in *Escherichia coli*. *Journal of Food Science*, **60**, 551-553.

[42] Trudgill, P.W. (1994) Microbial metabolism and transformation of selected monoterpenes. In: *Biochemistry of Microbial Degradation* (Ratledge, C. ed) pp 33-61, Kluwer Academic Publishers, Dordrecht, The Netherlands.

[43] van der Werf, M.J., de Bont, J.A.M., Leak, D.J. (1997) Opportunities in microbial transformation of monoterpenes. *Advances in Biochemical Engineering/Biotechnology*, **55**, 147-177.

[44] Best, D.J., Floyd, N.C., Magalhaes, A., Rhodes, P.M. (1987) Initial steps in the degradation of alpha-pinene by *Pseudomonas fluorescens* NCIMB 11671. *Biocatalysis*, **1**, 147-159.

[45] Compagno, C., Tura, A., Ranzi, B.M., Martegani, E. (1993) Bioconversion of lactose whey to fructose diphosphate with recombinant *Saccharomyces cerevisiae* cells. *Biotechnology and Bioengineering*, **42**, 398.

[46] Kumagai, M.H., Sverlow, G.G., Dellacioppa, G., Grill, L.K. (1993) Conversion of starch to ethanol in a recombinant *Saccharomyces cerevisiae* strain expressing rice alpha amylase from a novel *Pichia pastoris* alcohol oxidase promoter. *Bio/Technology*, **11**, 606-610.

[47] Hirata, D., Aoki, S., Watanabe, K., Tsukioka, M., Suzuki, T. (1992) Stable overproduction of isoamyl alcohol by *Saccharomyces cerevisiae* with chromosome- integrated multicopy LEU4 genes. *Bioscience, Biotechnology and Biochemistry*, **56**, 1682-1683.

[48] Miyagawa, K., Kanzaku, N., Kimura, H., Sumino, Y., Akiyama, S., Nakao, Y. (1989) Increased inosine production by a *Bacillus subtilis* xanthine-requiring mutant derived by insertional activation of the IMP dehydrogenase gene. *Bio/Technology*, **7**, 821-824.

[49] Berger, R.G. (1995) *Aroma Biotechnology* pp 51-77, Springer-Verlag, Berlin.

[50] Fell, D.A. (1998) Increasing the flux in metabolic pathways: a metabolic control analysis perspective. *Biotechnology and Bioengineering* **58**, 121-124.

[51] Nielsen, J. (1998). Metabolic engineering: techniques for analysis of targets for genetic mutations. *Biotechnology and Bioengineering*, **58**, 125-132.

[52] Tan, C.-T. (1995) Physical chemistry in flavour products preparation, ACS Symposium Series **610**, Flavor Technology, 1-19, American Chemical Society, Washington DC.

[53] Karlsson, H.O.E., Trägardh, G. (1996) Applications of pervaporation in food processing, *Trends in Food Science and Technology*, 78-83.

[54] Baudot, A., Marin, M. (1997) Pervaporation of aroma compounds: comparison of membrane performances with vapour-liquid equilibria and engineering aspects of process improvement, *Transactions of International Chemical Engineering*, **75**, Part C,117-142.

[55] Modey, W.K., Mulholland, D. A., Raynor, M. W. (1996) Analytical supercritical fluid extraction of natural products, *Phytochemical Analysis*, **7**, 1-15.

[56] Seidlitz, H., Lack, E. (1996) Separation of essential oils & flavours from spices and herbs. Market potential and economics, *Chemical Industry Digest*, **9**, 91-97.

[57] Kapfer, G. F., Krüger, H., Nitz, S., Drawert, F. (1990) Gewinnung von fermentativ hergestellten Aromen durch Adsorption und CO_2 Hochdruckextraktion, *ZFL Lebensmitteltechnik*, **41**,51-55.

[58] Sherman, S.R., Trampe, R.D., Bush, D.M. *et al.* (1996) Compilation and correlation of limiting activity coefficients of nonelectrolytes in water. *Industrial and Engineering Chemistry Research*, **35**, 1044-48.

[59] Schmid, C., Steinbrecher, R., Ziegler, H. (1992) Partition coefficients of plant cuticles for monoterpenes. *Trees*, **6**, 32-36.

[60] Weidenhamer, J.D., Macias, F.A., Fischer, N. *et al.* (1993) Just how insoluble are monoterpenes? *Journal of Chemical Ecology*, **19**, 1799-1807.

[61] Turner, L.H., Chiew, Y.C., Ahlert, R.C. *et al.* (1996) Measuring vapor-liquid equilibrium for aqueous-organic systems: review and a new technique. *AIChE Journal*, **42**, 1772-88.

[62] Lyman, W.J., Reehl, W.F., Rosenblatt, D.H. (1990) *Handbook of chemical property estimation methods*, American Chemical Society, Washington DC.

[63] Meylan, W.M., Howard, P.H., Boethling R.S. (1996) Improved method for estimating water solubility from octanol/water partition coefficient. *Environmental Toxicology and Chemistry*, **15**, 100-106.

[64] Meylan, W.M., Howard, P.H. (1995) Atom/fragment contribution method for estimating octanol-water partition coefficients. *Journal of Pharmaceutical Sciences*, **84**, 83-92.

[65] Voutsas, E.C., Tassios, D.P. (1996). Analysis of the UNIFAC-type group-contribution models at the highly dilute region. 1. Limitations of the combinatorial and residual expressions. *Industrial and Engineering Chemistry Research*, **36**, 4965-4972.

[66] Carrea, G., (1984) Biocatalysis in water-organic solvent two-phase systems. *Trends in Biotechnology*, **2**, 102-106.

[67] Larroche, C., Creuly, C., Gros, J.-B. (1992) 2-Heptanone production by spores of *Penicillium roquefortii* in a water-organic solvent two-phase system. *Biocatalysis*, **5**, 163-173.

[68] Woodley, J.M., Lilly, M.D., (1992) Process engineering of two-liquid phase biocatalysis. In: *Biocatalysis in Non-Conventional Media* (Tramper, J., Vermüe, M.H., Beeftink, H.H., von Stockar, U., eds.), Elsevier, Amsterdam, 147-154.

[69] van Keulen, F., Correia, C.N., da Fonseca, M.M.R., Solvent selection for the biotransformation of terpenes by *Pseudomonas putida*. *Journal of Molecular Catalysis B: Enzymatic* (in press).

[70] van der Werf, M.J., Overkamp, K.M., Swarts, H.J., de Bont, J. (1998) Purification and characterization of limonene-1,2-epoxide hydrolase from *Rhodococcus erythropolis* DCL14; an enzyme showing sequential substrate degradation. *J. Bacteriology* (submitted).

[71] Fichan, I., Larroche, C., Gros, J.-B. Water solubility, vapor pressure and activity coefficients of terpenes and terpenoids. *Journal of Chemical Engineer Data* (submitted).

[72] Rosenberg, M., (1984) Bacterial adherence to hydrocarbons: a useful technique for studying cell surface hydrophibicity. *FEMS Microbiology Letters*, **22**, 289-295.

[73] Krings, U., Berger, R. G. (1995) Porous polymers for fixed bed adsorption of aroma compounds in fermentation processes. *Biotechnology Techniques*, **9**, 19-24.

[74] Souchon, I., Rojas, J. A., Voilley, A., Grevillot, G. (1996) Trapping of aromatic compounds by adsorption on hydrophobic sorbents. *Separation Science and Technology*, **31**, 2473-2491.

[75] Krings, U., Kelch, M., Berger, R. G. (1993) Adsorbents for the recovery of aroma compounds in fermentation processes. *Journal of Chemical Technology and Biotechnology*, **58**, 293-299.

[76] As cited in: Itsuno, S., Uchikoshi, K., Kto, K. (1990) Novel method for halomethylation of cross-linked polystyrenes. *Journal of the American Chemical Society*, **112**, 8187-8188.

[77] Farall, M. J., Fréchet, J. M. J. (1976) Bromination and Lithiation: Two important steps in the functionalisation of polystyrene resins. *Journal of Organic Chemistry*, **41**, 3877-3882.

[78] Hodge, P., Hunt, B. J., Shakhshier, I. H. (1985) Preparation of cross-linked polymers using acenaphthylene and the chemical modification of these polymers, *Polymer*, **26**, 1701-1707.

[79] Dumont, P. J., Fritz, J. S. (1995) Effect of resin sulfonation on the retention of polar organic compounds in solid-phase extraction, *Journal of Chromatography A*, **691**, 123-131.

[80] Masqué, N., Galià, M., Marcé, R. M., Borull, F. (1997) Solid-phase extraction of phenols and pesticides in water with a modified polymeric resin, *Analyst*, **122**, 425-428.

[81] Fritz, J. S., Dumont, P. J., Schmidt, L. W. (1995), Methods and Materials for solid-phase extraction, *Journal of Chromatography A*, **691**, 133-140.

[82] Thyssen, K. (1985) Determination of chloroacetanilides and chloroanilines in sulphanilamides and waste water by HPLC, *Journal of Chromatography A*, **319**, 99-104.

[83] Witholt, B., de Smet, M.-J., Kingma, J., van Beilen, J.B., Kok, M., Lageveen, R.G., Eggink, G (1990) Bioconversions of aliphatic compounds by Pseudomonas olevorans in multiphase bioreactors: background and economic potential. *Trends in Biotechnology*, **8**, 46-52.

[84] Krings, U., Berger, R.G. (1998) Biotechnological production of flavours and fragrances, *Applied Microbiology and Biotechnology*, **49**, 1-8.

[85] Speelmans, G., Bijlsma, A., Eggink, G. (1998) Limonene bioconversion to high concentrations of a single stable product, perillic acid, by a solvent-resistant *Pseudomonas putida* strain. *Applied Microbiology and Biotechnology*, accepted for publication

[86] Wubbolts, M.G., Favre-Bulle, O., Witholt, B. (1996) Biosynthesis of synthons in two-liquid-phase media. *Biotechnology and Bioengineering*, **52**, 301-308.

LIPID DERIVED FLAVOURS

WOLFGANG FITZ, JOSEF KERLER, AND HUGO WEENEN
Quest International, Huizerstraatweg 28, 1411 GP Naarden
The Netherlands

1. Introduction

Lipid derived flavour substances play an important role in the aroma of all foods. They can be enzymatically generated, or formed by the process of autoxidation or photooxidation from lipids. Primary oxidation products are prone to secondary oxidation, often resulting in further C-C cleavage.

Lipid derived flavour substances include a wide variety of molecules, including acids, esters, lactones, ketones and aldehydes. Both saturated and unsaturated aliphatic species occur, with the unsaturated ones generally having low to very low odour thresholds and more impact on the overall odour.

Acids are generally the result of hydrolysis of lipids, and can react further to form esters. Lactones are mainly of enzymatic (microbial) origin, but aldehydes and ketones can be both of enzymatic (lipoxygenase) or of thermal (autoxidation) origin.

The autoxidation of lipids is well recognised as a major factor responsible for the limited shelf-life of foods. It is of great importance to food manufacturers to come up with methods to prevent or at least slow down lipid oxidation and thus the development of off-flavours. On the other hand, autoxidation can also lead to the formation of important flavour components which positively contribute to the overall flavour of a food product.

Both enzymatic oxidation and autoxidation are used by the flavour industry to produce desirable aromas, for application in a wide variety of products. Patents suggest the use of lipoxygenases by the flavour and food industry, to produce green notes and mushroom notes, and autoxidation processes to produce fatty notes. In addition micro-organisms are used to produce lactones and methyl ketones. Both autoxidation processes and enzyme catalysed processes have been reported for the generation of carotenoid derived products.

K.A.D. Swift (ed.), Current Topics in Flavours and Fragrances, 171–214.

2. Non-Enzymatic Lipid Oxidation

Since Criegee discovered in 1939 that the primary product of the autoxidation of cyclohexene is the corresponding allylic hydroperoxide [1], the autoxidation of alkenes in general, and of unsaturated fatty acids in particular, has been extensively studied and is the subject of several monographs and reviews [2-6].

This chapter does not aim to give a representative overview of the vast amount of literature on non-enzymatic lipid oxidation, but rather to focus on some key aspects, which are essential in understanding flavour formation from fat. Examples of applications in today's food industry are also given.

2.1 AUTOXIDATION

The autoxidation of lipids proceeds via a free radical chain reaction as shown in Figure 1.

Initiation:

In_2	→	$2\ In^{\cdot}$	(1)
$In^{\cdot} + RH$	→	$InH + R^{\cdot}$	(2)

Propagation:

$R^{\cdot} + O_2$	→	$RO_2^{\cdot}$	(3)
$RO_2^{\cdot} + RH$	→	$RO_2H + R^{\cdot}$	(4)

Termination:

$R^{\cdot} + RO_2^{\cdot}$	→	RO_2R	(5)
$2RO_2^{\cdot}$	→	RO_4R	(6)
RO_4R	→	$R_2C{=}O + R_2CHOH + O_2$	(7)

Figure 1. General mechanism of autoxidation

Chain initiation (2) in the absence of added inhibitors is usually ascribed to the formation of active oxygen species through the decomposition of peroxidic impurities in the starting material. This is catalysed by transition metals (in particular copper and iron), light or irradiation, and even metal traces in glass walls [7,8].

The first step in propagation, namely the addition of oxygen to an alkyl radical (3), is very rapid, and in most cases diffusion-controlled. At a partial oxygen pressure above 200 torr, the reaction of the alkyl peroxy radical with a substrate molecule (4) limits the propagation rate. This makes hydrogen abstraction from lipids selective for the most weakly bound hydrogen, thus the ease of dioxygenation strongly depends on structural features such as the number and position of double bonds. Conjugated 1,4-

pentadienyl systems are preferred over isolated double bonds. Relative reaction rates for oleic acid, linoleic acid, and linolenic acid are respectively 1, 12, and 25 [9]. In the case of linoleic acid, a hydrogen atom is abstracted from the bisallylic C-11 position, with the formation of a pentadienyl radical. This radical reacts with oxygen to give the 9- and 13-hydroperoxides [10,11]. Cis-trans products are obtained mainly, but trans-trans products are formed as well. The ratio between cis-trans and trans-trans products is dependent on the availability of H-atom donors such as phenolic antioxidants [12]. In the case of linolenic acid, hydrogen abstraction takes place at C-11 and C-14, giving rise to the conjugated dienes with the hydroperoxy group at the C-9, C-13, C-12- and C-16 positions. In the case of oleic acid, the hydrogen abstraction takes place at the C-8 and the C-11 positions. The intermediate allylic radicals react with oxygen to yield the 8-, 9,-, 10-, and 11-allylic hydroperoxides [13,14].

At normal oxygen pressures termination occurs exclusively by the self reaction of two alkylperoxy radicals to give tetroxides (6), which in the case of primary and secondary alkylperoxy radicals undergo disproportionation (7) [15].

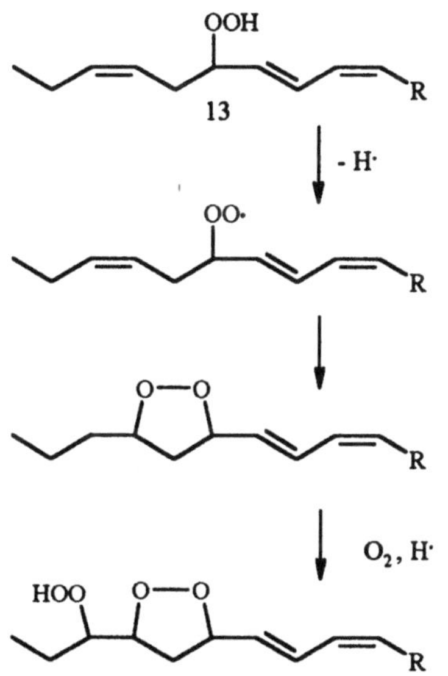

Figure 2. Formation of 5-membered hydroperoxy epidioxides [16]

The unsaturated hydroperoxides which are obtained as the primary products of autoxidation, can further oxidise to yield dihydroperoxides. Five-membered hydroperoxy epidioxides are formed from monohydroperoxides in significant amounts, if a double bond is present in the β-position of the hydroperoxy group, as is the case with the 12- and 13-hydroperoxides of autoxidised linolenic acid (Figure 2) [6,16]. The higher oxidised species can equally serve as precursors in the formation of volatiles.

The autoxidation of triglycerides is about one order of magnitudes lower than the autoxidation of the corresponding fatty acids [17]. This might be due to steric hindrance but also to the restricted oxygen solubility in the oil phase as compared to aggregates formed by free acids or methyl esters. The 2-position in triglycerides is as susceptible to oxidation as the more exposed 1-, and 3-position, indicating that the reduced oxidisability of triglycerides cannot solely be attributed to steric hindrance [18,19].

2.2 PHOTOOXIDATION

Photooxidation of unsaturated fatty acids can take place, if natural sensitisers such as hemoglobin, chlorophyll, riboflavin, or synthetic sensitisers like rose bengal and methylene blue are present. Photooxidation is about three orders of magnitude faster than autoxidation, but it may be accompanied by autoxidation, if the conditions allow for the breakdown of the formed hydroperoxides. For reviews of photooxidation, see [20,21]. There are two types of photooxidation:

In the type-I mechanism a sensitiser forms an activated triplet-complex with the substrate which will subsequently react with triplet oxygen to finally give hydroperoxides. A typical catalyst is riboflavin. This is a radical process, and the hydroperoxide products formed in type-I reactions are the same as obtained by autoxidation. As opposed to autoxidation, however, the type-I photooxidation has no induction period. With regard to the production of flavour blocks, this translates to the production of similar volatile patterns as in an autoxidative process. It is interesting to note, however, that a type-I process will be much faster, thus making such a process rather efficient.

An example of a commercial process involving photooxidation is the following: a synthetic butter flavour whose milk flavour and butter flavour were enriched was prepared from milk fat in a two step procedure: the first step involved the lipase-catalysed hydrolysis of milk fat. This was followed by ultraviolet irradiation of the fat hydrolysate at 45°C to yield a flavour block for dairy and bakery applications [22].

The type-II mechanism involves singlet oxygen, which is formed by a photochemically excited sensitiser from triplet oxygen. Singlet oxygen then reacts with double bonds in an ene-reaction to give a trans allylic hydroperoxide. No radicals are involved, and the obtained hydroperoxide pattern significantly differs from the one obtained in both autooxidation and in type-1 photooxidation [7]. In the case of oleic acid, only the 9- and the 10-hydroperoxides are formed. In the case of fatty acids containing a 1,4-pentadienyl system, both the nonconjugated and the conjugated hydroperoxide products are formed. With linoleic acid, the 9-, 10-, 12-, and 13-hydroperoxides are obtained. Singlet oxygen oxidation of linolenic acid gives the 9-, 10-, 12-, 13-, 15-, and 16-hydroperoxides. The reaction rates of different fatty acids mainly reflect statistics due to the different numbers of double bonds present. Relative reaction rates of 1.0, 1.7 and 2.3 are observed for oleic acid, linoleic acid and linolenic acid, respectively [23].

Under photooxidation conditions, the primary monohydroperoxides can be further oxidised to dihydroperoxides and to six-membered hydroperoxy epidioxides (Figure 3) [6,16].

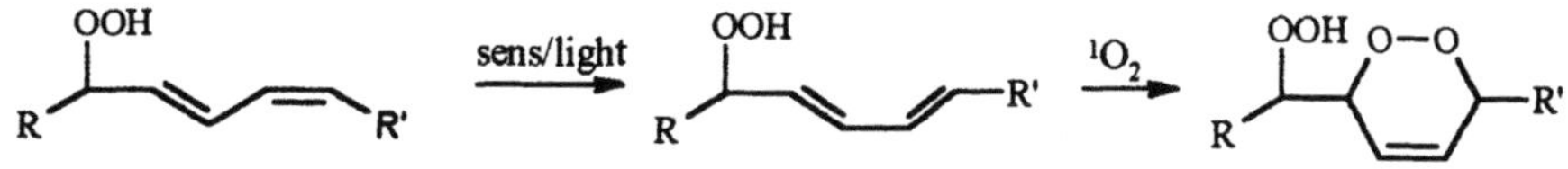

Figure 3. Formation of 6-membered hydroperoxy epidioxides [6]

The role of photooxidation in normal foods is demonstrated by the occurrence of 3-methyl-2,4-nonanedione in a variety of foods (see Table 2). 3-Methyl-2,4-nonanedione generation is best explained by the photooxidation of furan fatty acids (Figure 4) [24-26].

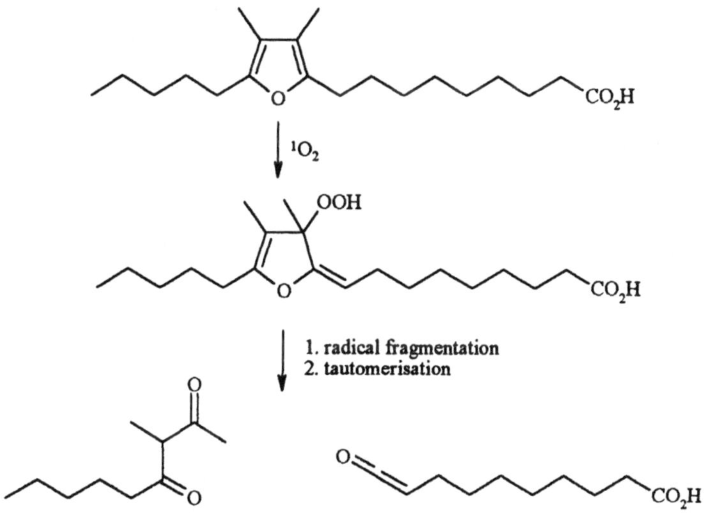

Figure 4. Suggested mechanism for the formation of 3-methyl-2,4-nonanedione [25,26]

2.3 BREAKDOWN OF HYDROPEROXIDES

2.3.1 Acid-Catalysed Breakdown of Hydroperoxides

Under acidic conditions allylic hydroperoxides are cleaved in a clean reaction via a heterolytic, Hock-type mechanism [6,27]. Examples are, treatment of the allylic hydroperoxide with acid-washed Fuller's earth [28], with BF_3-Et_2O in ether [29] or with 5% HCl in methanol [30]. Cleavage occurs exclusively on the olefinic side. In the case of the 9-hydroperoxide of methyl linoleate for example, trans-2-nonenal and methyl 9-oxononate were obtained (Figure 5). In an analogous manner, 2,6-nonadienal is formed from linolenic acid as the starting material.

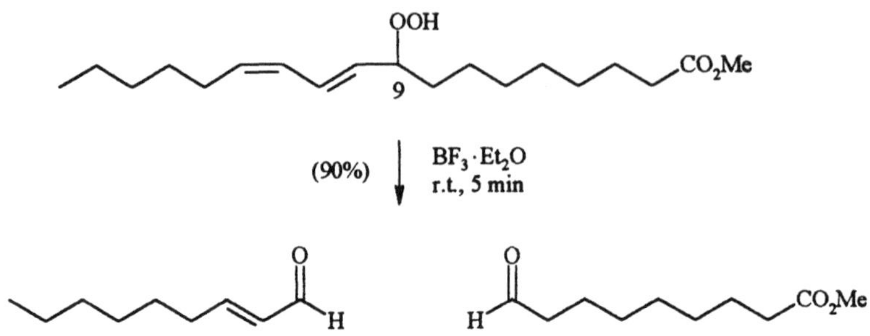

Figure 5. Heterolytic cleavage of hydroperoxides [29]

2.3.2. Thermal Breakdown of Hydroperoxides

The thermal decomposition of hydroperoxides involves a complex set of reaction pathways, and gives rise to a complicated mixture of non-volatile and volatile products. As non-volatile products are less accessible to routine analytical techniques, considerably less is known about them than about the volatiles. In studies of model systems (autoxidised 2,5-undecadiene, 3,6-nonadiene, and 4,7-heptadiene) both an ether bridged dimer and a peroxy bridged dimer have been identified as oxidation products, among others [31-33]. In the case of fatty acids, non-volatile products include dimerised and polymerised fatty acids, keto(di)ene acids, ketoepoxy acids and hydroxy(di)ene acids. In many cases non volatiles will actually be the major products of hydroperoxide decomposition. These are reviewed elsewhere [5,34]. In the following, only reactions which are relevant to the formation of volatile flavour compounds are discussed.

The thermal decomposition of monohydroperoxides starts with the homolytic cleavage of the O-O bond to give an alkoxy radical and a hydroxy radical (Figure 6). The alkoxy radical intermediate undergoes β-scission yielding an oxo compound and an alkyl or alkenyl radical. It is generally believed that cleavage can occur on either side of the alkoxy carbon: cleavage A in the scheme results in the formation of one fragment with an α,β-unsaturated aldehyde group and an alkyl radical fragment, while cleavage B gives a fragment with a saturated aldehyde group and a fragment with a vinyl radical. The alkenyl radical may further combine with a hydroxy radical to give an aldehyde, or with a hydrogen radical from a hydrogen atom donor to give an alkene. In an analogous manner, an alkyl radical is converted into an alcohol, or an alkane.

A lot has been speculated, but little is actually known about how the two cleavage pathways A and B depend on various reaction conditions (see [14] for an excellent update of the discussion). It has been suggested that A should be favoured over B under thermodynamically controlled conditions, as A gives a conjugated oxoene product [27]. A similar outcome is predicted, arguing that an alkyl radical (the second product in A) is more stable compared to a vinyl radical (the second product in B). However, temperature, oxygen pressure, the oxidation state of the lipid and the presence of antioxidants all seem to have an influence, and the situation is even more complicated, because in many cases the formation of shorter aldehydes may not only be ascribed to the β-cleavage of hydroperoxides, but is also due to secondary reactions.

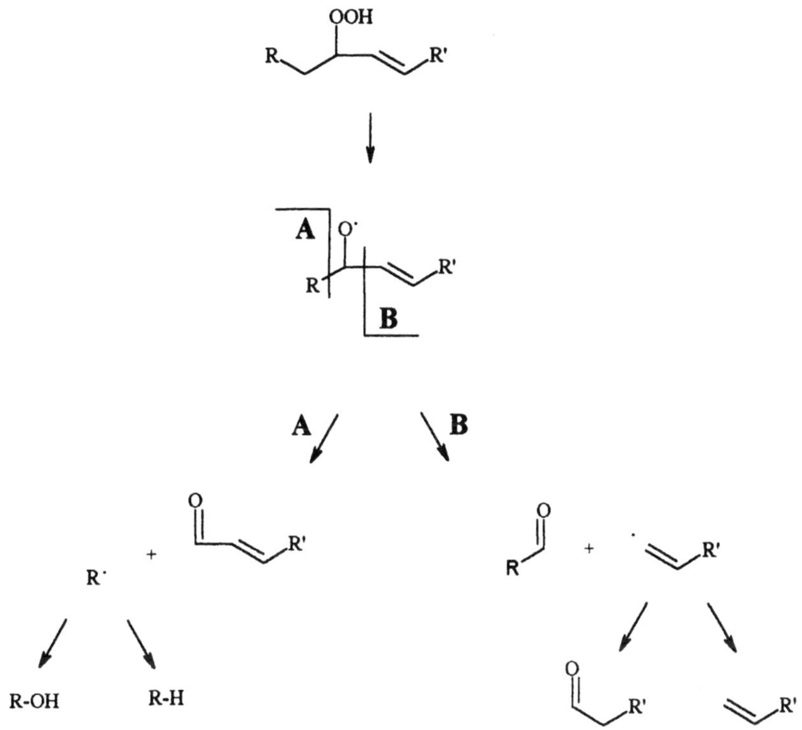

Figure 6. Homolytic breakdown of hydroperoxides

Some unsaturated aldehydes were shown to be more prone to autoxidation than the starting unsaturated fatty acids [35]. For example, 2,4-decadienal, the major cleavage product of the 9-hydroperoxide of linoleic acid, disappeared in the presence of oxygen, and 2-octenal and hexanal were formed, indicating that both double bonds in the aldehyde can undergo oxidative cleavage. Peroxide intermediates have been suggested in this process (see Figure 7, route I) [35-38]. In addition, a non-oxidative, retroaldol-type mechanism has been suggested for the formation of short chain aldehydes from unsaturated aldehyde precursors (see Figure 7, route II) [39].

The mechanism for the formation of 2,4-nonadienal, however, has not yet been elucidated. The fact that it cannot be simply explained by β- or Hock-cleavage of the usual hydroperoxide intermediates underlines the need for more fundamental research on the secondary oxidation products of unsaturated aldehydes.

In conclusion, long-chain aldehydes are preferably formed under breakdown conditions such as high temperatures, short reaction times, exclusion of oxygen, and the presence of antioxidants. The formation of short chain aldehydes on the other hand is favoured at moderate temperatures, prolonged reaction times, presence of metals and high oxygen pressure [40-44].

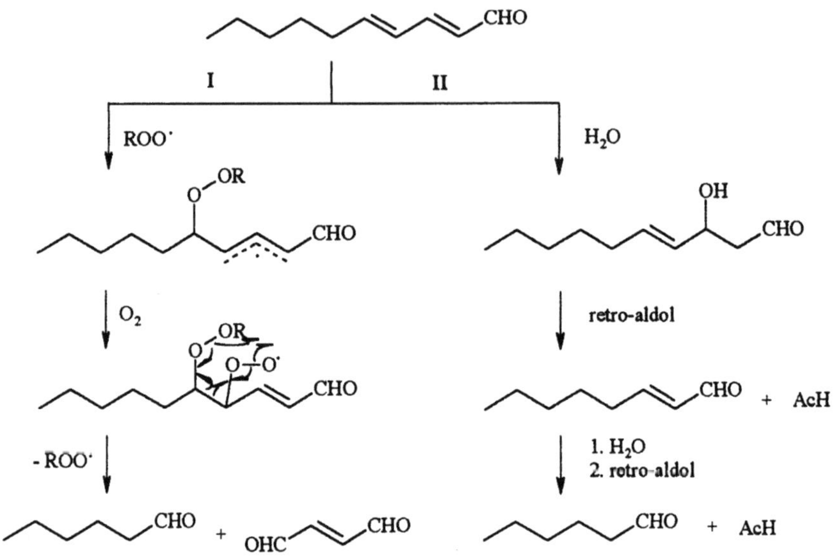

Figure 7. Suggested mechanisms for the formation of hexanal from 2,4-decadienal [35,39]

Dihydroperoxides are cleaved via the same pathways as the monohydroperoxides, but products with short chain lengths are formed, including dialdehydes. Epoxy hydroperoxides also fragment according to the mechanisms shown for monohydroperoxides to yield epoxy aldehydes (Figure 8, route A) [45,46]. Another pathway has been suggested for the formation of epoxy aldehydes involving the epoxidation of unsaturated aldehydes, (Figure 8, route B) [46].

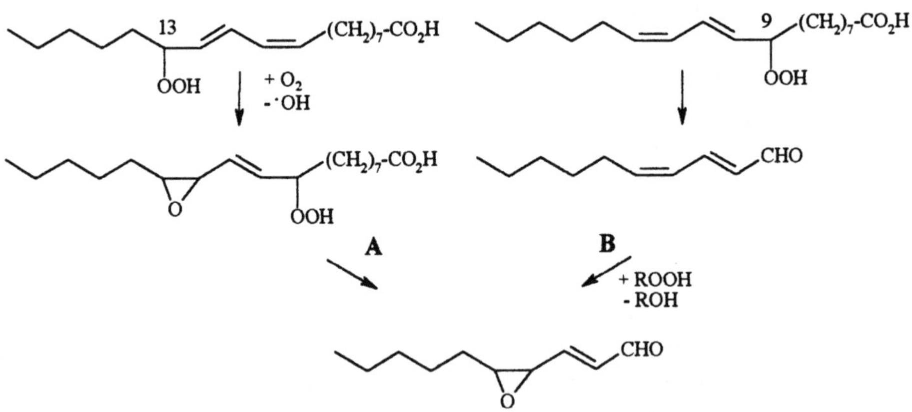

Figure 8. Suggested mechanisms for the formation of 4,5-epoxy-2-decenal [45,46]

Five-membered hydroperoxy epidioxides fragment via (i) β-cleavage analogous to monohydroperoxides, (ii) cleavage of the dioxolane ring and (iii) cleavage between the dioxolane ring and the double bond [47]. Figure 9 shows the products which were obtained upon pyrolysing at 210°C methyl 9-hydroperoxy-10,12-epidioxy-13-octadecenoate, a photooxidation product of methyl linoleate [48].

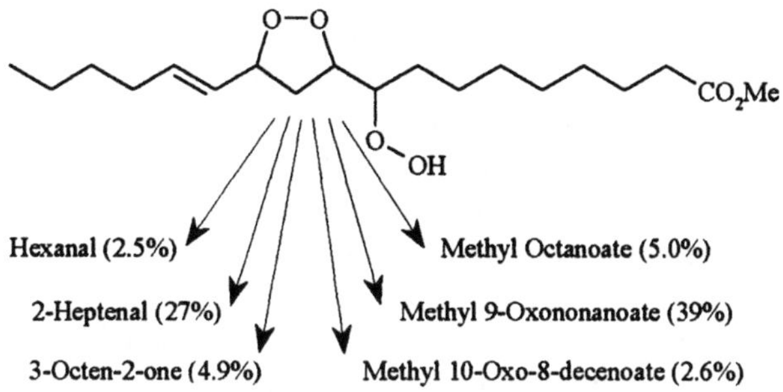

Figure 9. Pyrolysis of methyl 9-hydroperoxy-10,12-epidioxy-13-octadecenoate [48]

2.4. APPLICATIONS

2.4.1. Margarine's and Spreads

Today a wide range of margarine's and spreads are available, suitable for various applications and varying in fat content and composition, melting behaviour and taste. The water content in very low fat spreads can be as high as 80% which drastically reduces the buttery flavour which is associated with the fat phase. All in all, spreads are a challenging target from a flavour point of view.

Several approaches are used to manufacture flavour blocks which in combination lend a balanced butter flavour to margarine and low-fat spreads [49]. One approach relies on a mild non-enzymatic oxidation of butter oil. When a butter oil/water emulsion in the presence of α-tocopherol is stirred at elevated temperatures for 16 h, a balanced butter block containing aldehydes, lactones and methyl ketones is obtained [50,51]. Another approach involves releasing fatty acids from butter fat via chemical or lipase-catalysed lipolysis yielding mainly fatty acids with an even-number of carbon atoms in their chain [52]. Thirdly, a Maillard-type approach involving the moderate heating of butter or butter oil with milk powder as the protein source gives a sweet, caramel-like butter flavour which is relatively high in maltol [53,54].

Flavours obtained by a suitable combination of the above three blocks can further be enhanced by adding volatile top-notes. For example, a sour cream application requires higher levels of acetic acid, acetoin, and diacetyl, while a sweet cream character is obtained by increasing the levels of δ-lactones, vanillin and dimethyl sulphide [55]. Cheesy notes are obtained by providing fatty acids and methyl ketones, and for ghee-like flavours the emphasis will be on maltol, phenols and esters [56]. The required top-notes can be generated by chemical synthesis or by enzyme treatment of natural food materials (e.g. diacetyl from starter distillate, a by-product in the manufacture of cheese) or by the extraction of natural raw materials (e.g. dimethyl sulfide from spearmint oil, δ-lactones from coconut oil) [57].

2.4.2. Savoury

Using an approach related to the above mentioned mild non-enzymatic oxidation of butter oil, a variety of vegetable oils and animal fats were mildly oxidised to give enhanced lipid oxidation blocks for a variety of savoury end-uses. Using α-tocopherol concentrations of 0.1% or more gave the best results. For example, the oxidation of chicken fat gave a complex mixture of volatiles which was rich in 2,4-decadienal and could serve as an ingredient in chicken flavours [58].

Upon heating fatty acids at 300 °C in an air stream, coupled with the immediate trapping of the released volatiles into a cold trap, flavouring preparations for an end-use in savoury products were obtained. Using this procedure with oleic acid as the starting material, a mixture of volatiles containing high levels of saturated and monounsaturated aldehydes was obtained, which had an overall impact described as 'grilled and roasted beef'. Similarly, linoleic acid gave a product with a chicken-type flavour impression [59].

2.4.3. Tobacco

Due to their attractive aroma qualities and their low odour thresholds, carotenoid-derived aroma compounds such as ionones, β-damascenone, theaspiranes and vitispiranes are important flavour ingredients [60].

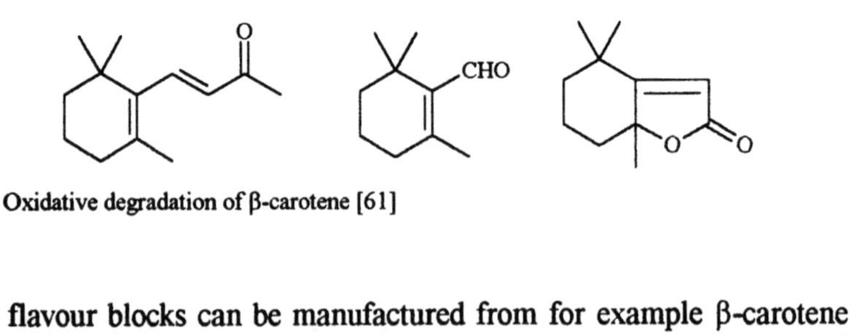

Figure 10. Oxidative degradation of β-carotene [61]

Powerful flavour blocks can be manufactured from for example β-carotene either via autoxidation at elevated temperatures and air pressures, or via photooxidation [61]. With β-carotene as the raw material, a floral and woody block is obtained which is useful for tobacco flavours. The manufacturing protocol relies on the possibility to direct the fragmentation towards the formation of significant amounts of small molecular weight products including C_{13}-norisoprenoids such as β-ionone by applying a high pressure (see Figure 10) [62].

3. Enzymatic Lipid Oxidation

The enzymes involved in the lipoxygenase pathway produce primary volatiles similar to the ones obtained via autoxidation or photooxidation. In plants, only two fatty acids which occur at significant levels, linoleic and linolenic acids, serve as precursors for a range of aldehydes and alcohols which in turn contribute to the overall aroma of a living plant or a fresh plant foodstuff. On the other hand, storage of fresh non-heated plant material, and processing procedures which result in the cracking of the cell structures may result in uncontrolled enzymatic reactions and often generate off-flavours (see Table 3) [63]. Over the past 60 years, enzymes involved in the lipoxygenase pathway have been found to occur in many plant materials (see [64-68] for reviews). The lipoxygenase pathway starts with the liberation of free fatty acids by the action of lipases, generally upon disrupture of the cell membranes (Figure 11). Lipoxygenases then react with free polyunsaturated fatty acids containing a cis.cis-1,4-pentadienyl system to give the fatty acid hydroperoxides, usually in a regioselective and a stereoselective fashion (see [66] for an overview). The hydroperoxides can be further converted into a range of volatile products by the action of hydroperoxide lyases. The lyase products can be further metabolised by the action of double bond isomerases and dehydrogenases. Overall, the specificities of both lipoxygenase and hydroperoxide lyase lead to a certain volatile pattern which is specific of a certain fruit, vegetable or other plant material (see section 4).

3.1. LIPOXYGENASE

As has been shown for a soybean lipoxygenase isozyme (soybean LOX-1), the dioxygenation requires iron and does not involve a haem-type cofactor (see [64.69] for a detailed discussion of the mechanism of lipoxygenase action). Although considerably less mechanistic insight is available with regard to lipoxygenases from other plants, the significant sequence homologies with soybean LOX-1 make it reasonable to assume that they operate in a similar fashion [70].

In its active form lipoxygenase contains a high-spin Fe(III) centre which specifically abstracts a hydrogen from the substrate. Two mechanisms have been proposed to account for this reaction. The first proposal involves a free radical intermediate that reacts directly with dioxygen (route A in Figure 12) [71-73]. The second proposal suggests the formation of a γ-organoiron intermediate, followed by the insertion of dioxygen in the σ-organometallic bond (route B in Figure 12) [74,75]. Although there is a substantial body of evidence supporting the notion of radical intermediates [64,73], it has not been possible to rule out the organoiron route.

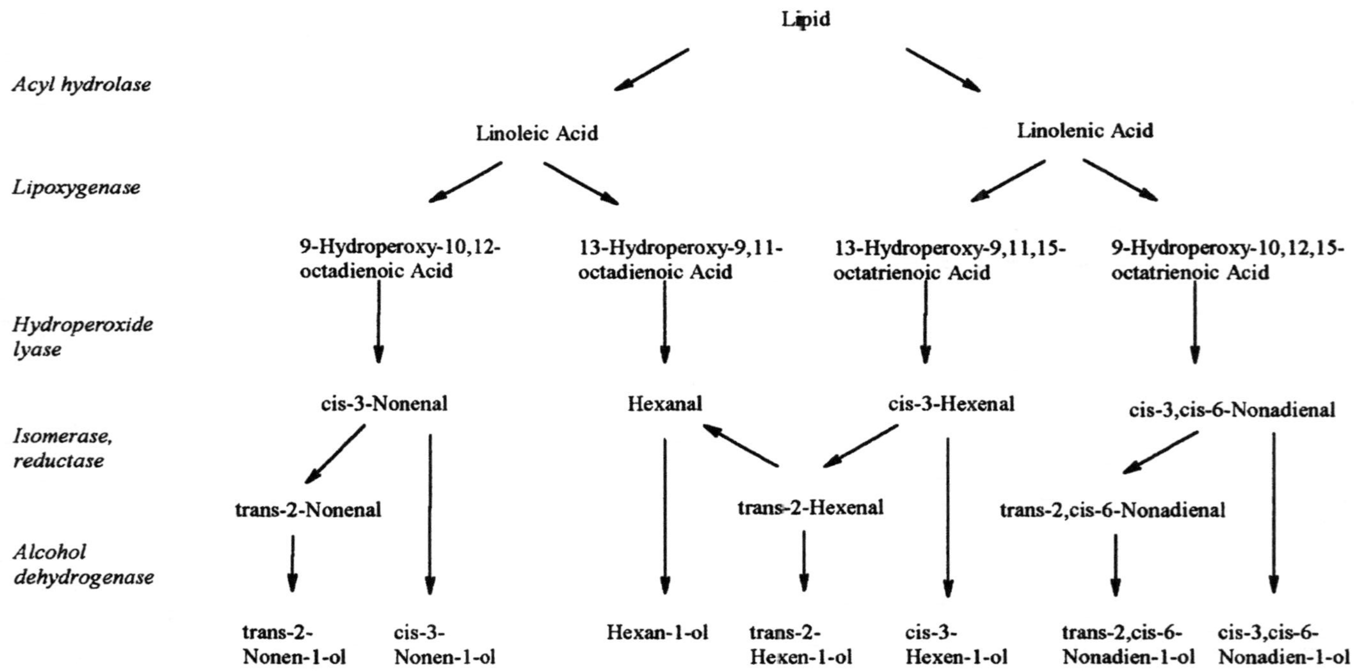

Figure 11. Enzyme-mediated lipid oxidation.

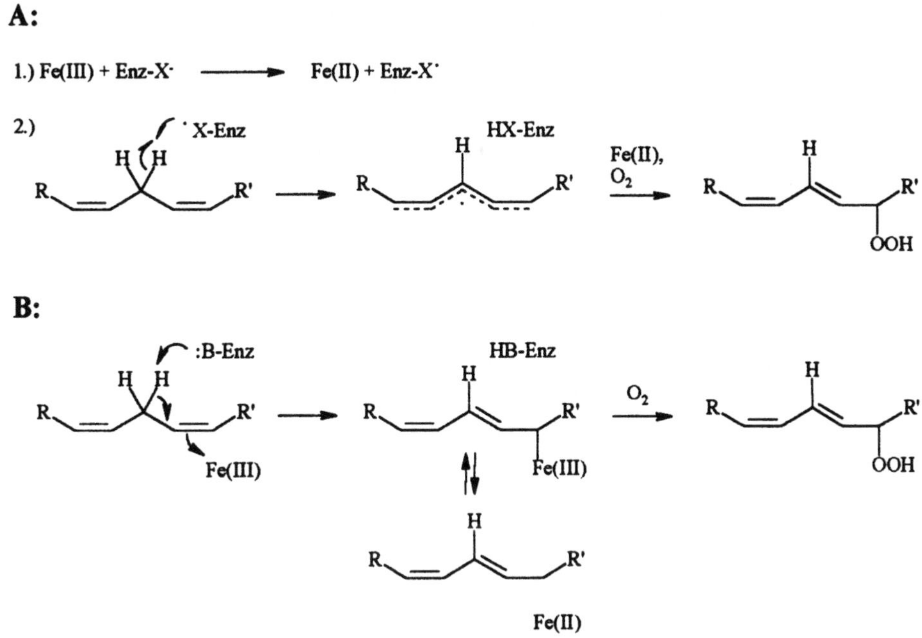

Figure 12. Two mechanisms proposed for the lipoxygenase reaction [71-75]

In terms of their substrate specificities, lipoxygenases from plant sources are most active with free fatty acids, but some lipoxygenase species are reported to react also with fatty acid glycerides and phospholipids [76]. In general, lipoxygenases from plants require a 1,4-diene unit in the ω-6-position of their substrate; longer chain fatty acids (>C18) are usually oxidised at considerably lower rates. In the case of the typical lipoxygenase substrates linoleic and linolenic acids, either C-13 or C-9 hydroperoxides are formed by lipoxygenases from plant origin.

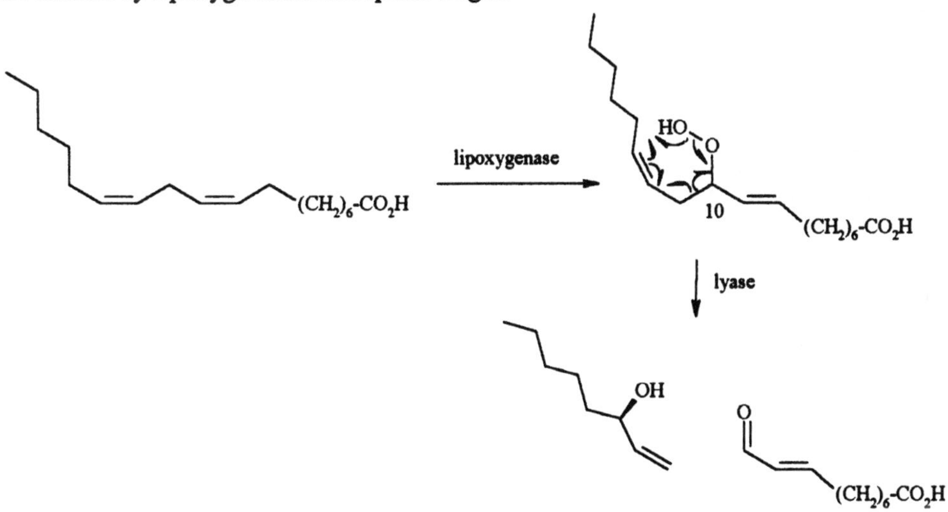

Figure 13. Formation of (3R)-1-octen-3-ol in mushrooms [77,78]

Fungi contain an unique lipoxygenase which, starting from linoleic acid, produces a 10-hydroperoxide which is converted by an equally specific lyase to form 1-octen-3-ol, a key character compound in mushroom flavour (Figure 13) [77-79].

Table 1 gives an overview of the site of the oxygen insertion for lipoxygenases from different plant materials.

Table 1. Product specificities of lipoxygenases from plants

Lipoxygenase from (experiments done at)	Product specificity of lipoxygenase		References
	% 13-HPO	% 9-HPO	
Soybean seed (pH 7)	40	60	[80-87]
Soybean seed (pH 9)	90	10	
- purified LOX-1 (pH 9.5)	97.5	2.5	
- purified LOX-2 (pH 6)	25	75	
Cucumber (pH 5.5)	25	75	[88]
Maize (pH 7.4)	9	91	[65,89]
Maize (pH 6.6)	14	86	
Tomato (pH 5.5)	15	85	[90,91]
Alfalfa (pH 6.8)	50	50	[92]
Potato (pH 5.5)	5	95	[93,94]
Pea (pH 6.3)	42	58	[91,95,96]
- purified LOX-1 (pH 6.8)	39	61	
- purified LOX-2 (pH 6.8)	89	11	
Bell pepper (pH 7.5)	-	100	[13]
Peanut (pH 6.0)	100	-	[97]
Cauliflower (pH 7.5)	2	98	[13]
Tea leaves (pH 6.3)	84	16	[98,99]
Cotton (pH 6.5)	94	6	[100]
Flax (pH 6.5)	80	20	[101]
Wheat germ (pH 6.8)	15	85	[91]
Corn germ	7	93	[80]
Oat (pH 7.0)	10	90	[102]
Barley seeds (pH 7.5)	4	96	[103-105]
- purified LOX-1 (pH 7.5)	4	96	
- purified LOX-2 (pH 7)	90	10	
Rice (pH 7)	15	85	[106-108]
- purified LOX-1 (pH 4.5)	48	52	
- purified LOX-2 (pH 5.5)	49	51	
- purified LOX-3 (pH 7)	17	83	
Apple (pH 7.5)	85	15	[109]
Pear (pH 7.5)	5	95	[109]
Grape (pH 8)	80	20	[110]
Gooseberry (pH 6.5)	50	50	[111]
Strawberry (pH 6.5)	80	20	[111]
Special enzyme systems:			
Mushrooms (pH 6.5)	exclusive formation of 10-hydroperoxide		[77,78]

3.1.1. *Soybean Lipoxygenase*
Soybeans contain at least three lipoxygenase isozymes, one of which (LOX-1) has a pH optimum of 9-10 and produces the 13-hydroperoxide in a selective manner. LOX-2 and LOX-3 are most active at pH 6-7, and give mixtures of positional hydroperoxide isomers. Crystalline soybean lipoxygenase (LOX-1) is commercially available, and LOX-1 has become a well-studied model enzyme [69]. Soybean LOX-1 is a soluble iron-containing non haem enzyme with a molecular weight of 94 kDa. It has been sequenced [112], and a crystal structure of the enzyme is available [113].

Using crystalline soybean LOX-1, linoleic acid has been dioxygenated under 4 atm O_2 pressure to give (13S)-13-hydroperoxy-9,11-octadecadienoic acid in 80% yield, at a linoleic acid concentration of 28 g/L [114]. Lipoxygenase has been shown to oxidise substrates in organic co-solvents [115,116]. For example, nearly quantitative conversions of linoleic acid (at a concentration of 3g/L) have been achieved with soybean LOX-1 in microemulsions, containing an organic solvent, water, and a combination of the non-ionic detergents Nonidet P-40 and Triton X-35. Toluene, isooctane, and diisopropyl ether were the best solvents [117].

Various solid supports have been used to immobilise lipoxygenase [118,119], sometimes in combination with an organic co-solvent [120]. Using soybean LOX immobilised on a carbonyldiimidazole-activated matrix, the 13-hydroperoxide of linoleic acid was produced in an isooctane-water two-phasic system in 80% yield, employing a linoleic acid concentration of 2 g/L [121].

Another way to get around the use of expensive pure enzyme, is to use cheap soy flour as a source of the enzyme. At pH 9-10 LOX-1 is the only active soybean isozyme in the crude preparation, so that the 13-hydroperoxides of linoleic or linolenic acids can be selectively prepared [122]. An additional advantage of the basic reaction conditions is the increased solubility of the fatty acid substrates. Using this approach, hydroperoxide concentrations of up to 80 g/L have been obtained, and the process has been commercialised [123,124].

3.1.2. C-9 Specific Lipoxygenases
For preparative purposes, crude extracts of tomato, potato or barley materials can be employed for the preparation of the 9-hydroperoxides of linoleic and linolenic acids. For example, the 9-hydroperoxide of linoleic acid was prepared in 69% yield using a crude tomato extract, at a linoleic acid concentration of 0.3 g/L [90]. With a crude enzyme preparation from potato the 5-hydroperoxide of arachidonic acid was synthesised on preparative scale [125]; with linoleic and linolenic acids the 9-hydroperoxides are obtained [93,94].

Higher substrate concentrations (up to 14 g/L) have been achieved with an extract of barley seeds: linoleic and linolenic acids were converted into the corresponding 9-hydroperoxides which, on triphenylphosphine reduction and methylation, yielded (9S)-hydroxyoctadeca-10,12-dienoic acid methyl ester and (9S)-hydroxyoctadeca-10,12,15-

trienoic acid methyl ester, respectively, in 39-44% yield and > 98 e.e. [103,126]. Overall, the volume efficiencies which have so far been obtained in C-9-dioxygenations have to our knowledge not yet allowed for the commercialisation of a cost-effective process.

3.2. HYDROPEROXIDE LYASE

In spite of hydroperoxide lyase playing such an essential role in fruit aroma biosynthesis and its applications in the food industry, the enzyme is not well studied. The main difficulties are that hydroperoxide lyases are membrane-bound enzymes and that they are present in plant tissues in small amounts only. Hydroperoxide lyases can be classified into two groups which either cleave (13S)-hydroperoxides or (9S)-hydroperoxides. The former gives a C_6-aldehyde, and the latter a C_9-aldehyde product. C-13 specific lyases have been identified in alfalfa seedlings [127], tomato fruits [128], green bell pepper fruits [129], watermelon seedlings [130], tea leaves [131,132], and tobacco cells [133]. C-9 specific lyases were identified in pear fruits [134,135], whilst both C-9 and C-13 lyase activities were found in cucumber [136-138], soybean seedlings [65,139] and kidney bean [128].

The hydroperoxide lyase from tea leaves was purified to homogeneity and was found to have a molecular weight of 55 kDa. The enzyme was inhibited by anti-oxidants such as BHA and nordihydroguaiaretic acid, which are known as potent lipoxygenase inhibitors, and was also rapidly inactivated by higher concentrations of its substrate, the 13-hydroperoxide of linoleic acid [140,141]. Two hydroperoxide lyase isozymes from green bell pepper have also been purified to homogeneity and were found to be haem proteins. Both isozymes are trimers of 55 kDa subunits and are inhibited by lipophilic anti-oxidants such as α-tocopherol and nordihydroguaiaretic acid [142]. Recently, a hydroperoxide lyase was purified from tomato leaves and was reported to be a trimer of 73 kDa units. The enzyme was irreversibly inhibited by its substrate, the 13-hydroperoxide of linoleic acid, but also by five other hydroperoxide isomers which were not degraded by the enzyme [143].

3.3. APPLICATIONS

The potential of lipoxygenases and hydroperoxide lyases for the *in vitro* production of natural, label-friendly flavour ingredients [144] has long been recognised [64], and significant achievements have been made over the recent years.

3.3.1. C_6-Aldehydes and Alcohols

A three step process leading to natural C_6-aldehydes and alcohols has been commercialised using soy flour extract as a source of C-13-specific lipoxygenase, followed by treating the 13-hydroperoxides with plant homogenates (e.g. alfalfa sprouts, green bell pepper) which are relatively rich in hydroperoxide lyase. By pH tuning, the conversion could be directed to the production of either cis-3-hexenal or of

trans-2-hexenal. Finally, cis-3-hexenol (leaf alcohol) could be obtained using the reductive power of yeast [123,145,146].
Using a similar approach but alleviating problems resulting from the intrinsic instability of hydroperoxide lyase, an improved process was recently reported which produces cis-3-hexenol in relatively high yield. The procedure relies on the cloning of hydroperoxide lyase from banana and the overexpression of the enzyme in bakers yeast [147].

3.3.2. Mushroom Flavours
The 1-octen-3-ol content in mushrooms can be increased by incubating a mushroom homogenate with linoleic acid in the presence of oxygen; the mixture was not worked up but was directly spray-dried [148].

The submerged culture of edible fungi has been directed towards the production of fungal mycelium as a source of flavour. In general, good yields of fungal biomass are obtained, but the biomass is low in the desired product, 1-octen-3-ol [149,150]. When linoleic acid was exogenously added to growing cultures of *Pleurotus pulmonarius*, the production of 1-octen-3-ol was sevenfold enhanced [151].

3.3.3. Various
Lipoxygenases cannot only be used to produce hydroperoxides as precursors of volatiles in their own right, but can also mediate co-oxidations. Various examples have been reported, including a manufacturing process of ionones (via the soybean lipoxygenase dioxygenation of linoleic acid in the presence of carotenoids) [152], and of nootkatone (via the soybean lipoxygenase dioxygenation of linoleic acid in the presence of valencene) [153].

Lipoxygenase has also been used as a mediator in the oxidation of phenolic compounds. Of interest to the flavour industry are the lipoxygenase-catalysed oxidations of isoeugenol and coniferyl benzoate from Siam resin, resulting in the formation of vanillin in 4-15% yield [154].

4. Lipids as a Source of Positive and Negative Flavours in Foods

Lipids occur in all living cells and, thus, in nearly all foodstuffs. They have metabolic and structural functions, and they are powerful precursors of both desirable and undesirable aroma compounds. In addition, lipids strongly affect flavour release by restraining the escape of fat soluble aroma substances into the air. They also contribute to the taste of foods through mouthfeel. For example, butter and chocolate lipids impart a characteristic 'cooling' effect, and in milk they generate a long lasting rich and creamy impression.

Unsaturated fatty acids which are part of triacylglycerides and phospholipids are susceptible to autoxidation and, above all, in their free form also to enzymatic oxidation (see 2 and 3). These reactions lead to a multiplicity of aroma-active compounds, such as aldehydes, ketones, alcohols, acids, esters, and hydrocarbons. The minor components of lipids, e.g. hydroxy- and oxo fatty acids, furanoid fatty acids, carotenoids, and sterols can also act as flavour precursors, giving rise after oxidation to lactones and unsaturated ketones. Additionally, short chain free fatty acids which are formed by lipolysis of triacylglycerides contribute to the flavour of foodstuffs as well.

Aroma compounds which are derived from lipids have been detected in nearly all foods [155]. Because of their low odour thresholds (see Table 4) they often contribute significantly to the overall flavour impression of many foodstuffs, such as fresh milk, buttermilk, butter, cream, yoghurt, vegetable oils, cheese, French fries, fish, cooked meat as well as fruits and vegetables [13,14,156]. Unsaturated aldehydes impart a green, fatty, deep-fried or cucumber-like odour to these foods. On the other hand, the same aroma compounds can also be responsible for off-flavours, such as the rancid flavour of fats, peanuts, coconut, chocolate, the oxidised/metallic flavour of milk and buttermilk, the warmed-over flavour of meat, the cardboard flavour of butter, the reversion flavour of vegetable oils, the hay-like flavour of dry parsley, the green, fatty/trainy flavour of fish as well as off-flavours in beer, dried potatoes and baked goods. There are probably many other examples.

In this chapter attention is focused on those lipid derived aroma compounds, whose contribution to the (off-)flavour of a foodstuff has been assessed by either gc/olfactometry or by quantitative data. Their occurrence and aroma generation is briefly reviewed and their odour and taste thresholds (in oil and water) are summarised.

4.1. LIPIDS AS PRECURSORS OF DESIRABLE FLAVOURS

In fat containing foods, autoxidation of unsaturated fatty acids which are free or bound to glycerol results in a great number of low-molecular weight volatile compounds (mechanism see section 2). These are responsible for the fatty, green, fresh, deep-fried, rancid, fruity, cocos-like, fishy, mushroom-like or cucumber-like odour notes of these foodstuffs. In vegetables and fruits, however, aroma compounds derived from lipid oxidation are formed enzymatically. Linoleic and linolenic acid, which serve as precursors in plants, are part of phospholipids of cell membranes. In vegetables, they are split into volatile compounds by enzymatic reactions during the homogenisation, whereas the aromagenesis in fruits occurs in intact cells during the post harvest ripening process. Table 2 gives an overview of important lipid derived aroma compounds, their precursors and the foodstuffs, in which they were identified.

Six groups of lipid derived aroma compounds are reported to contribute to the flavour of foods. They are aldehydes, ketones, alcohols, acids, esters, and lactones.

4.1.1. Aldehydes

Aliphatic and especially unsaturated aldehydes are the most important group of lipid derived aroma-active components, since they are the most abundant volatile breakdown products of lipids and exhibit low odour thresholds (see Table 4). They are mainly formed from linoleic and linolenic acid by autoxidation or enzymatic oxidation. Other precursors are oleic acid, and arachidonic acid in meat, and eicosapentaenoic as well as docosapenta(hexa)enoic acid in fish.

The deep-fried aroma of French fries, chicken and puff-pastry can be attributed to (E,E)-2,4-decadienal. It is also responsible for the fatty odour impression of chocolate, tea, sesame and popcorn (Table 2). With a few exceptions, (E,E)-2,4-nonadienal and 4,5-epoxy-(E)-2-decenal occur in the same foodstuffs as 2,4-decadienal, suggesting similar formation pathways. The generation of 4,5-epoxy-(E)-2-decenal in thermally treated fats has been elucidated by Gassenmeier and Schieberle [46] (see Figure 8). The formation pathway of 2,4-nonadienal, however, is still unknown (see also 2.3).

Aldehydes are also responsible for the species specific character of meat [215,216]. In comparison to chicken and pork, their abundance is much lower in beef, with (E)-2-nonenal, (E,E)-2,4-decadienal and 12-methyltridecanal being the most important contributors to the green, fatty and animalic odour notes of cooked beef [168,169,217]. In fish, aldehydes are mainly derived from polyunsaturated C_{20} and C_{22} fatty acids, which are characteristic for fish lipids. Oxidation of these lipids leads to a range of unsaturated aldehydes, such as (Z)-3-hexenal, (Z)-4-heptenal, (E,Z)-2,6-nonadienal, and (E,E)-2,4-nonadienal.

They are responsible for the green, cucumber-like and fatty odour notes of cooked salmon, trout or cod [162]. In addition, propionaldehyde causes a fresh impression in boiled salmon and trout.

The formation of (E,Z)-2,6-nonadienal from (Z,Z,Z,Z,Z)-5,8,11,14,17-eicosapentaenoic acid (I) is shown in Figure 14 (according to [218]). The reaction starts with a lipoxygenase catalysed oxygenation at C12, followed by a lyase catalysed β-cleavage of the hydroperoxyde II, which results in (Z,Z)-3,6-nonadienal (III). Isomerisation of III then leads to (E,Z)-2,6-nonadienal (IV). IV is further degraded to (Z)-4-heptenal (VI) through water addition and subsequent retro-aldol cleavage of 3-hydroxy-6-nonenal (V) [219].

In vegetables and fruits, the so-called leaf aldehydes (E)-2-hexenal and (Z)-3-hexenal as well as hexanal are mainly responsible for the green odour notes (see also Figure 11). They contribute to the aroma of most fruits and vegetables, e.g. tomato, olive oil, apple, apricot, orange (juice), banana, straw-, blue- and raspberry (Table 2). The aroma of cucumber is characterised by (E,Z)-2,6-nonadienal and (E)-2-nonenal [183,220]. The latter compounds are also responsible for the cucumber-like and green-fatty odour of fish, tea, buttermilk, chocolate, meat, cheese, French fries, sesame and soybean oil (Table 2).

Table 2. Occurrence of important lipid derived compounds in foods and their potential precursors

Compound	Precursor	Detected in[c]
I. aldehydes		
propanal	18-HPO[b] of eicosapentaenoic acid, 20-HPO of docosa-penta(hexa)enoic acid, 16-HPO of linolenic acid	fish [157,158], yoghurt [172], cucumber [183], olive oil [211]
hexanal	13-HPO of linoleic acid, 2,4-decadienal	apple [174], tomato [159]; olive oil [160,212], cheese [161], fish [162], yoghurt [172], chocolate [175], cucumber [183], apricot [190], orange (juice) [195,205], pork [204]
(E)-2-hexenal	15-HPO of linolenic acid, (Z)-3-hexenal	tomato [159], olive oil [160,212], apple [174], apricot [190], banana [198]
(Z)-3-hexenal	15-HPO of linolenic acid	tomato [159], olive oil [160,212], strawberry [163], butter [164], fish [162], tea [180], raspberry [199], orange juice [205], parsley leaves [214]
(Z)-4-heptenal	(E,Z)-2,6-nonadienal or (Z,Z)-11,15-octadecadienoic acid	fish [162], butter [164], chocolate [175], tea [180]
octanal	11-HPO of oleic acid	chicken [165], beef [167-169], yoghurt [172], orange (juice) [195,205], white wine [209], olive oil [212]
(E)-2-octenal	9- HPO of linoleic acid, 2,4-decadienal	chocolate [175], French fries [176], pork [204], beef [206]
nonanal	10-HPO of oleic acid	chicken [166], beef [167,168,206], cheese [170], chocolate [175], French fries [176]
(E)-2-nonenal	9-HPO of linoleic acid, 12-HPO of arachidonic acid	fish [162], chicken [165,166], beef [167,168], cheese [170], chocolate [175], French fries [176], tea [180], cucumber [183], sesame [184], soybean oil [200], pork [204], olive oil [211,212]
(Z)-2-nonenal	9-HPO of linoleic acid, 12-HPO of arachidonic acid	olive oil [160,212], beef [167-169], cheese [170], puff-pastry [46], chocolate [175], French fries [176,213], popcorn [185], soybean oil [200]
(Z)-3-nonenal	10-HPO of linoleic acid	French fries [176]
(E,E)-2,4-nonadienal	2,4-decadienal (hypothetical)	fish [162], butter [164,203], chicken [165,166], beef [167-169,206], puff-pastry [46], chocolate [175], French fries [176], tea [180], sesame [184], popcorn [185], olive oil [212]
(Z,Z)-3,6-nonadienal	9-HPO of linolenic acid, 14-HPO of eicosapentanoic acid	fish [162], butter [203]
(E,Z)-2,6-nonadienal	9-HPO of linolenic acid, 14-HPO of eicosapentanoic acid	fish [162], chocolate [175], tea [180], buttermilk [181], cucumber [183]
(E)-2-decenal	9-HPO of oleic acid	chicken [166], French fries [176]
(E,E)-2,4-decadienal	9-HPO of linoleic acid	fish [162], butter [164,203], chicken [165,166], beef [167-169,206], puff-pastry [46], chocolate [175], French fries [176,213], tea [180], sesame [184], popcorn [185], olive oil [212], parsley leaves [214]
(E,Z)-2,4-decadienal	9-HPO of linoleic acid	olive oil [160,212], chicken [165,166], beef [167-169], puff-pastry [46], French fries [176,213], pork [204]
4,5-epoxy-(E)-2-decenal	9-HPO of linoleic acid via 2,4-decadienal, 13-HPO of	olive oil [160,212], cheese [170], puff-pastry [46], French fries [176,213], tea

	linoleic acid via 12,13-epoxy-9-hydroperoxy-10-octadecenoic acid	[180], buttermilk [181], sesame [184], popcorn [185], soybean oil [200], orange juice [205]
2-undecenal	8-HPO of oleic acid	chicken [166], buttermilk [181], pork [204]
12-methyltridecanal	plasmalogens	beef [168,169]
II. ketones		
2-heptanone	3-oxooctanoic acid	yoghurt [172], blue cheese [178], pork [204]
2-nonanone	3-oxodecanoic acid	cheese [161,178], yoghurt [172], puff-pastry [46],
2-undecanone	3-oxododecanoic acid	cheese [161,178], yoghurt [172], beef [206]
1-hexen-3-one	13-HPO of linolenic acid	butter [164]
1-octen-3-one	10-HPO of linoleic acid, 12-HPO arachidonic acid	camembert cheese [161], fish [162], butter [164], chicken [165,166], beef [167-169], puff-pastry [46], chocolate [175], French fries [176], tea [180], sesame [184], mushroom [197], orange juice [205], olive oil [211,212], parsley leaves [214]
(Z)-1,5-octadien-3-one	10-HPO of linolenic acid	fish [162], butter [164], French fries [176], tea [180], mushroom [197], orange juice [205], olive oil [212], parsley leaves [214]
3-methyl-2,4-nonan-dione	10,13-epoxy-11,12-dimethyloctadeca-10,12-dienoic acid	tea [180]
(E)-β-damascenone	9-O-glycoside of 3-hydroxy-7,8-didehydro-β-ionol	cheese [161,170], apple (juice) [173], tea [180], beer [182], popcorn [185], tomato [186], coffee [202], white wine [207,209,210], olive oil [212]
α-ionone	α-carotene	carrot [187], raspberry [188], tea [189], apricot [192]
β-ionone	β-carotene	tomato [159], raspberry [188], tea [189], apricot [190], peach [193], orange [194], beef [206]
III. alcohols		
(Z)-3-hexenol	15-HPO of linolenic acid, via (Z)-3-hexenal	olive oil [160,212], apple [174], white wine [209], parsley leaves [214]
2-heptanol	3-oxooctanoic acid, via 2-heptanone	camembert cheese [161]
1-octen-3-ol	10-HPO of linoleic acid	camembert cheese [161], mushroom [197]
(Z)-1,5-octadien-3-ol	10-HPO of linolenic acid	camembert cheese [161]
IV. acids		
butyric acid	triglycerides	cheese [161,170,179], butter [164,203], yoghurt [172], puff-pastry [46], buttermilk [181]
hexanoic acid	triglycerides	cheese [161,179], yoghurt [172], coconut [196]

V. esters		
(Z)-3-hexenyl acetate	15-HPO of linolenic acid via (Z)-3-hexenol	olive oil [160,211,212], apricot [190], peach [193], parsley leaves [214]
ethyl butanoate	butyric acid	strawberry [163], cheese [171], beer [182], apple [191], orange (juice) [194,195,205], banana [201], white wine [207-209], olive oil [211,212]
ethyl hexanoate	hexanoic acid	cheese [161,171], beer [182], apple [191], orange (juice) [194,205], white wine [207-209]
VI. lactones		
δ-octalactone	5-hydroxyoctanoic acid	camembert cheese [161], butter [164], milk [177], buttermilk [181], apricot [190], peach [193], coconut [196]
γ-decalactone	oleic acid, via (R)-12-hydroxyoleic acid	yoghurt [172], chocolate [175], buttermilk [181], French fries [213], apricot [190], peach [193]
δ-decalactone	linoleic acid, via (R) or (S)-13-hydroxy-9,11-octadeca-dienoic acid	cheese [161,170,171], butter [164,203], yoghurt [172], puff-pastry [46], chocolate [175], milk [177], tea [180], buttermilk [181], French fries [213], apricot [190], peach [193], coconut [196]
(Z)-6-dodecen-γ-lactone	linoleic acid, oleic acid	cheese [161,170], butter [164], puff-pastry [46], white wine [209]

Footnotes to table 2.

[a]: The aroma compounds were evaluated based on gc/o or quantitative analysis.

[b]: HPO: hydroperoxide.

[c]: References are given in brackets.

4.1.2. Ketones

Aroma-active ketones are also very important lipid derived flavour substances. Methyl ketones, such as 2-heptanone, 2-nonanone, and 2-undecanone are formed during ripening of cheese in a side pathway of the β-oxidation of triacylglycerides [221], which is catalysed by *Penicillium* and *Aspergillus* species as well as by *Ascomycetes, Phycomycetes* and Fungi *imperfecti*. They are responsible for the typical aroma of mould-ripened cheese [161,178] and also contribute to the flavour of yoghurt [172]. The formation of methyl ketones by moulds is shown in Figure 15. During the conventional β-oxidation cycle, various fungi are capable of deacylating the β-ketoacyl-CoA to β-ketoacid and CoA-SH by a thiohydrolase, which apparently exceeds in these micro-organisms the activity of the β-ketothiolase. A subsequent decarboxylation, which is catalysed by a β-ketoacid decarboxylase [222] yields methyl ketones and carbon dioxide.

Vinyl ketones, such as 1-octen-3-one and (Z)-1,5-octadien-3-one are generated by oxidation of their corresponding alcohols which, themselves, are enzymatic cleavage products of the 10-hydroperoxides of, respectively, linoleic and linolenic acid (see Figure 13). 1-Octen-3-one, a character impact compound of mushrooms, is also formed as a minor product of autoxidation, or a major product of the photooxidation of acyllipids in some fermented and processed foods (Table 2). (Z)-1,5-Octadien-3-one is also a key compound in mushrooms [197] and is responsible for the geranium-like odour of tea [180], parsley [214] and boiled fish [162]. Both vinyl ketones have very low odour thresholds (see Table 4).

3-Methyl-2,4-nonandione contributes to the flavour of green and black tea [180]. It is formed by photooxidation of furan fatty acids (see Figure 4), which occurs in the green parts of some plants and vegetables (e.g. chives, cabbage, potato, spinach) [24], in vegetable oils [25] and butter [26] as well as in fish and liver [223].

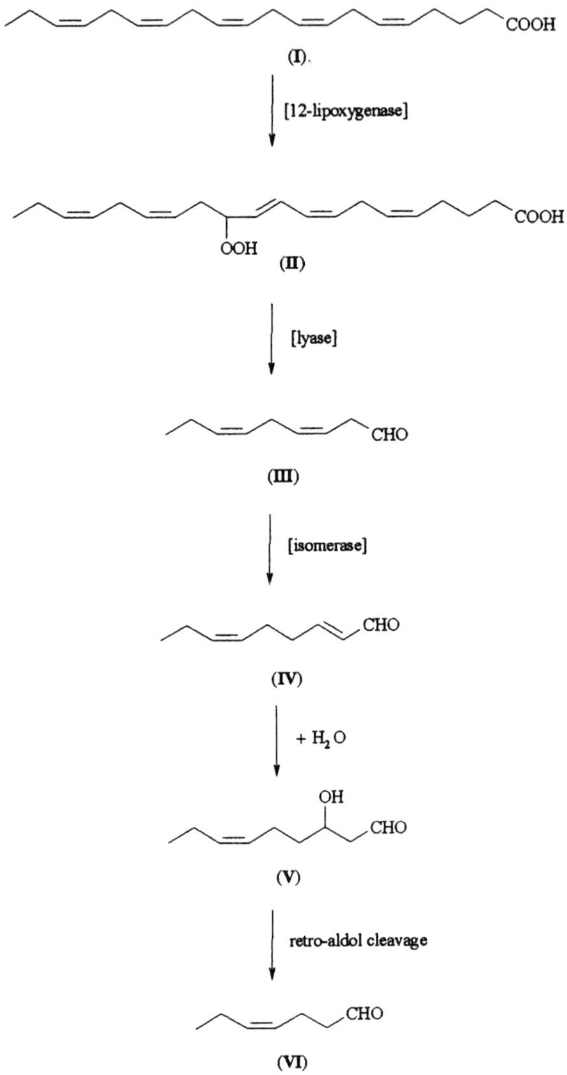

Figure 14. Formation of (E,Z)-2,6-nonadienal (IV) and (Z)-4-heptenal (VI) from (Z,Z,Z,Z,Z)-5,8,11,14,17-eicosapentaenoic acid (I) (according to [218,219])

Thermal and enzymatic conversions of carotenoids lead to a number of important flavour substances, such as (E)-β-damascenone, α- and β-ionone. (E)-β-Damascenone is responsible for the floral character of cooked apples [173], coffee [202], beer [182], wine [207,209,210] and popcorn [185], whereas α- and β-ionone have a higher impact in apricot [190,192], peach [193], carrot [187], raspberry [188] and orange [194]. α- and β-Ionone are autoxidation and photooxidation products of α- and β-carotene (see Figure 10).

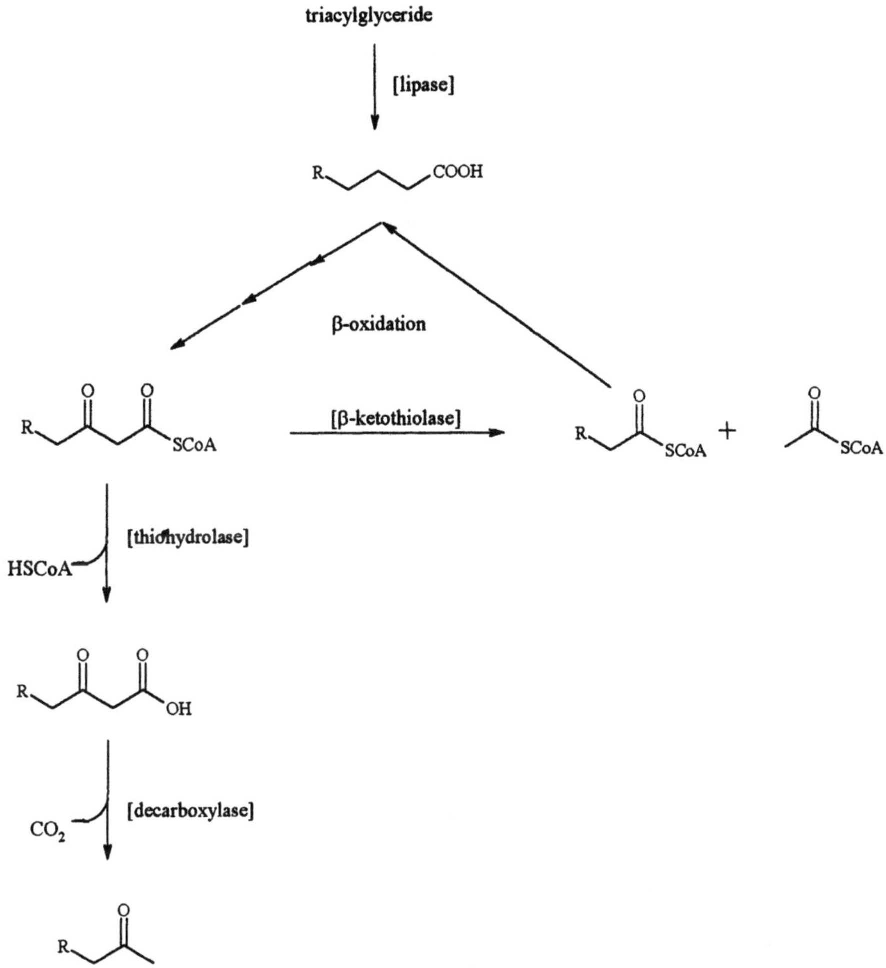

Figure 15. Formation of methyl ketones from triacylglycerides by moulds (according to [221]

Although the knowledge of the biochemistry of carotenoid catabolism is still very limited, there is some evidence that three steps are involved in the formation of (E)-β-damascenone in vegetables and fruits: (i) oxidative cleavage of carotenoids, (ii) enzymatic transformations (e.g. reduction), and (iii) acid catalysed conversions (Figure 16, according to [224]). Neoxanthin (I) is assumed to be enzymatically oxidised to grasshopper ketone (II) (cleavage of the 9,10 double bond) [225,226]. II is then enzymatically reduced to the allenic triol III, which has been detected in plants in the form of its glycoside. After hydrolysis of the glycosidic bond, III has been shown to be an effective precursor of β-damascenone (IV) [227]. The first step in Figure 16 has been proposed to be catalysed by a iron-containing dioxygenase system, which is also suggested to be involved in the biosynthesis of α- and β- ionone from α- and β-carotene [224].

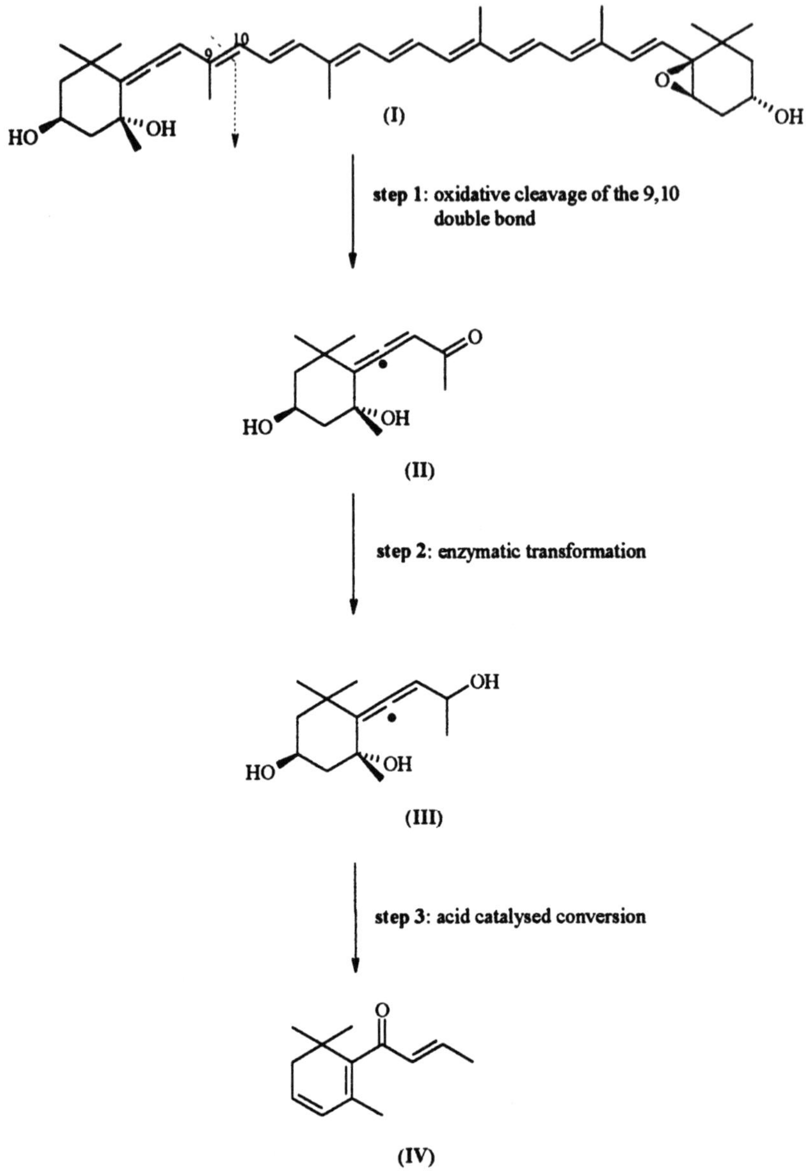

Figure 16. General steps for the conversion of neoxanthin (I) into β–damascenone (IV) (according to [224])

A related mechanism has been proposed for the formation of (E)-β-damascenone in apples, starting from 3-hydroxy-7,8-didehydro-β-ionol (no.I in Figure 17) [173]. The acetylenic diol I (glycosidically bound to α-L-arabinofuranosyl-(1,6)-β-D-glucopyranoside at the OH-group at C_9) was identified as the most abundant of eight polar β-damascenone precursors in apples [173], as well as in rose petals (bound to glucose at the OH-group at C_9) [228]. After hydrolysis of the glycoside, the ionol I undergoes acid catalysed dehydration, followed by a Meyer-Schuster rearrangement. This rearrangement involves the 1,3 shift of the OH-group of the secondary α-acetylenic alcohol II to β-damascenone (IV) via the allenic alcohol III (Figure 17).

4.1.3. Alcohols

Alcohols generally have a lower impact on the flavour of foodstuffs than the corresponding aldehydes, since their thresholds are significantly higher (see Table 4). Unsaturated alcohols, however, are quite potent odorants. 1-Octene-3-ol and (Z)-1,5-octadien-3-ol contribute to the aroma of mushrooms [197] and camembert cheese [161]. These vinyl alcohols are formed from linoleic and linolenic acid, catalysed by lipoxygenase and lyase enzymes (see Figure 13).

Alkenols such as (Z)-3-hexenol are derived from the corresponding aldehyde, which is reduced by the action of an alcohol dehydrogenase (see Figure 11). (Z)-3-hexenol has a green, leaf-like odour and contributes after oxidation to the flavour of olive oil [160,212], parsley [214], apple [174] and white wine [209]. Alkanols such as 2-heptanol which imparts a sweetish character to camembert cheese [161] are also formed by reduction of the corresponding ketone.

4.1.4. Acids and Esters

Acids derived from the lipolysis of triacylglycerides are important contributors to the flavour of milk products, e.g. cheese [161,170,179], butter [164,203] and yoghurt [172]. Esters are responsible for the fruity odour of many fruits, but were also found to have an impact on the aroma of e.g. white wine [207-209], beer [182], olive oil [160,211,212], parsley [214] and cheese [161,171] (see also Table 2).

4.1.5. Lactones

Lactones are often used as flavouring substances because of their broadly appreciated odour qualities. Since they are chiral components, their enantiomeric composition in foods is of interest for two reasons: (i) it is characteristic for each kind of fruits and milk products, and (ii) R and S enantiomers often differ in odour quality and threshold. For example, R(+)-γ- and δ-decalactone exhibit a fruity, peach-like odour and contribute to the flavour of e.g. apricot [190] and peach [193]. R(+)-γ-Decalactone is also a key aroma compound in buttermilk [181], yoghurt [172] and chocolate [175]. In contrast to its R(+) enantiomer, S(-)-δ-decalactone has a coconut-like odour quality. It is a key odourant in milk and milk products such as bovine milk [177], butter [203] and puff-pastries prepared with butter [46].

Various micro-organisms and enzymes were found to be involved in the biosynthesis of γ- and δ-lactones from fatty acids [229-231]. Albrecht *et al.* [232] studied the biosynthesis of R(+)-γ- and δ-decalactone using the lactone producing yeast *Sporobolomyces odorus*. Based on experiments with deuterium labelled precursors such as [9,10,12,13-$^2H_4$]-linoleic acid and [2,2-$^2H_2$]-(E)-3,4-epoxydecanoic acid, they proposed biosynthetic pathways for R(+)-

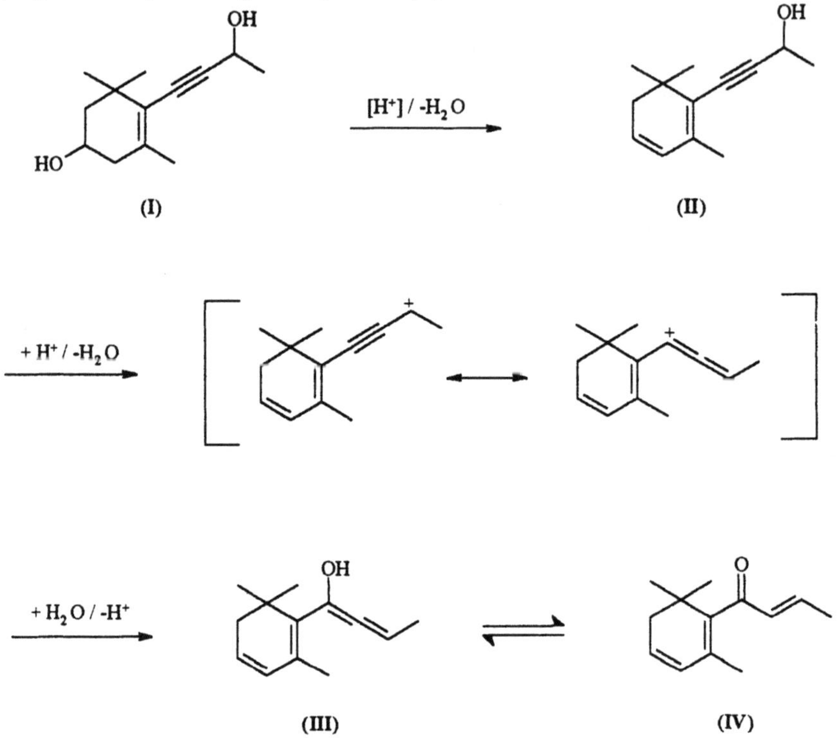

Figure 17. Proposed mechanistic pathway for the acid-catalysed formation of β-damascenone (IV) from 3-hydroxy-7,8-didehydro-b-ionol (I)(according to [173])

γ- and δ-decalactone, which are shown in Figures 18 and 19. The biosynthesis of R(+)-δ-decalactone starts with the enzymatic oxygenation of linoleic acid to (S)-13-hydroperoxy-(Z,E)-9,11-octadecadienoic acid (I) (Figure 18). I is converted via 13-oxo-(Z,E)-9,11-octadecadienoic acid (II) to (R)-13-hydroxy-(Z,E)-9,11-octadecadienoic acid (III). The enzymatic reduction of the oxo-group of II results in III with an enantiomeric purity of >98%. β-Oxidation, hydration and cyclisation of III yields R(+)-δ-decalactone (IV).

The proposed biosynthetic route for R(+)-γ-decalactone (III) from linoleic acid (Figure 19) is based on the observation that both deuterium labelled (E)-3-decenoic acid (I) and (E)-3,4-epoxydecanoic acid (II) were converted into deuterium labelled R(+)-γ-decalactone (III) [233], whereas I has been identified as an intermediate of the β-oxidation of linoleic acid. However, this route seems to be rather hypothetical, because

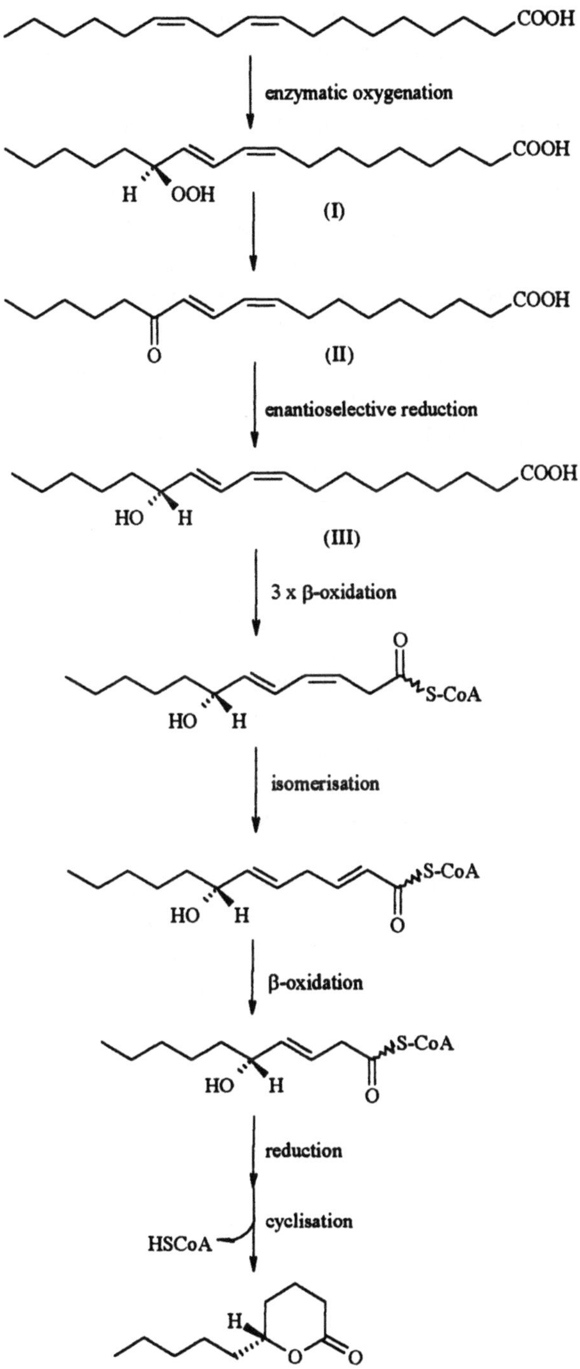

Figure 18. Proposed biosynthetic pathway of R(+)-δ-decalactone (IV) in *Sporobolomyces odorus* (according to [232])

of two reasons: (i) only trace amounts of labelled γ-decalactone were found after administration of [9,10,12,13-$^{2}H_4$]-linoleic acid to cultures of *Sporobolomyces odorus*, and (ii) γ-decalactone was formed from racemic ethyl-(E)-3,4-epoxydecanoate only with a purity of 38% enantiomeric excess.

Further investigations revealed that oleic acid is a good precursor for R(+)-γ-decalactone [234]. This is based on the observation that [9,10-$^2H_2$]-oleic acid was converted to [2-^{2}H]-R(+)-γ-decalactone by the yeast *Sporobolomyces odorus* in relatively high yields and with a enantiomeric purity of >99%. The formation pathway is shown in Figure 20. The key step of this conversion is the enantioselective hydroxylation of [9,10-$^2H_2$]-oleic acid (I) at C_{12}, which is suggested to be catalysed by a (R)-12-hydroxylase. β-Oxidation of (R)-12-hydroxy-octadecenoic acid (II) leads to [5,6-$^2H_2$]-(R)-8-hydroxy-5-tetradecenoic acid-CoA (III), which upon reduction, isomerisation and β-oxidation is further converted into [2-$^2H_1$]-(R)-4-hydroxydecanoic acid-CoA (IV). Finally, a thiolase catalyses the cyclisation of III to (R)- γ-decalactone (V). The authors also showed that under the same conditions another biochemical pathway, which starts with the transformation of oleic acid into linoleic acid catalysed by a Δ^{12}-desaturase leads to (R)-(Z)-6-dodecen-γ-lactone.

In contrast to the biosynthesis of R(+)-δ-decalactone in fruits, S(-)-δ-decalactone in milk is suggested to be derived from (S)-δ-hydroxydecanoic acid, which is bound to the α-position of glycerol [235]. This hypothesis is based on the observation that free lactones are absent in freshly secreted milk, but their formation is induced by heating or storage [177]. δ-Hydroxyalkanoic acids were probably formed as products of fatty acid metabolism in the mammary gland. Infusions of 1-^{14}C-decanoic acid into the lactating gland of goats revealed that the labelling could be detected in high yields in δ-lactones [177]. This implies the existence of an enzyme which catalyses the δ-hydroxylation of fatty acids. However, further studies by Swenson and Dimick [236] have failed to prove this hypothesis. The positive correlation of the abundance of lactones with those of short-chain (C_4-C_{14}) saturated fatty acids in milk, as well as labelling experiments with 1-^{14}C-acetate [237] provide additional information for further research in this area.

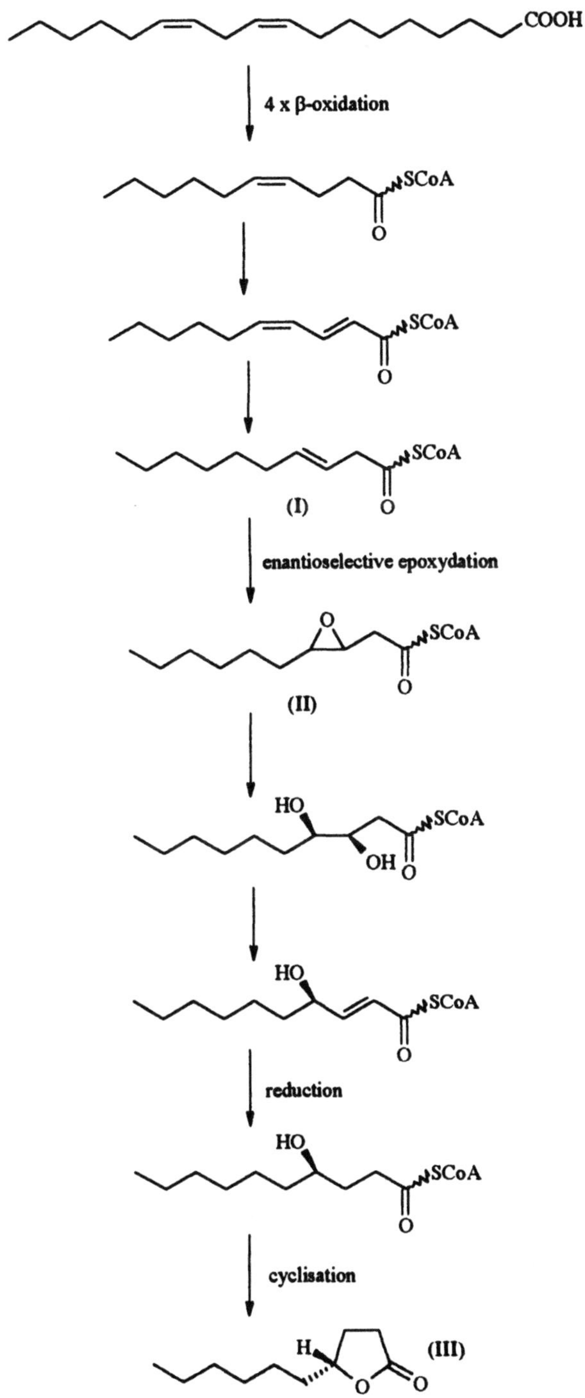

Figure 19. Proposed biosynthetic pathway of R(+)-γ-decalactone (III) from linoleic acid in *Sporobolomyces odorus* (according to [232])

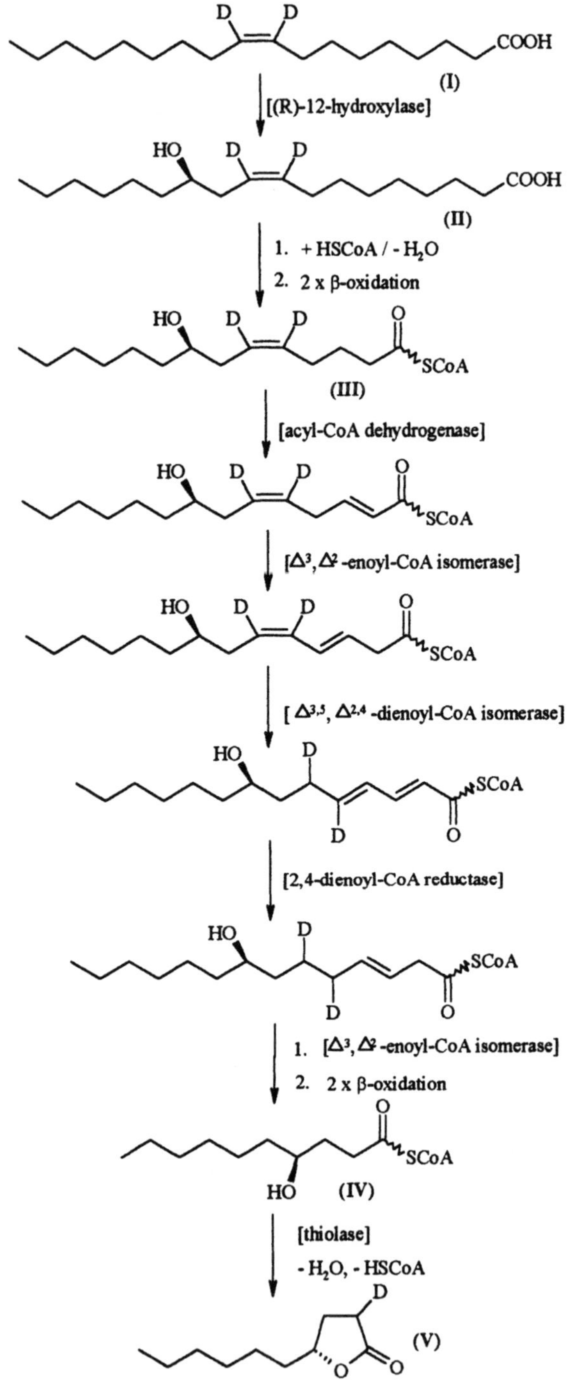

Figure 20. Proposed biosynthesis of [2-2H]-(R)-γ-decalactone from [9,10-2H2]-oleic acid by *Sporobolomyces odorus* (according to [234])

4.2. LIPIDS AS PRECURSORS OF UNDESIRABLE FLAVOURS

Lipid oxidation is often associated with flavour defects in foods. Most of the lipid derived off-flavours occur in fat containing processed foods, and are due to autoxidation of unsaturated fatty acids. In contrast to triacylglycerides, phospholipids contain high amounts of polyunsaturated fatty acids, and are therefore quite susceptible to autoxidation. They have been suggested, for example, as precursors of the undesirable aroma compounds responsible for the warmed-over flavour of cooked meat [238]. Furthermore, desirable aroma compounds of foods can easily be attacked by free radicals generated during autoxidation. Their decomposition has a great impact on aroma deterioration of foods. Examples of such compounds, which often have antioxidant properties, are 2-methyl-3-furanthiol, 2-furfurylthiol, 3-mercapto-2-pentanone, furaneol and sotolon. Their breakdown contributes to meat flavour deterioration during the storage of cooked beef and chicken [165,239].

Table 3 shows some examples of off-flavours in food, which are caused by lipid oxidation. The term 'warmed-over flavour' has been used to describe an off-flavour which develops rapidly, when cooked meat is stored at 4°C and is re-heated before consumption [240]. The high rate of lipid oxidation in heated meat probably has the following reasons: (i) heating of meat generates radicals and, thus, promotes the formation of hydroperoxides from phospholipids, which are released due to the destruction of cellular structures [241]; and (ii) cooking denatures the haem molecule and increases the concentration of non-haem iron, which is believed to catalyse the formation of radicals from hydroperoxides during the storage of meat [242,243]. Hexanal which was found to be the major off-flavour compound [244-246], can also be considered as a marker substance for warmed-over flavour development [244]. In addition, 4,5-epoxy-(E)-2-decenal and 1-octen-3-one were found to be responsible for the metallic odour of reheated meat [246,247].

The flavour of boiled fish depends on the freshness of the raw material. Substantial oxidative changes occur during frozen storage (-10°C to -15°C) of raw fish, which results in a green-fatty/trainy and fishy off-flavour after boiling of e.g. trout and salmon [158,248]. By evaluating the aroma of boiled trout and salmon before and after frozen storage at -13°C for 17 and 14 weeks, the authors found that (Z)-3-hexenal and (Z,Z)-3,6-nonadienal were mainly responsible for the flavour defects. Based on its higher stability, (Z)-3-hexenal was recommended as marker substance for such oxidative flavour changes in fish [248]. McGill *et al.* [249] reported the identification of compounds responsible for the 'cold-storage odour' of frozen stored, boiled cod. Based on quantitative data, (Z)-4-heptenal was considered as primarily responsible for this off-odour. Another fishy off-flavour which is also based on lipid oxidation was studied by Badings [250] in butter. It develops during cold storage of butter and was found to be due to (Z)-4-heptenal, 1-octen-3-one, (E,Z)-2,6-nonadienal, hexanal and (E)-2-nonenal.

Table 3. Examples of lipid derived off- flavours in foods and the responsible compounds

Type of off-flavour	Responsible compounds [a]
warmed-over flavour of meat	hexanal [244,246,247], 4,5-epoxy-(E)-2-decenal [246,247], 1-octen-3-one [246,247]
off-flavour in fish	(Z)-3-hexenal [158,248,249], (Z,Z)-3,6-nonadienal [158,248], (E,Z)-2,6-nonadienal [158], (Z)-4-heptenal [249]
fishy flavour of butter	(Z)-4-heptenal [250], 1-octen-3-one [250], (E,Z)-2,6-nonadienal [250], hexanal [250], (E)-2-nonenal [250]
cardboard flavour of butter	(E)-2-nonenal [251], (Z)-2-nonenal [251]
metallic and grassy flavour of milk	1-octen-3-one [252,253], (Z)-1,5-octadien-3-one [252,253], (E,Z)-2,6-nonadienal [252]
bean-like flavour of dry-milk	(E)-6-nonenal [254]
fruity flavour of milk	ethylbutyrate [255], ethylhexanoate [255]
metallic flavour of buttermilk	(E,Z)-2,6-nonadienol [256]
reversion flavour of soybean oil	3-methyl-2,4-nonanedione [200]
hay-like flavour of dry parsley and spinach	3-methyl-2,4-nonanedione [257,258]
hay-like flavour of peas	3,5-octadien-2-one [259], hexanal [259]
stale beer flavour	(E)-2-nonenal [260,261]
floral flavour of stored carrots	α/β-ionone [262], β-ionone-5,6-epoxide [262]
rancid flavour of oat meal	hexanal [263], caproic acid [263]
perfume rancidity of coconut oil	methyl ketones, e.g. heptan-2-one [264]

[a]: References are given in brackets.

Storage of butter oil at room temperature in the dark leads to an off-flavour, which is described as cardboard-like. Based on quantitative analysis and organoleptic evaluations, Widder and Grosch [251] showed that this odour defect was due to increasing concentrations of (E)- and (Z)-2-nonenal in butter oil during storage. It is worth noting that after 90 days of storage at 35°C, the cardboard odour was already perceptible, although the odour activity values (OAV: ratio of concentration to odour threshold in oil) of (E)- and (Z)-2-nonenal were found to be 0.7 and 0.4. The authors could recombine the off-odour by adding a mixture of the two nonenals at a concentration of 0.5x of their odour thresholds (in butter oil) to fresh butter oil. This is a good example of the additive or synergistic effect of two different aroma compounds and also underlines the need for a systematic evaluation of off-flavours, e.g. according to the approach which was recently described by Maarse and Grosch [265]. Precursor studies revealed that (E)- and (Z)-2-nonenal were formed in butter oil by autoxidation of palmitoleic acid [266].

In milk, off-flavours which are due to autoxidation of milk fat, were characterised as oxidised, metallic and grassy [252]. The authors identified 1-octen-3-one and (E,Z)-2,6-nonadienal as the major cause of the metallic and the grassy taint in milk. Both 1-octen-3-one and (Z)-1,5-octadien-3-one were reported to be responsible for the metallic off-odour of butterfat [253]. A beany off-odour was detected in spray-dried milk, which was prepared during hot weather. This taint was attributed to the formation of (E)-6-

nonenal [254], which is suggested to be formed by ozonolysis of 8,15- and 9,15-isolinoleic acid [259].

The typical mild, buttery aroma of fresh sour-cream buttermilk is considerably more unstable than the aroma of fermented (sweet-cream) buttermilk. This is due to different processing conditions. Sour-cream buttermilk is prepared by fermenting cream of about 40% fat, whereas fermented buttermilk contains only about 1% fat at the start of fermentation. Although sour-cream buttermilk is preferred by the consumer, a metallic off-flavour develops already after 4 days at 8°C [181,267]. Quantitative instrumental and sensory studies revealed that (E,Z)-2,6-nonadienal was the key contributor to the metallic flavour defect of sour-cream buttermilk [256]. In a second study, Heiler and Schieberle [268] found that α-linolenic acid is enzymatically oxidised by lipoxygenases present in the starter cultures of buttermilk to its 9-hydroperoxide, which upon cleavage under the acidic conditions of buttermilk yields (E,Z)-2,6-nonadienal. The aldehyde is finally converted into the potent (E,Z)-2,6-nonadienol by the reducing activity of the starter cultures. The odour thresholds of the unsaturated alcohol were found to be quite low (0.2 ppt in water and 1.3 ppt in buttermilk) [256].

The so-called reversion flavour of soybean oil can be characterised by beany, grassy and hay-like off-notes [269]. According to the author, the flavour defect develops quite rapidly when soybean oil is exposed to daylight, although the concentration of peroxides is still low. Guth and Grosch [200] showed that 3-methyl-2,4-nonandione was mainly responsible for the light-induced off-flavour of soybean-oil. In a further study, they identified two furanoid fatty acids, 10,13-epoxy-11,12-dimethyloctadeca-10,12-dienoic acid and 12,15-epoxy-13,14-dimethyleicosa-12,14-dienoic acid, as precursors of this potent off-flavour compound [25]. When soybean oil was exposed to sunlight at room temperature, the two precursors were completely degraded within two days, whereas linoleic and linolenic acid were not affected [270]. 3-Methyl-2,4-nonandione was also found to be responsible for hay-like off-flavours of dry parsley [257] and dry spinach [258].

Storage of beer results in a complex pattern of staling. Meilgaard [261], for example, characterised ageing beers as increasing astringency combined with a transient development of a papery/cardboard character. Such complex aroma changes imply that not a single, but several changes in aroma compounds contribute to staling. However, oxygen is considered by many brewers as the most important factor responsible for beer staling. Hashimoto [271] reported the contribution of volatile aldehydes to the oxidised flavour of beer, as was shown by the observation that the addition of carbonyl reagents, e.g. 2,4-dinitrophenylhydrazine, metabisulfite and hydroxylamines to aged beer restored the fresh flavour. The cardboard flavour of beer has later been attributed to (E)-2-nonenal [260,261]. (E)-2-nonenal is also reported to contribute to the stale character of beer in concentrations below its odour threshold of 0.3-0.5 ppb by acting additively or synergistically [261].

The oxidation of carotenoids can also lead to the formation of off-flavours, e.g. the floral aroma defect in carrots. Ayres *et al.* [262] identified α-and β-ionone as well as β-ionone-5,6-epoxide as major contributors to the flavour defect in stored carrots. Another example of ketones which act as off-flavour compounds are methyl ketones, which are formed from fatty acids with less than 15 C-atoms by various fungi (see Figure 15). They contribute to the so-called 'perfume rancidity' [272] of mould-ripened cheeses as well as of coconut and coconut oil [264].

4.3. ODOUR THRESHOLDS OF LIPID DERIVED AROMA COMPOUNDS

Table 4 lists odour (nasal perception) and taste (retronasal perception) threshold values (in oil and water) of important lipid derived compounds of foodstuffs. The odour thresholds are strongly dependent on the nature of the solvent (oil/water), which is mainly due to differences in solubility, as reflected by the Henry constants k_h ($k_h = p/N$; p = partial pressure of the compound above the solution, N = mole fraction of the compound in solution) of a substance in water and oil [273]. With the exception of butyric and hexanoic acid, all compounds have considerably lower threshold values in water than in oil. This is due to the better solubility of the odorants in oil, because most are rather hydrophobic substances. This applies, above all, to aldehydes and ketones with long carbon chains and/or high molecular weight.

Odour and taste thresholds can vary considerably, depending on the group of assessors. This makes it difficult to compare some of the values given in Table 4. With a few exceptions, the taste thresholds (in oil and water) of the lipid derived aroma compounds are significantly lower than their odour thresholds. With regard to aldehydes, cis-configurated compounds have considerably lower threshold values than trans-configured and saturated substances. Very low threshold values were found for e.g. (Z)-3-hexenal, (Z)-4-heptenal, (Z)-2-nonenal, (Z,Z)-3,6-nonadienal, (E,Z)-2,6-nonadienal, and (E,Z)-2,4-decadienal. The introduction of an epoxy-group also decreases the flavour threshold, as shown by the lower values of 4,5-epoxy-(E)-2-decenal as compared to (E,E)-2,4-decadienal. The differences in the threshold values of alkanals, (E)-2-alkenals and (E,E)-2,4-alkadienals, however, is relatively small.

The odour and taste thresholds of vinyl ketones, some norisoprenoid compounds (β-damascenone, α- and β-ionone) and 3-methyl-2,4-nonandione are in the range of the most potent aldehydes or even below. According to Table 4, (Z)-1,5-octadien-3-one is the most potent lipid derived aroma compound. Alcohols generally have significantly higher threshold values than the corresponding aldehydes or ketones.

The threshold values of acids are relatively high and depend on the pH of the media, since only the undissociated molecule is volatile and can cause an odour [274]. Esters and lactones show considerably lower threshold values than acids, above all in water. Ethyl butanoate and ethyl hexanoate have the lowest thresholds within these two groups.

Table 4. Odour thresholds (in oil and water) of important lipid derived compounds

Compound	Threshold Value [μg/kg] [a]			
	In oil		In water	
	Odour	Taste	Odour	Taste
I. aldehydes				
propanal	3600 [275]	1600 [275]	23 [276]	7 [162]
hexanal	300 [200]	73 [200]	4.5 [277]	10.5 [248]
(E)-2-hexenal	424 [212]	257 [212]	50 [278]	
(Z)-3-hexenal	13.5 [200]	2.8[200]	0.25 [186]	0.03 [248]
(Z)-4-heptenal	10 [279]	1.6[279]	0.2 [280]	0.06 [248]
octanal	56 [212]	56 [212]	0.7 [281]	5 [276]
(E)-2-octenal	98 [164]	61 [164]	4.0 [282]	
nonanal	1000 [277]	200 [250]	1.0 [277]	
(E)-2-nonenal	900 [200]	66 [200]	0.25 [162]	0.08 [248]
(Z)-2-nonenal	4.5 [200]	0.6 [200]		
(Z)-3-nonenal	250 [279]	35 [279]		
(E,E)-2,4-nonadienal	2500 [275]	40 [164]	0.1 [162]	0.06 [248]
(Z,Z)-3,6-nonadienal			0.2 [162]	0.05 [248]
(E,Z)-2,6-nonadienal	3.8 [200]	1.4 [200]	0.01 [162]	0.02 [248]
(E)-2-decenal	33800 [275]	150 [250]	0.35 [276]	230 [283]
(E,E)-2,4-decadienal	180 [200]	41 [200]	0.2 [169]	0.05 [217]
(E,Z)-2,4-decadienal	4 [213]	4 [213]	0.04 [167]	
4,5-epoxy-(E)-2-decenal	1.3 [200]	3 [200]	0.12 [246]	0.015[284]
12-methyltridecanal			0.1 [217]	
II. ketones				
2-heptanone	1500 [171]	1500 [171]	29 [285]	
2-nonanone	103 [276]			
2-undecanone	450 [286]			
1-octen-3-one	10 [200]	0.3 [200]	0.05 [162]	0.01 [248]
(Z)-1,5-octadien-3-one	0.45 [200]	0.03 [200]	0.0012 [253]	0.0004 [248]
3-methylnonan-2,4-dione	22.5 [200]	1.5 [200]	0.03 [257]	0.02 [287]
(E)-β-damascenone	11 [212]	3.7 [212]	0.002 [186]	0.001 [287]
α-ionone			0.4 [288]	
β-ionone			0.007 [186]	
III. alcohols				
(Z)-3-hexenol	1100 [212]	364 [212]	38.9 [214]	30 [292]
2-heptanol			410 [286]	
1-octen-3-ol			1 [281]	
IV. acids				
butyric acid	135 [203]		1000 [169]	6200 [289]
hexanoic acid	5400 [203]		5000 [278]	5400 [290]
V. esters				
(Z)-3-hexenyl acetate	200 [212]	750 [160]	7.8 [214]	12.1 [214]
ethyl butanoate	28 [171]	3.5 [171]	1 [182]	0.1 [292]
ethyl hexanoate	40 [171]	20 [171]	5 [182]	0.5 [292]
VI. lactones				
δ-octalactone			90 [286]	

γ-decalactone	320 [213]	385 [213]	5 [278]	88 [280]
δ-decalactone	400 [171]	1600 [171]	100 [291]	125 [276]
(Z)-6-dodecen-γ-lactone	250 [203]			

[a]: References are given in brackets.

5. References.

1. Criegee R., Pilz H., Flygare H. (1939) *Chem. Ber.*, **72**, 1799.
2. Sheldon R.A., Kochi J.K. (1981) *Metal-Catalysed Oxidations of Organic Compounds* Academic Press, New York, pp 18.
3. Frankel E.N. (1980) *Progr. Lipid Res.*, **19**, 1.
4. Frankel E.N. (1982) *Progr. Lipid Res.*, **22**, 1.
5. Frankel E.N. (1991) *J. Sci. Food Agric.*, **54**, 495.
6. Porter N.A. (1992) In *Organic Peroxides* (Ando W., ed.) Wiley, Chichester, pp 101.
7. Korycka-Dahl M.B., Richardson T. (1978) *CRC Crit. Rev. Food Sci. Nutr.*, **12**, 209.
8. Niki E. (1993) In *Atmospheric Oxidation and Antioxidants, vol. 3* (Scott G., ed.) Elsevier, Amsterdam, pp 1.
9. Grosch W. (1975) *Z. Lebensm. Unters. Forsch.*, **157**, 70.
10. Porter N.A., Wolf R.A., Yarbro E.M., Weenen H. (1979) *Biochem. Biophys. Res. Comm.*, **89**, 1058.
11. Porter N.A., Weber B.A., Weenen H., Khan J.A. (1980) *J. Am. Chem. Soc.*, **102**, 5597.
12. Weenen H., Porter N.A. (1982) *J. Am. Chem. Soc.*, **104**, 5216.
13. Grosch W. (1982) In *Food Flavours, Part A. Introduction* (Morton I.D., MacLeod A.J., eds.) Elsevier, Amsterdam, pp 325.
14. Grosch W. (1987) In *Autoxidation of Unsaturated Lipids* (Chan H.W.S., ed.) Academic Press, London, pp 95.
15. Russell G.A. (1957) *J. Am. Chem. Soc.*, **79**, 3871.
16. Coxon D.T., Price K.R., Chan H.W.S. (1981) *Chem. Phys. Lipids*, **28**, 365.
17. Labuza T.P., Tsuyuki H., Karel M. (1969) *J. Am. Oil Chem. Soc.*, **46**, 409.
18. Neff W.E., Frankel E.N., Miyashita K. (1990) *Lipids*, **25**, 33.
19. Frankel E.N., Neff W.E., Miyashita K. (1990) *Lipids*, **25**, 40.
20. Chan H.W.S. (1977) *J. Am. Oil Chem. Soc.*, **54**, 100.
21. Min D.B., Lee S.H., Lee E.C. (1989) *In Flavour Chemistry of Lipid Foods* (Min D.B., Smouse T.H., eds.), AOCS, Champaign, IL, pp 57.
22. Ganmyo K., Kawai N., Nakamura T. (1996) *European Patent* EP 0 766 925 A2, 1996, Snow Brand Milk Products Co., Ltd.
23. Terao J., Matsushita S. (1977) *J. Am. Oil Chem. Soc.*, **54**, 234.
24. Hannemann K., Puchta V., Simon E., Ziegler H., Ziegler G., Spiteller G. (1989) *Lipids*, **24**, 296.
25. Guth H., Grosch W. (1991) *Fat Sci. Technol.*, **93**, 249.
26. Guth H., Grosch W. (1992) *Z. Lebensm. Unters. Forsch.*, **194**, 360.
27. Ohloff G. (1973) In *Fette als funktionelle Bestandteile von Lebensmitteln* (Solms J., ed.) Forster Verlag, Zuerich, pp 119.
28. Kimoto W.I., Gaddis A.M. (1974) *J. Am. Oil Chem. Soc.*, **51**, 307.
29. Gardner H.W., Plattner R.D. (1984) *Lipids*, **19**, 294.
30. Frankel E.N., Neff W.E., Selke E. (1984) *Lipids*, **19**, 790.
31. Muizebelt W.J., Hubert J.C., Venderbosch R.A.M. (1994) *Progr. Org. Coat.*, **24**, 263.
32. Muizebelt W.J., Nielen M.W.F. (1996) *J. Mass Spectrom.*, **31**, 545.
33. Hubert J.C., Venderbosch R.A.M., Muizebelt W.J., Klaasen R.P., Zabel, K.H. (1997) *Progr. Org. Coat.*, **31**, 331.
34. Frankel E.N. (1985) *Progr. Lipid Res.*, **23**, 197.
35. Schieberle P., Grosch W. (1981) *J. Am. Oil Chem. Soc.*, **58**, 602.
36. Lillard D.A., Day E.A. (1964) *J. Am. Oil Chem. Soc.*, **41**: 549.
37. Swoboda P.A.T., Lea C.H. (1965) *J. Sci. Food Agric.*, **16**, 680.
38. Matthews R.F., Scanlan R.A., Libbey J.M. (1971) *J. Am. Oil Chem. Soc.*, **48**, 745.

39. Josephson D.B., Lindsay R.C. (1987) *J. Food Sci.*, **52**, 1186,1218.
40. Terao J., Ogawa T., Matsushita S. (1975) *Agric. Biol. Chem.*, **39**, 397.
41. Grosch W. (1976) *Z. Lebensm. Unters. Forsch.*, **160**, 371.
42. Grosch W., Laskawy G., Fischer K.-H. (1977) *Z. Lebensm. Unters. Forsch.*, **163**, 203.
43. Frankel E.N., Neff W.E., Selke E. (1981) *Lipids*, **16**, 279.
44. Frankel E.N., Neff W.E., Selke E., Brooks D.D. (1987) *Lipids* 1, **22**, 322.
45. Swoboda P.A.T., Peers K.E. (1978) *J. Sci. Food Agric.*, **29**, 803.
46. Gassenmeier K., Schieberle P. (1994) *J. Am. Oil Chem. Soc.*, **71**, 1315.
47. Neff W.E., Frankel, E.N., Selke E., Weisleder D. (1983) *Lipids*, **18**, 868.
48. Frankel E.N., Neff W.E., Selke E., Weisleder D. (1982) *Lipids*, **17**, 11.
49. Schutte L. (accepted for publication) In *Flavor Chemistry: 30 Years of Progress* (Teranishi R., Wick E.L., Hornstein I., eds.) Plenum, New York.
50. Haring P.G.M. (1988) *European Patent* EP 0 298 552 B1, Unilever NV.
51. Haring P.G.M. (1989) *European Patent* EP 0 377 239 A1, Unilever NV.
52. Huang H.T., Dooley J.G. (1976) *Biotech. Bioengineering*, **18**, 909.
53. Doornbos T.A., van der Heijden A., van der Kamp J.W. (1986) *European Patent* EP 173,401, Unilever BV.
54. Izzo H. (1986) *US Patent* 4,684,532.
55. Veringa H.A., van den Berg G., Stadhouders J. (1976) *Milchwissenschaft*, **31**, 658.
56. Urbach G., Gordon M.H. (1994) In *Fats in Food Products* Blackie Academic & Professional, Glasgow, pp 347.
57. Hulsker F.H., de Jong K., Turksma H. (1991) *US Patent* 5,130,157.
58. Simmons J., Visser J. (1990), *European Patent* EP 0 463 660 A1, Quest International B.V.
59. Chen T.-K. (1989) *European Patent* EP 0 295 509 A1, Nestle S.A.
60. Enzell C. (1985) *Pure Appl. Chem.*, **57**, 693.
61. Heij J.T. de, Dort J.M. van (1992) In *Progress in Flavour Precursor Studies* (Schreier P., ed.) Allured Publishing, Carol Stream IL, pp 409.
62. Dort J.M. van, Heij J.T. de, Renes H. (1992) *US patent* 5,084,292, Unilever BF.
63. Wilson L. (1996) In *Lipoxygenase and Lipoxygenase Pathway Enzymes* (Piazza G.J., ed.) AOCS Press, Champaign, Illinois, pp 209.
64. Veldink G.A., Vliegenthart J.F.G., Boldingh J. (1977) *Prog. Chem. Fats other Lipids*, **15**, 131.
65. Gardner H.W. (1991) *Biochim. Biophys. Acta*, **1084**, 221.
66. Gardner H.W. (1996) *J. Am. Oil Chem. Soc.*, **73**, 1347.
67. Vick B.A. (1993) In *Lipid Metabolism in Plants* (Moore Jr. T.S., ed) CRC Press, Boca Raton, pp 167.
68. Hatanaka A. (1996) *Food Rev. Int.*, **12**, 303.
69. Prigge S.T., Boyington J.C., Gaffney B.J., Amzel L.M. (1996) In *Lipoxygenase and Lipoxygenase Pathway Enzymes* (Piazza G.J., ed.) AOCS Press, Champaign, Illinois, pp 1.
70. Shibata D. (1996) In *Lipoxygenase and Lipoxygenase Pathway Enzymes* (Piazza G.J., ed.) AOCS Press, Champaign, Illinois, pp 39.
71. Groot J.J.M.C. de, Veldink G.A.; Vliegenthart J.F.G., Boldingh J., Wever R., van Gelder B.F. (1975) *Biochim. Biophys. Acta*, **377**, 71.
72. Petersson L., Slappeudel S., Feiters M.C., Vliegenthart J.F.G. (1987) *Biochim. Biophys. Acta*, **913**, 228.
73. Nelson M.J. (1996) In *Lipoxygenase and Lipoxygenase Pathway Enzymes* (Piazza G.J., ed.), AOCS Press, Champaign, Illinois, pp 80.
74. Corey E.J., Nagata R. (1987) *J. Am. Chem. Soc.*, **109**, 8107.
75. Corey E.J. (1987) *Pure Appl. Chem.*, **59**, 269.
76. Chen A.O., Whitaker J.R. (1986) *J. Agric. Food Chem.*, **34**, 203.
77. Wurzenberger M., Grosch W. (1983) *Z. Lebensm. Unters. Forsch.*, **176**, 16.
78. Wurzenberger M., Grosch W. (1986) *Lipids*, **21**, 261.
79. Mau J.-L., Beelman B., Ziegler G.R. (1992) *J. Food Science*, **57**, 704.
80. Hamberg M. (1971) *Anal. Biochem.*, **43**, 515.
81. Axelrod B., Cheesbrough T.M., Laalso S. (1981) *Methods Enzymol.*, **71**, 441.
82. Theorell H., Holman R.T., Akeson A. (1947) *Acta Chem. Scand.*, **1**, 571.
83. Christopher J.P., Pistorius E.K, Axelrod B. (1970) *Biochim. Biophys. Acta*, **198**, 12.
84. Christopher J.P., Pistorius E.K, Axelrod B. (1972) *Biochim. Biophys. Acta*, **284**, 54.
85. Os C.P.A. van, Vente M., Vliegenthart J.F.G. (1979) *Biochim. Biophys. Acta*, **574**, 103.

86. Andre J.C.; Funk M.O.(1986) *Anal. Biochem.*, **158**, 316.
87. Gardner H.W. (1989) *Biochim. Biophys. Acta*, **1001**, 274.
88. Gardner H.W., Phillips D.R. (1976) *Biochim. Biophys. Acta*, **431**, 278.
89. Yamamoto A., Fujii Y., Yasumoto K., Mitsuda H. (1980) *Agric. Biol. Chem.*, **44**, 443.
90. Matthew J.A., Chan H.W.S., Galliard T. (1977) *Lipids*, **12**, 324.
91. Kuehn H., Wiesner R., Lankin V.Z., Nekrasov A., Alder L., Schewe T. (1987) *Anal.Biochem.*, **160**, 24.
92. Chang C.C., Esselman W.J., Clagett C.O. (1971) *Lipids*, **6**, 1971.
93. Galliard T., Phillips D.R. (1971) *Biochem. J.*, **124**, 431.
94. Nikolaev V., Reddanna P., Whelan J., Hildenbrandt G., Reddy C.C. (1990) *Biochem. Biophys. Res. Commun.*, **170**, 491.
95. Eriksson C.E.; Svensson S.G. (1970) *Biochim. Biophys. Acta*, **198**, 449.
96. Arens D., Seilmeier W., Weber F., Kloos G., Grosch W. (1973) *Biochim. Biophys. Acta*, **327**, 295.
97. St.Angelo A.J., Dupuy H.P., Ory R.L. (1972) *Lipids*, **7**, 793.
98. Kajiwara T., Nagata N., Hatanaka A., Naoshima Y. (1980) *Agric. Biol. Chem.*, **44**, 437.
99. Sekiya S., Numa S., Kajiwara T., Hatanaka A. (1976) *Agric, Biol. Chem.*, **40**, 185.
100. Vick B.A., Zimmerman D.C. (1981) *Plant. Physiol.* **67**, 92.
101. Zimmerman D.C., Vick,B.A. (1970) *Lipids*, 392.
102. Heimann W., Dresen P., Klaiber V. (1973) *Z. Lebensm. Unters. Forsch.*, **153**, 1.
103. Aarle P.G.M. van, Barse M.M.J. de, Veldink G.A., Vliegenthart J.F.G. (1991) FEBS Lett.,**280**, 159.
104. Yabuuchi S., Amaha M. (1975) *Phytochemistry*, **14**, 2569.
105. Yabuuchi S. (1976) *Agr. Biol. Chem.*, **40**, 1987.
106. Ida S., Masaki Y., Morita Y. (1983) *Agric. Biol. Chem.*, **47**, 637.
107. Ohta H., Ida S., Mikami B., Morita Y. (1986) *Agric. Biol. Chem.*, **50**, 3165.
108. Ohta H., Ida, S., Mikami B., Morita Y. (1986) *Plant Cell Physiol.*, **27**, 911.
109. Grosch W. Z. (1977) *Lebensm. Unters. Forsch.*,**163**, 4.
110. Cayrel A., Crouzet J., Chan H.W.S., Price K.R. (1983) *Am. J. Enol. Vitic.*, **34**, 77.
111. Kim I.-S., Grosch W. (1978) *Z. Lebensm. Unters. Forsch.*, **167**, 324.
112. Shibata D., Steczko J., Dixon J.E., Hermodson M., Yazdanparast R., Axelrod B. (1987) *J. Biol. Chem.*, **262**, 10080.
113. Boyington J.C., Gaffney B.J., Amzel L.M. (1993) *Science*, **260**, 1482.
114. Iacazio G., Langrand G., Baratti J., Buono G., Triantaphylides C. (1990) *J. Org. Chem.*, **55**, 1690.
115. Emken E.A. (1973) *US patent* 3729379.
116. Drouet P., Thomas D., Legoy M.D. (1994) *Tetrahedron Lett.*, **35**, 3923.
117. Piazza G.J. (1992) *Biotechnol. Lett.*, **14**, 1153.
118. Battu S., Cook-Moreau J., Beneytout J.L. (1994) *Biochim. Biophys. Acta*, **1211**, 270.
119. Butovich I.A., Mogilevich T.V., Ogiy S.A., Kutniaya M.U., Kukhar V.P. (1994) *Bioorg. Khim.*, **20**, 1095.
120. Parra-Diaz D., Brower D.P., Medina, M.B., Piazza G.J. (1993) *Biotechnol. Appl. Biochem.*, **18**, 359.
121. Piazza G.J., Brower D.P., Parra-Diaz D. (1994) *Biotechnol. Appl. Biochem.*, **19**, 243.
122. Elshof M.B.W., Janssen M., Veldink G.A., Vliegenthart J.F.G. (1996) *Recl. Trav. Chim. Pays-Bas*, **115**, 499.
123. Muller B.L., Dean C., Whitehead I.M. (1995) In *Bioflavour 95* (Etievant P., Schreier P., eds.), INRA, Paris, pp 339.
124. Whitehead I.M., Muller B.L., Dean C. (1995) *Cereal Foods World*, **40**, 193.
125. Corey E.J., Albright J.O., Barton A.E., Hashimoto S. (1980) *J. Am. Chem. Soc.*, **102**, 1435.
126. Martini D., Buono G., Montillet J.-L., Iacazio G. (1996) *Tetrahedron Asymmetry*, **7**, 1489.
127. Sekiya J., Kajiwara T., Hatanaka A. (1979) *Agric. Biol. Chem.*, **43**, 969.
128. Matthew A.J., Galliard T. (1978) *Phytochemistry*, **17**, 1043.
129. Shibata Y., Matsui K., Kajiwara T., Hatanaka A. (1995) *Plant Cell Physiol.*, **36**, 147.
130. Vick B.A., Zimmerman D.C. (1976) *Plant Physiol.*, **57**, 780.
131. Hatanaka A., Kajiwara T., Koda T. (1979) *Agric. Biol. Chem.*, **43**, 2115.
132. Hatanaka A., Kajiwara T., Sekiya J., Inouye S. (1982) *Phytochemistry*, **21**, 13.
133. Sekija J., Tanigawa S., Kajiwara T., Hatanaka A. (1984) *Phytochemistry*, **23**, 2439.
134. Kim I.-S., Grosch W. (1981) *J. Agric. Food Chem.*, **29**, 1220.
135. Kajiwara T., Sekiya J., Odake Y., Hatanaka A. (1977) *Agric. Biol. Chem.*, **41**, 1481.
136. Galliard T., Phillips D.R., Reynolds J. (1976) *Biochim. Biophys. Acta*, **441**, 181.
137. Galliard T., Phillips D.R. (1976) *Biochim. Biophys. Acta*, **431**, 278.

138. Matsui K., Shibata Y., Kajiwara T., Hatanaka A. (1989) *Z. Naturforsch.*, **44c**, 883.
139. Matoba T., Hidaka H., Narita H., Kitamura K., Kaizuma N., Kito M. (1985) *J. Agric.Food Chem.*, **33**, 852.
140. Matsui K., Toyota H., Kajiwara T., Hatanaka A. (1991) *Phytochemistry*, **30**, 2109.
141. Matsui K., Kajiwara T., Hatanaka A. (1992) *J. Agric. Food Chem.*, **40**, 175.
142. Shibata Y., Matsui K., Kajiwara T., Hatanaka A. (1995) *Biochem. Biophys. Res. Commun.*, **207**, 438.
143. Fauconnier M.-L., Perez A.G., Sanz C.; Marlier M. (1997) *J. Agric. Food Chem.*, **45**, 4232.
144. Whitehead I.M. (1998) *Food Technology*, **52**, 40.
145. Pascal B. (1989) *French patent* 2 652 587, Pernod Ricard S.A.
146. Muller B.L., Gautier A.E., Dean C., Kuhn J.-C. (1993) *Intl. Patent Applic.* WO93/24644, Firmenich S.A.
147. Muheim A., Haeusler A., Schilling B., Lerch K. (1997) In *Flavours and Fragrances* (Swift K.A.D., ed.) The Royal Society of Chemistry, Cambridge, pp 11.
148. Kibler L.A., Kratky Z., Tandy J.S. (1988) *U.S. Patent* 5114734, Nestec S.A.
149. Schindler, F. (1989) *US patent* 4,810,504, Huels AG.
150. Hadar Y., Dosoretz C.G. (1991) *Trends in Food Science & Technology*, 215.
151. Belinky P.A., Masaphy S., Levanon D., Hadar Y., Dosoretz C.G. (1994) *Appl. Microbiol. Biotechnol.*, **40**, 629.
152. Ropert F., Dumont B., Belin J.M. (1995) In *Bioflavour 95* (Etievant P., Schreier P., eds.) INRA, Paris, pp 275.
153. Muller B.L., Dean C., Schmidt C., Kuhn J.-C. (1997) *Intl. Patent Applic.* WO97/22575, Firmenich S.A.
154. Markus P.A., Peters A.L.J., Roos, R. (1991) *European Patent* 542284, Quest International B.V.
155. Maarse H., Visscher C.A. (1991) In *Volatile Compounds in Food. Qualitative nad Quantitative Data. Supplement 2*, TNO-CIVO Food Analysis Institute, Zeist, The Netherlands.
156. Litman I., Numrych S. (1978) *ACS Symp. Series*, **75**, 17.
157. Milo C., Grosch W. (1995) *J. Agric. Food Chem.*, **43**, 459.
158. Milo C., Grosch W. (1996) *J. Agric. Food Chem.*, **44**, 2366.
159. Buttery R.G., Teranishi R., Ling L.C., Flath R.A., Stern D.J. (1988) *J. Agric. Food Chem.*, **36**, 1247.
160. Guth H., Grosch W. (1993) *J. Am. Oil Chem. Soc.*, **70(5)**, 513.
161. Kubickova J., Grosch W. (1997) *Int. Dairy Journal*, **7**, 65.
162. Milo C. (1995) *PhD Thesis*, Technical University of Munich.
163. Schieberle P., Hofmann T. (1997) *J. Agric. Food Chem.*, **45**, 227.
164. Widder S. (1994) *PhD Thesis*, Technical University of Munich.
165. Kerler J., Grosch W. (1997) *Z. Lebensm. Unters. Forsch.*, **205**, 232.
166. Gasser U., Grosch W. (1990) *Z. Lebensm. Unters. Forsch.*, **190**, 3.
167. Kerler J. (1996) *PhD Thesis*, Technical University of Munich.
168. Kerscher R., Grosch W. (1997) *Z. Lebensm. Unters. Forsch.*, **204**, 3.
169. Guth H., Grosch W. (1994) *J. Agric. Food Chem.*, **42**, 2862.
170. Milo C., Reineccius G.A. (1997) *J. Agric. Food Chem.*, **45**, 3590.
171. Preininger M., Grosch W. (1994) *Lebensm. Wiss. Technol.*, **27**, 237.
172. Ott A., Fay L.B., Chaintreau A. (1997) *J. Agric. Food Chem.*, **45**, 850.
173. Roberts D.D., Acree T.E. (1995) In *Fruit Flavors, Biogenesis, Characterization, and Authentication* (Rouseff R.L., and Leahy M.M., eds.) ACS Sym. Ser. 596, Washington, DC, 190.
174. Acree T.E., McLellan M.R. (1994) In *Juice Technology Workshop, Cornell University, October* 1993, New York State Agricultural Experiment Station, Downing D.L., 38.
175. Schnermann P., Schieberle P. (1997) *J. Agric. Food Chem.*, **45**, 867.
176. Wagner R., Grosch W. (1997) *Lebensm. Wiss. Technol.*, **30**, 164.
177. Dimick P.S., Walker N.J., Patton S. (1969) *J. Agric. Food Chem.*, **17**, 649.
178. De Frutos M., Sonz J., Martinez-Castro I. (1991) *J. Agric. Food Chem.*, **39**, 524.
179. Ha J.K., Lindsay R.C. (1991) *J. Dairy Sci.*, **74**, 3267.
180. Guth H., Grosch W. (1993) *Flavour Fragrance J.*, **8**, 173.
181. Heiler C., Schieberle P. (1996) *Lebensm. Wiss. Technol.*, **29**, 460.
182. Schieberle P. (1991) *Z. Lebensm. Unters. Forsch.*, **193**, 558.
183. Forss D.A., Dunstone E.H., Ramshaw E.H., Stark W. (1962) *J. Food Sci.*, **27**, 90.
184. Schieberle P. (1996) *Food Chemistry*, **55(2)**, 145.
185. Schieberle P. (1991) *J. Agric. Food Chem.*, **39**, 1141.
186. Buttery R.G., Teranishi R., Ling L.C., Turnbaugh J.G. (1990) *J. Agric. Food Chem.*, **38**, 336.

187. Linko R.R., Kallio H., Pyysalo T., Rainio K. (1978) *Z. Lebensm. Unters. Forsch.*, **166**, 208.
188. Vereshchagin P.V., Bezzubov A.A. (1981) *Appl. Biochem. Microbiol.*, **17**, 53.
189. Kawakami M., Yamanishi T. (1983) *Agric. Biol. Chem.*, **47**, 2077.
190. Guichard E., Souty M. (1988) *Z. Lebensm. Unters. Forsch.*, **186**, 301.
191. Berger R.G. (1984) *J. Sci. Food Agric.*, **35**, 1318.
192. Kim Y.-H., Kwag J.-J., Kwon Y.-J., Yang K.-K. (1990) *Korean J. Food Sci. Technol.*, **22**, 543.
193. Kakiuchi et al. (1991) *J. Jpn. Soc. Cold. Preserv. Food*, **17**, 14.
194. Schreier P., Drawert F., Junker A., Mick W. (1977) *Z. Lebensm. Unters. Forsch.*, **164**, 188.
195. Moshonas M.G., Shaw P.E. (1989) *J. Agric. Food Chem.*, **37**, 157.
196. Jayalekshmy A., Narayanan C.S., Mathew A.G. (1991) *J. Am. Oil Chem. Soc.*, **68**, 873.
197. Tressl R., Bahri D., Engel K.-H. (1982) *J. Agric. Food Chem.*, **30**, 89.
198. Tressl R., Drawert F., Heimann W. (1970) *Z. Lebensm. Unters. Forsch.*, **142(4)**, 249.
199. Winter M., Enggist P. (1971) *Helv. Chim. Acta*, **54**, 1891.
200. Guth H., Grosch W. (1990) *Lebensm. Wiss. Technol.* **23**, 513.
201. Berger R.G., Drawert F., Kollmannsberger H. (1986) *Z. Lebensm. Unters. Forsch.*, **183**, 169.
202. Semmelroch P., Laskawy G., Blank I., Grosch W. (1995) *Flavour Fragrance J.*, **10**, 1.
203. Schieberle P., Gassenmeier K., Guth H., Sen A., Grosch W. (1993) *Lebensm. Wiss. Technol.*, **26**, 347.
204. Ramarathnam N., Rubin L.J., Diosady L.L. (1991) *J. Agric. Food Chem.*, **39**, 344.
205. Hinterholzer A., Schieberle P. (1998) *Flavour Fragrance J.*, **13**, 49.
206. Gasser U., Grosch W. (1988) *Z. Lebensm. Unters. Forsch.*, **186**, 489.
207. Chrisholm M.G., Guiher L.A., Vonah T.M., Beaumont J.L. (1994) *Am. J. Enol. Vitic.*, **45**, 201.
208. Moio L., Schlich P., Etievant P. (1994) *Sci. Aliments.*, **14**, 601.
209. Guth H. (1997) *J. Agric. Food Chem.* **45**, 3022.
210. Berger R.G. (1995) In *Aroma Biotechnology* (Berger R.G., ed.) Springer Verlag, Berlin, 11.
211. Reiners J., Grosch W. (1997) *Lebensmittelchemie.*, **51(2)**, 44.
212. Reiners J., Grosch W. (1998) *J. Agric. Food Chem.*, **46**, 2754.
213. Wagner R., Grosch W. (1998) *J. Am. Oil Chem. Soc.*, in press.
214. Masanetz C., Grosch W. (1998) *Flavour Fragrance J.*, **13(2)**, 115.
215. Minor L.J., Pearson A.M., Dawson L.E., Schweigert B.S. (1965) *J. Food Sci.*, **30 (4)**, 686.
216. Rothe M., Kirova E., Schischkoff G. (1981) *Nahrung*, **25**, 543.
217. Grosch W, Zeiler-Hilgart G., Cerny C., Guth H. (1993) In *Progress in Flavour Precursor Studies* (Schreier P., Winterhalter P., eds.), Allured Publishing Corp, Carol Stream, 329.
218. Josephson D.B., Lindsay R.C. (1986) In *Biogeneration of Aromas* (Parliment T.H., Croteau R., eds.) ACS Symposium Series 317, 201.
219. Josephson D.B., Lindsay R.C. (1987) *J. Food Sci.*, **52**, 328.
220. Schwartz J.M., Grosch W. (1971) *Lipids*, **6**, 351.
221. Kinsella J.E., Hwang D.H. (1976) *Crit. Rev. Food Sci. Nutr.*, **8**, 191.
222. Franke W., Platzeck A., Eichhorn G. (1961) *Arch. Microbiol.*, **40**, 73.
223. Glass R.L., Krick T.P., Sand D.M., Olson D.L., Thorson R.L. (1977) *Lipids*, **12**, 828.
224. Winterhalter P. (1996) In *Biotechnology for Improved Foods and Flavors* (Takeoka G.R., Teranishi R., Williams P.J., eds.) American Chemical Society, Washington, D.C., 295.
225. Isoe S., Katsumura S., Sakan T. (1973) *Helv Chim. Acta*, **56**, 1514.
226. Williams P.J., Sefton M.A., Francis I.L. (1992) In *Flavor Precursors, Thermal and Enzymatic Conversions* (Teranishi R., Takeoka G.R., Güntert M., eds.) American Chemical Society, Washington, D.C., 74.
227. Ohloff G., Rautenstrauch V., Schulte-Elte K.H. (1973) *Helv. Chim. Acta*, **56**, 1503.
228. Straubinger M., Knapp H., Oka N., Watanabe N., Winterhalter P. (1997) *J. Agric. Food Chem.*, **45**, 4053.
229. Forss D.A. (1973) In *Progress in the Chemistry of Fats and other Lipids* (Holman R.T., ed.) Pergamon Press, Oxford, New York, 177.
230. Maga J.A. (1976) *Crit. Rev, Food Sci. Nutr.*, **9**, 1.
231. Welsh F.W., Murray W.D., Williams R.E. (1989) *Crit. Rev. Biotechnol.*, **9**, 105.
232. Albrecht W., Heidlas J., Schwarz M., Tressl R. (1992) In *Flavor Precursors, Thermal and Enzymatic Conversions* (Teranishi R., Takeoka G.R., and Güntert M., eds.) ACS Symp. Ser. 490, American Chemical Society, Washington, 46.
233. Albrecht W., Tressl R. (1990) *Z. Naturforsch.*, **45**, 207.
234. Haffner T., Tressl R. (1996) *J. Agric. Food Chem.*, **44**, 1218.

235. Kinsella J.E., Patton S., and Dimick P.S. (1967) *J. Am. Oil Chem. Soc.*, **44**, 202.
236. Swenson P.E., Dimick P.S. (1971) *Biochem. J.*, **125**, 1139.
237. Walker N.J., Patton S., Dimick P.S. (1968) *Biochim. Biophys. Acta*, **152**, 445.
238. Igene J.O., Pearson A.M. (1979) *J. Food Sci.*, **44**, 1285.
239. Kerscher R., Grosch W. (1998) *J. Agric. Food Chem.*, **46**, 1954.
240. Tims M.J., Watts B.M. (1958) *Food Technol.*, **12**, 240.
241. Asgar A., Gray J.I., Buckley D.J., Pearson A.M., Booren A.M. (1988) *Food Technol.*, **42(6)**, 102.
242. Igene J.O., King J.A., Pearson A.M., Gray J.I. (1979) *J. Agric. Food Chem.*, **27**, 838.
243. Gray J.I., Pearson A.M. (1991) In *Flavor of Meat and Meat Products* (Shahidi F.,ed.) Blackie Academic & Professional, London, 116.
244. St.Angelo A.J., Vercellotti J.R., Legendre M.G., Vinnett C.H., Kuan J.W. James Jr. C., Dupuy H.P. (1987) *J. Food Sci.*, **52**, 1163.
245. Konopka U.C., Grosch W. (1991) *Z. Lebensm. Unters. Forsch.*, **193**,. 123.
246. Kerler J., Grosch W. (1996) *J. Food Sci.*, **61(6)**, 1271.
247. Konopka U.C., Grosch W. (1995) *Z. Lebensm. Unters. Forsch.*, **201**, 339.
248. Milo C., Grosch W. (1993) *J. Agric. Food Chem.*, **41**, 2076.
249. McGill A.S., Hardy R., Gunstone F.D. (1977) *J. Sci. Food Agric.*, **28**, 200.
250. Badings H.T. (1970) *Ned. Melk Zuiveltijdschr.*, **24**, 147.
251. Widder S., Grosch W. (1994) *Z. Lebensm. Unters. Forsch.*, **198**, 297.
252. Hammond E.G., Hill F.D. (1964) *J. Am. Oil Chem. Soc.*, **41**, 180.
253. Swoboda P.A.T., Peers K.E. (1977) *J. Sci. Food Agr.*, **28**, 1019.
254. Parks O.W., Wong N.P., Allen C.A., Schwartz D.P. (1969) *J. Dairy Sci.*, **51**, 953.
255. Witter L.D. (1961) *J. Dairy Sci.*, **44**, 983.
256. Heiler C., Schieberle P. (1997) *Int. Dairy J.*, **7**, 659.
257. Masanetz C., Grosch W. (1998) *Z. Lebensm. Unters. Forsch.*, **206**, 114.
258. Masanetz C., Guth H., Grosch W. (1998) *Z. Lebensm. Unters. Forsch.*, **206**, 108.
259. Belitz D.-H., Grosch W. (1992) *Lehrbuch der Lebensmittelchemie, 4th edition*, SpringerVerlag, Berlin, 308.
260. Jamieson A.M., van Gheluwe J.E.A. (1970) *Proc. Am. Soc. Brew. Chem.*, **28**, 192.
261. Meilgaard M.C. (1972) *Brew. Dig.*, **47(4)**, 48.
262. Ayres J.E., Fishwick J.J., Land D.G., Swain T. (1964) *Nature*, **203**, 81.
263. Guth H., Grosch W. (1993) *Z. Lebensm. Unters. Forsch.*, **196**, 22.
264. Kellard B., Busfield D.M., Kinderlerer J.L. (1985) *J. Sci. Food Agric.*, **36**, 415.
265. Maarse H., Grosch W. (1996) In *Food Taints and Off-Flavours* (Saxby M.J., ed.) Blackie Academic & Professional, Glasgow, 72.
266. Widder S., Grosch W. (1997) *Nahrung*, **41(1)**, 42.
267. Huβ M., Kessler H.G. (1989) *Deutsche Molkereizeitung*, **49**, 1570.
268. Heiler C., Schieberle P. (1997) *Int. Dairy J.*, **7**, 667.
269. Smouse T.H. (1985) In *Flavor Chemistry of Fats and Oils* (Min D.B., Smouse T.H.,eds.) American Oil Chemists' Society, 85.
270. Guth H., Grosch W. (1997) *J. Am. Oil Chem Soc.*, **74**, 323.
271. Hashimoto N. (1966) *Rep. Res. Lab. Kirin Brewing Co., Yokohama*, **9**, 1.
272. Thaler H., Geist G. (1939) *Biochem Z.*, **302**, 369.
273. Nelson P.E., Hoff J.E. (1968) *J. Food Sci.*, **33**, 479.
274. Day E.A. (1966) *Advan. Chem. Ser.*, **56**, 94.
275. Meijboom P.W. (1964) *J. Am. Oil Chem. Soc.*, **41**, 326.
276. Leffingwell J.C., Leffingwell D. (1991) *Perfum Flavor*, **16**, 2.
277. Guadagni D.G., Buttery R.G., Turnbaugh J.G. (1972) *J. Sci. Food Agric.*, **23**, 1435.
278. Larsen M., Poll L. (1992) *Z. Lebensm. Unters. Forsch.*, **195**, 120.
279. Meijboom P.W., Jongenotter G.A. (1981) *J. Am. Oil Chem. Soc.*, **58**, 680.
280. Ohloff G. (1978) *Prog. Chem. Org. Nat. Prod.*, **35**, 431.
281. Buttery R.G., Turnbaugh J.G., Ling L.C. (1988) *J. Agric. Food Chem.*, **36**, 1006.
282. Eriksson C.E., Lundgren B., Vallentin K. (1976) *Chem. Senses Flavor*, **2**, 3.
283. Boelens M.H., Gemert J. (1987) *Perfum Flavor*, **12**, 31.
284. Konopka U.C. (1993) *PhD Thesis*, Technical University of Munich.
285. Schnabel K.-O., Belitz H.-D., von Ranson C. (1988) *Z. Lebensm. Unters. Forsch.*, **187**, 215.
286. Siek T.J., Albin I.A., Sather L.A., Lindsay R.C. (1971) *J. Dairy Sci.*, **54**, 1.

287. Guth H., Grosch W. (1993) In *Progress in Flavour Precursor Studies* (Schreier P., Winterhalter P., eds.), Allured Publishing Corp, Carol Stream, 401.

288. Teranishi R., Buttery R.G. (1987) In *Flavour Science and Technology* (Martens M., Dalen G.A., Russwurm H., eds.) John Wiley & Sons, Chichester, UK, 513.

289. Siek T.J., Albin I.A., Sather L.A., Lindsay R.C. (1969) *J. Food Sci.*, **34**, 265.

290. Patton S. (1964) *J. Food Sci.*, **29**, 679.

291. Engle K.H., Flath R.A., Buttery R.G., Mon T.R., Ramming D.W., Teranishi R. (1988) *J. Agric. Food Chem.*, **36**, 549.

292. Guth H. (1996) *Habilitationsschrift*, Technical University of Munich.

SAFETY AND LEGISLATION OF FLAVOURINGS

DR F. GRUNDSCHOBER
I.O.F.I / I.F.R.A.
8 Rue Charles-Humbert
Geneva 1205
Switzerland

1. Introduction

The purpose of food legislation is to protect the health of consumers, to prevent fraud and to assure fair competition on the market place. Food flavourings are part of the food supply and have therefore to be regulated adequately. It is however important to note that flavourings are regulated differently than food additives.

There exist several administrative, scientific and legal reasons to explain and to justify this difference :

- The number of food additives in each functional group is small and one additive may be sufficient to obtain the desired effect, whereas several thousand flavouring substances are known and up to hundred different substances may be used to realise a specific flavour.

- Food additives are generally used in many food categories. Few flavouring substances are used in many food categories ; certain flavour notes are very specific and many flavouring substances will only be consumed in a limited number of foods. Also the foods in a given food category will not all have the same flavour.

- The use levels of food additives are relatively high, often over 1000 ppm and most over 100 ppm, whereas the majority of flavouring substances are used at a level of about 10 ppm, with many finding usage at levels lower than 1 ppm.

- The world-wide volume of use for food additives is large, often 100 or sometimes up to 1000 tons/year whereas flavouring substances are used in low volumes, e.g. 80% are used at less than 50 kg/year.

- The consumer is normally not aware of the presence of food additives in the food they are consuming daily, whereas consumers are aware of the presence of flavourings and choose food on the basis of their flavour. Consumers are likely to

K.A.D. Swift (ed.), Current Topics in Flavours and Fragrances, 215–229.

choose differently flavoured foods and this variation prevents them from consuming the same substance every day through a whole lifetime.

- The overuse of a food additive will remain unknown to consumers whereas overdosing of flavourings makes the food unpalatable. Flavouring substances are generally self-limiting in use.

- Food additives rarely occur naturally in foods ; they are added to the food product by the manufacturer, whereas many foods are inherently flavoured. These are sources of flavouring substances that are part of the human diet. Most flavouring substances are consumed mainly as a result of their natural occurrence.

2. Systems of regulation

Food additives are regulated in many countries by positive lists. A positive list is a list of authorised substances, any substance not listed is automatically prohibited. Positive lists are established by governmental agencies and the inclusion of a substance in a positive list implies that the agency accepts responsibility that the substance is indeed safe under the conditions of use. So far, it has not been possible to agree on complete positive lists of flavouring substances, due to the large number of substances involved. In addition, it was considered unnecessary to authorise the use of flavouring substances which are already present naturally in foods.

Several systems of flavour regulation can be distinguished. A systematic analysis of flavour regulations leads to the following:

- Premarket approval (in Poland, Czech Republic, Russia). There exists no published detailed regulation and the notifier of a flavouring may risk approval being refused, creating insecurity and distortion of competition. This system should be replaced by detailed regulations.

- Negative list (in Canada, African countries) Substances which may present a health risk are banned, the use of all other substances is free. This system is easy to enforce but it is no longer adequate for industrialised countries.

- GRAS-System (in the USA). A category of substances "generally recognised as safe" by experts qualified by training and experience, is exempted from the food additive definition. The Flavour Extract Manufacturers' Association of the USA – FEMA, has organised a panel of experts and is publishing the results of the GRAS evaluation by this group. The GRAS evaluations of FEMA are recognised by the US FDA and by many national and international authorities. GRAS reflects GMP of the flavour industry.
- Mixed system (in Germany, the Netherlands, Spain, Italy). Natural and nature-identical flavouring substances are exempted from the food additive definition.

They are regulated by a restrictive list, artificial flavouring substances by a positive list.

- Partial positive list (in Japan, Taiwan, Korea). Individual substances and groups of chemically related substances are listed. Members of these groups can be used freely.

- Inventory (in EU Member States after October 1998). All the substances listed in an official inventory or register can be used.

- Positive list (in EU after 2004). Only substances especially authorised can be used. This system has still to be elaborated. During the next years, the EU Register will have to be transformed into the EU positive list. Once a satisfactory list is prepared, it is likely that it will become a world-wide list.

3. Definitions

All flavour regulations are based on definitions. Definitions are important for the classification of flavouring substances and the label declaration. There exist differences between the definitions of the USA and of the EU.

The term "natural flavour" or "natural flavouring" means the essential oil, oleoresin, essence or extractive, protein hydrolysate, distillate, or any product of roasting, heating or enzymolysis, which contains the flavouring constituents derived from a spice, fruit or fruit juice, vegetable or vegetable juice, edible yeast, herb, bark, bud, root, leaf or similar plant material, meat, seafood, poultry, eggs, dairy products, or fermentation products thereof, whose significant function in food is flavouring rather than nutritional (US CFR 21 §101.22(a)(3)). The term "artificial flavour or artificial flavouring" means any substance which is not derived from a natural source.

The definitions of the European Flavour Directive 88/388/EEC (EEC 1988) distinguish natural, nature-identical, artificial flavouring substance, flavouring preparation, process flavouring and smoke flavouring.

Natural flavouring substances (Art 1,2(b)(i)) and preparations (Art 1,2(c)) are obtained by appropriate physical processes (including distillation and solvent extraction) or enzymatic or microbiological processes from material of vegetable or animal origin either in the raw state or after processing for human consumption by traditional food-preparation processes (including drying, torrefaction and fermentation).

A nature-identical flavouring substance (Art 1,2(b)(ii)) is obtained by chemical synthesis or isolated by a chemical processe(s). It is chemically identical to a substance naturally present in material of vegetable or animal origin as described in Art 1,2(b)(i).

An artificial flavouring substance is obtained by chemical synthesis but is not chemically identical to a substance naturally present in material of vegetable or animal origin as described in Art 1,2(b)(i).

Process flavouring means a product which is obtained according to good manufacturing practices by heating to a temperature not exceeding 180°C for a period not exceeding 15 minutes a mixture of ingredients, not necessarily themselves having flavouring properties, of which at least one contains nitrogen (amino) and another is a reducing sugar; (Art 1,2(d)). It has to be noted that a revision of this definition is presently under discussion.

Smoke flavouring means a smoke extract used in traditional foodstuffs smoking processes (Art 1,2(e)).

The Directive does not apply to substances which have exclusively a sweet, sour or salty taste. This means that bitter substances are covered as well as for example acids which also have a flavour value.

A comparison of the US and EU definitions for natural shows that both give weight to the natural source of the material and to the manufacturing processes (which comprises protein hydrolysis, roasting, heating or enzymolysis in the USA and distillation, solvent extraction, enzymatic or microbiological processes, drying, torrefaction and fermentation in Europe). These definitions are very close. There is however a difference between natural flavourings in the USA and Europe. For example most European process flavourings are considered natural in the US and some natural hydrolysates from the USA are not considered natural in Europe. There may also exist differences in the interpretations. More weight is given in the USA to the natural source and in Europe to the manufacturing process.

4. Labelling

4.1. DECLARATION ON THE LABEL OF FLAVOURINGS

Flavourings are industrial products sold to food manufacturers. The label has to show all the information needed by food manufacturers to label their products correctly.

In the USA, the label of flavouring has to show at least:

- The correct designation of the content.
- The net weight or volume.
- A distinguishing batch number.
- The name and addresses of the manufacturer.

Article 9 of the European Flavourings Directive 88/388 (EEC 1998) requires the following information on the label:

- Name and address of the manufacturer or seller established in the EU.
- Sales description "flavouring" or a more specific name.
- The statement for "foodstuffs" or a more specific reference to the foodstuff for which the flavouring is intended.
- A list in descending order of weight of the categories of flavouring substances and preparations present in the flavouring.
- The names and EEC numbers of other substances mixed with the flavouring e.g. additives for the flavouring e.g. carrier solvents, antioxidants, etc.
- An indication of the maximum quantity of each limited substance (see maximum limits of Annex II or of food additives), or appropriate information to comply with such restrictions (recommended dosage).
- An indication identifying the consignment.
- The nominal quantity.

4.2. DECLARATION ON FOOD LABELS

The flavour declaration on the ingredient list on the label of finished foods in the USA uses the terms "natural", "artificial" or "natural and artificial flavouring".

The use of flavourings is also declared together with the name of the food. If the food is expected to contain a characterising ingredient, e.g. strawberries, but does not contain sufficient fruits to characterise the food, the declaration has to be "natural strawberry flavoured shortcake" or "strawberry flavoured shortcake".

If the natural flavour used is not derived from the product whose flavour is simulated, the food should be labelled with the name of the flavour or as "artificially flavoured".

If the food contains both a characterising flavour and other natural flavour, the name of the food shall be followed by the words "with other natural flavour" (WONF).

If the food contains any artificial flavour which simulates, resembles or reinforces the characterising flavour, the name of the food shall be accompanied by the words "artificial" or "artificial flavoured" e.g. "artificial vanilla", "artificially flavoured strawberry" (US CFR 21 §101.22 (i)(1)).

The Flavouringy Directive (88/388/EEC) requires the declaration "flavouring". The word "natural" can only be used if the flavouring part of the flavouring contains exclusively natural flavouring substances and/or preparations.

If the sales description of the flavouring contains a reference to a foodstuff or a flavouring source, the word "natural" may not be used unless the flavouring has been isolated solely or almost solely from the foodstuff or the flavouring source concerned (Art 9,2). The term "artificial" is no longer used in Europe or in other parts of the world. The future of labelling goes in the direction of the term "flavour" or "flavouring"; the adjective "natural" may only be used under restricted conditions.

5. Suitable Flavouring Substances

The inventories of flavour manufacturers contain over a thousand different substances which can be combined to yield hundreds of thousands of different flavourings. The choice of the flavouring substances used depends on the legal requirements of the country where the flavouring is to be used. It is a normal tendency of manufacturers to simplify their stocks and to produce flavourings which can be used in as many countries as possible. Since the ultimate aim of a global list of usable substances has not yet been achieved, flavourings are today compounded either according to the GRAS system or to the mixed system.

5.1. THE GRAS SYSTEM

The statutory definition of the term "food additive" in the US Federal Food, Drug and Cosmetic Act exempts any substance "generally recognised, among experts qualified by scientific training and experience to evaluate its safety, as having been adequately shown through scientific procedures (or, in the case of a substance used in food prior to January 1, 1958, through either scientific procedures or experience based on common use in food) to be safe under the conditions of its intended used". This provision has come to be designated by the acronym "GRAS". It should be emphasised that recognition of safety rests on the opinion of qualified experts, not limited to those in the regulatory agency, and relates specifically to the conditions of use of the substance in food.

Starting in 1960, the Flavour and Extract Manufacturers' Association (FEMA) organised a panel of scientists, not affiliated with the flavour industry but with a special competence in toxicology, pharmacology and biochemistry to evaluate the "GRAS status" of flavouring substances (Hall and Oser 1961). The Food and Drug Administration was kept informed of this activity of FEMA. Subsequent to the release in 1965 of the first GRAS list of some 1.100 flavouring substances and their use levels (Hall and Oser 1965), the FDA temporarily waived bulk labelling requirements and adopted virtually the entire list (21 CFR 121.1163 and 121.1164; now 21 CFR

172.510, 182.10. 182.20, 182.40, 182.50 for natural flavouring substances and 21 CFR 172.515, 182.60 for synthetic flavouring substances).

It may be noted that "scientific procedures" as defined in the original regulations included "not only original animal, analytical, and other scientific studies, but also an unprejudiced compilation of reliable information, both favourable and unfavourable, drawn from the scientific literature (21 CFR 121.1h)". It was later proposed to modify this definition to include "those human, animal, analytical and other scientific studies, whether published or unpublished, appropriate to establish the safety of a substance (21 CFR 170.3)."

It is within the spirit of these definition that the panel has evaluated the safety of flavouring substances. The results of these studies have been reported in a series of papers in Food Technology (Hall and Oser 1965, 1970; Oser and Hall 1972; Oser and Ford 1973a,b, 1974, 1975, 1977, 1978, 1979; Oser, Ford and Bernard 1984; Oser *et al.* 1985; Burdock *et al.* 1990; Smith and Ford 1993, Smith *et al.* 1996, and Newberne *et al.* 1998). The publications GRAS 3 to GRAS 18 cover substances with FEMA numbers 2001 to 3905 and report names, synonyms and average maximum use levels in foods. The panel employed these use levels as a basis for its judgement that the substances are generally recognised as safe for their intended uses. These figures should not be viewed to have the meaning of limits or restrictions (Hall and Oser 1965).

5.2. THE MIXED SYSTEM

Awaiting the completion of a positive list of flavouring substances, this system is used in Germany, the Netherlands, Spain, Italy and other countries. The system is characterised by a restrictive list of active principles of natural flavourings. The list of Annex II of the European Flavour Directive 88/388 is now legally binding in many countries (see Table 1). Even in countries where these restrictions have no force of law they are used in the interest of GMP. The use of nature-identical substances according to GMP is free, whereas artificial flavouring substances can only be used if authorised by a positive list.

The application of this system requires that documented evidence on natural occurrence of flavouring substances is available to distinguish nature-identical from artificial flavouring substances. This information can be found in the scientific literature and is summarised in the review 'Volatile Compounds in Food' published by CIVO-TNO in the Netherlands and also in an inventory prepared by IOFI as part of its Code of Practice.

The system also requires that a positive list of evaluated artificial flavouring substances is respected. Such a list has been published by IOFI in its Code of Practice. Several lists also exist in national flavour regulations.

The inconvenience of national positive lists is that they are all different. The first step to overcome this problem is the publication of an official register or inventory. The EU will publish an European Register before the end of 1998. It is then planned to transform this Register into a Positive list by the year 2004. In the meantime, it will be not possible for Member States to prohibit the use of substances listed in the Register.

It should be noted that the Mixed system is not only used for the horizontal authorisation of flavouring substances, but is also required if vertical regulations of food products restrict the use of certain categories of flavouring substances.

Table 1. †. Maximum limits for certain undesirable substances present in foodstuffs as consumed as a result of the use of flavourings. ‡. Maximum limits for certain substances obtained from flavourings and other food ingredients with flavouring properties present in foodstuffs as consumed in which flavourings have been used

Substance †	Foodstuffs	Beverages	
3,4-Benzopyrene	0.03μg/kg	0.03μg/kg	

Substance‡	Foodstuffs mg/kg	Beverages mg/kg	Exceptions and/or special restrictions
Agaric acid*	20	20	100mg/kg in alcoholic beverages and foodstuffs containing mushrooms.
Aloin*	0.1	0.1	50mg/kg in alcoholic beverages.
Beta asarone*	0.1	0.1	1mg/kg in alcoholic beverages and seasonings used in snack food.
Berberine*	0.1	0.1	10mg/kg in alcoholic beverages.
Coumarin*	2	2	10mg/kg in certain types of caramel confectionery.
Hydrocyanic acid*	1	1	50mg/kg in nougat, marzipan or its substitutes or similar products. 1 mg/%volume of alcohol in alcoholic beverages. 5mg/kg in canned stone fruit.
Hypericine*	0.1	0.1	10mg/kg in alcoholic beverages. 1mg/kg in confectionery
Pulegone*	25	100	250mg/kg in mint or peppermint-flavoured beverages. 350mg/kg in mint confectionery.
Quassine*	5	5	10mg/kg in confectionery in pastille form. 50mg/kg in alcoholic beverages.
Safrole & isosafrole*	1	1	2mg/kg in alcoholic beverages with not more than 25% volume of alcohol. 5mg/kg in alcoholic beverages with more than 25% volume of alcohol. 15mg/kg in foodstuffs containing mace and nutmeg.
Santonin*	0.1	0.1	1mg/kg in alcoholic beverages with more than 25% volume of alcohol.
Thuyone (alpha & beta)	0.5	0.5	5mg/kg in alcoholic beverages with not more than 25% volume of alcohol. 10mg/kg in alcoholic beverages with more than 25% volume of alcohol. 25mg/kg in foodstuffs containing preparations based on sage. 35mg/kg in bitters.

5.3. THE POSITIVE LIST SYSTEM

A complete positive list has not yet been established. The Japanese positive list comprises both of individual substances and groups of substances. It is planned that the EU Register will be transformed into a positive list by the year 2004. IOFI is contributing to this work and aims to establish a global positive list.

6. Safety Evaluation of Flavouring Substances

Several groups of experts have evaluated the safety of flavouring substances in the last 40 years. The first group to start was FEXPAN, the Flavour Expert Panel of the US Flavour and Extract Manufacturers Association FEMA.

6.1. FEXPAN

The principles and rationale adopted in evaluating the safety of flavouring substances as GRAS have been described in the FEMA GRAS publications and summarised in publications by H. Gerarde (1973), Oser and Hall (1977) and Woods and Doull (1991).

The evaluation of the safety of flavourings, as practised by the panel, involves the simultaneous application of a series of criteria, every one of which must be considered in arriving at a determination of safety-in-use. No single criterion is used to the exclusion of others. Instead, all must be mutually supportive and unanimously agreed upon for a conclusion on safety to be reached. These criteria are:

- Chemically identity of the substance.
- Structure and purity.
- Natural occurrence in food.
- Concentration in food and in the total diet.
- Toxicological evidence in animals and man.
- Metabolic fate in mammals.

The process of safety evaluation of flavouring substances involves a balanced appraisal of all available information relevant to the above criteria. Reasonable conclusions concerning safety can be arrived at even when animal data is lacking. A panel decision must be regarded as a value judgement taking into account all the relevant factors.

Lists of flavouring substances declared to be safe for their intended used in major food classes have been published periodically in order to expose the panel's judgements to scrutiny and possible criticism. Following publication, the absence of any sustainable challenge lends support to the GRAS status. In a few instances, the panel has withdrawn the GRAS status after review and consideration of new evidence.

The data used for the safety evaluation of GRAS substances have been published in the form of Scientific Literature Reviews (SLR), prepared by FEMA originally under contract with the FDA. The SLRs cover groups of substances with related chemical structures and therefore presumably, related metabolic and toxicological properties. The titles of these SLRs together with their order number are listed in Allured's Flavour and Fragrance Materials (1998). This publication contains for each GRAS substance an alphanumeric code relating to the SLR. The SLRs are available either paperbound or on microfiche from the Customer Service, National Technical Information Service (NTIS), 5285 Port Royal Road, Springfield, VA 22161; 703/487-4660.

6.2. COUNCIL OF EUROPE FLAVOUR EXPERTS

A Committee of Experts on flavouring substances of the Council of Europe, Partial Agreement in the Social and Public Health Field has evaluated the safety of defined flavouring substances. They have published a list of flavouring substances which may be added to foodstuffs without hazard to public health, and also a list of flavouring substances which may be added temporarily. These lists, published in the forth edition of the Blue Book Volume I (Council of Europe 1992) only comprise a limited number of substances. This list of source materials is presently under review. The third edition of the Blue Book (Council of Europe 1981) contains a list of botanical and zoological source materials for natural flavourings, subdivided into fruits and vegetables (N1), plants including herbs, spices and seasonings whose use are acceptable (N2), plants with a long history of use that are temporarily acceptable (N3) and plants which are used, but cannot be classified owing to insufficient information (N4). The use of certain source materials for flavourings may be limited by the presence of an active ingredient whose limit in the final product should not be exceeded. These limits for active ingredients have been used to form many national regulations, for Codex Guidelines and for the European Flavourings Directive.

6.3. JOINT FAO/WHO EXPERT COMMITTEE ON FOOD ADDITIVES – JECFA

JECFA is comprised of experts in safety assessment from a variety of scientific disciplines including toxicology, and food science. JECFA members are drawn from countries that participate in FAO and WHO activities and include government employees of national regulatory agencies and experts from academia. Membership in JECFA changes from year to year. JECFA publishes its decisions regularly.

As it is an independent expert committee it serves as the scientific advisory body to the Codex Alimentarius Commission on safety issues related to food additives and flavouring substances. JECFA conclusions are also used by individual countries for national regulations.

In its 17th report JECFA recognised (WHO 1974) that the safety evaluation of flavouring substances is different from the safety evaluation of food additives. In its

20th report (WHO 1976), JECFA considered the vast number of flavouring substances and agreed to establish an order of priority for their evaluation. In discussing the setting of priorities for its 22nd report (WHO 1978), JECFA recognised that it would be impossible to undertake full toxicological studies with all known substances within a reasonable period. Furthermore, to carry out such studies assigning equal importance to substances posing unequal risks would be a waste of effort.

It has been stated in the principles and guidelines adopted by JECFA for food safety evaluation, that in view of the very large number of substances used as food flavouring agents and also the fact that they are generally applied in low and self-limiting concentrations in foods, it is considered impractical and unreasonable to require that each food flavouring material be subjected to the same and extensive toxicological evaluation within a reasonable period.

Several approaches for the setting of priorities for flavouring substances were developed over the years. One approach is to give a low priority to nature-identical flavouring substances. It is based on the fact that the great majority of flavouring substances are consumed primarily from their natural occurrence in a wide variety of traditional foods.

The comparison of 'natural occurrence' to 'intentional addition' has been expressed as the consumption ratio (CR) (Stofberg and Kirschman, 1985). A CR of greater than one indicates food predominance (i.e. the flavouring substance is consumed at a higher level from foods than as an added substance) and further more, a CR greater than 10 indicates an insignificant contribution (< 10%) of the flavouring substance to the total intake (Stofberg and Grundschober, 1987). Stofberg and Grundschober calculated (1987) that out of 499 flavouring substances, 415 (83%) were food predominant (i.e. CR>1). They also calculated that for 309 (62%) flavouring substances, intentional addition made an insignificant contribution to the food supply (i.e. CR>10). The vast majority of flavouring substances are of minor safety concern and a high CR reduces any concern still further because any added use is trivial.

Another method for the setting of priorities for the safety review of food flavouring ingredients was proposed in the 33rd report of JECFA (WHO 1989). The method is based on the classification of flavouring substances into three structural classes using a Decision Tree with yes or no questions (Cramer *et al* 1978). This Decision Tree applies structure activity relationships to substances of know low or high toxicity, metabolism, physiological occurrence, and natural occurrence in food. The interrelationships and sequencing of the questions enable the Decision Tree to achieve substantially more discrimination among levels of toxicity than the use of structural alerts alone would permit. The Decision Tree leads to a classification in three classes which are further combined with the CR- and adjusted for existing toxicological data (Easterday *et al* 1992).

The setting of priorities is however not sufficient. A systematic codified expert evaluation of the exposure to flavouring substances is needed. An approach for the safety evaluation of flavouring substances based on the Decision Tree and on the principle of Threshold of Regulation (Rulis 1986) has been developed by Munro (1998). A database paper by Munro *et al* (1996) extends a concept proposed initially by Frawley (1967), that by using data from existing animal studies, a threshold below which there is no toxicological concern expressed in terms of a daily intake can be established for most substances. The database compared about 613 well tested substances with multiple NOELs (no effect levels). Using the structural classes of the Decision Tree, each substances was classified in one of the three classes of presumptive toxicity. The resulting cumulative distribution of NOELs for each structural class differed significantly, thus supporting the principle that chemical structure defines toxicity. The 5th centile NOEL for each class is 3, 0.91 and 0.15 mg/kg bw/d (bw/d is body weight/day). This leads, by using a safety factor of 100, to human exposure thresholds of 1800 (g/d for class I, 540 (g/d for class II and 90 (g/d for class III substances.

Human exposure threshold means a level of exposure below which any possibility of harm is minimal, if not absent. Exposure alone is not enough for a conclusion of safety. If combined with other considerations specific to the substance itself, such as metabolism or close structural relationships to substances of known toxicity, can justify a conclusion of safety in use. These human exposure thresholds are compared with the intakes of the substances. Intake estimations are based on the annual volume of use of the flavouring substances in specific geographical areas (e.g. the USA and Europe), divided by 10% of the population of this area to give a per capita daily intake for eaters only.

The evaluation sequence starts with the determination of the Decision Tree structural class, followed by a series of yes or no questions on

- Does it metabolise to innocuous products?
- Does intake exceed the exposure threshold?
- Does it metabolise to endogenous substances?
- Is there an adequate NOEL for the substance or for a substance of related structure?

This approach was used by the 44th JECFA (WHO 1995) and is now called the JECFA approach for the safety evaluation of flavouring substances (WHO 1996).

Several hundred substances have already been evaluated by JECFA using this approach and the conclusion was reached that the substances would not be expected to be of safety concern (WHO 1997, WHO 1998).

7. Future Developments

The GRAS substances are used in the USA and in many other parts of the world. Additional nature-identical flavouring substances are used in Europe and it is expected that a large register of flavouring substances in use in the European Union will be published by the end of 1998. Substances listed in this register will be used freely in the EU until 2004 at which point this register will have been transformed into a European positive list. It is likely that many more GRAS substances and European substances will be evaluated by JECFA and that they will become part of this European list. This list will be used as a global world-wide list of flavouring substances.

8. References

Allured (1998). 1998 *Allured's Flavour and Fragrance Materials.* Allured Publishing Corp. Carol Stream IL 60188-2787 USA (528 pages).

Burdock, G.A. Wagner, B.M. Smith, R.L., Munro, I.C., and Newberne, P.M. (1990) Recent progress in the consideration of flavouring ingredients under the Food Additives Amendment 15. GRAS Substances. *Food Technol.* **44**(2):78-82.

Council of Europe (1981) *Flavouring Substances and Natural Sources of Flavourings.* 3rd Ed. Council of Europe, Strasbourg, France.

Council of Europe (1992) *Flavouring Substances and Natural Sources of Flavourings.* Chemically-defined flavourings substances, 4th Ed. Vol. I. Council of Europe, Maisoneuve, France.

Cramer G.M., Ford R.A., and Hall R.L. (1978) Estimation of toxic hazard- A decision tree approach. *Food and Cosmetics Toxicology* **16**, 255-276.

Easterday O.D., Ford R.A., Hall R.L., Stofberg J., Cadby P. and Grundschober F. (1992). A flavour priority ranking system, acceptance and internationalisation. In: *Food Safety Assessment.* Edited by J.W. Finley, S.F. Robinson and D.J. Armstrong. American Chemical Society, ACS Symposium Series No. 484.

EEC (1988) Council Directive of 22 June 1988 on the approximation of the laws of the Member States relating to flavourings for use in foodstuffs and to source materials for their production (88/388(ECC)) *Official Journal of the European Communities* No. L 184/61-66.

Frawley J.P. (1967) Scientific evidence and common sense as a basis for food-packaging regulations. *Food and Cosmetics Toxicology* **5**, 293-308.

Gerarde H.W. (1973) Survey update: Determining safety factor for flavour chemicals. *Food Product Development* **7**, 82-85.

Hall, R.L. and Oser, B.L. (1965) Recent progress in the consideration of flavouring ingredients under the Food Additives Amendment. 3. GRAS substances, *Food Technol.* **19**(2):151.

Hall, R.L. and Oser, B.L. (1970) Recent progress in the consideration of flavouring ingredients under the Food Additives Amendment. 4. GRAS substances, *Food Technol.* **24**(5):25.

IOFI (1997) *IOFI Code of Practice as amended.* International Organisation of the Flavour Industry. Geneva.

Munro I.C., Ford R.A., Kennepohl E. and Sprenger J.G. (1996) Correlation of structural class with no-observed-effect levels: A proposal for establishing a threshold of concern. *Food and Chemical Toxicology* **34**, 829-867.

Munro I.C., Shubik P., and Hall R. (1998) Principles for the Safety Evaluation of Flavouring Substances. *Food and Chemical Toxicology* **36**, 529-540.

Newberene P., Smith R.L., Doull J., Goodwin J.I., Munro I.C., Portoghese P.S., Wagner B.M., Weil C.S. Woods L.A., Adams T.B., Hallagan J.B., Ford R.A. (1998) GRAS Flavouring Substances 18. *Food Technology*, **52**(9):65-90.

Oser, B.L. and Ford, R.A. (1973a) Recent progress in the consideration of flavouring ingredients under the Food Additives Amendment. 6. GRAS substances. *Food Technol.* **27**(1):64

Oser, B.L. and Ford, R.A. (1973b) Recent progress in the consideration of flavouring ingredients under the Food Additives Amendment. 7. GRAS substances. *Food Technol.* **27**(11):56

Oser, B.L. and Ford, R.A. (1974) Recent progress in the consideration of flavouring ingredients under the Food Additives Amendment. 8. GRAS substances. *Food Technol.* **287**(91):76

Oser, B.L. and Ford, R.A. (1975) Recent progress in the consideration of flavouring ingredients under the Food Additives Amendment. 9. GRAS substances. *Food Technol.* **29** (8):70

Oser, B.L. and Ford, R.A. (1977) Recent progress in the consideration of flavouring ingredients under the Food Additives Amendment. 10. GRAS substances. *Food Technol.* **30** (1):65

Oser, B.L. and Ford, R.A. (1978) Recent progress in the consideration of flavouring ingredients under the Food Additives Amendment. 11 GRAS substances. *Food Technol.* **32** (2):60

Oser, B.L. and Ford, R.A. (1979) Recent progress in the consideration of flavouring ingredients under the Food Additives Amendment. 12 GRAS substances. *Food Technol.* **33** (7):65

Oser, B.L., Ford, R.A. and Bernard, B.K. (1984) Recent progress in the consideration of flavouring ingredients under the Food Additives Amendment. 13. GRAS substances. *Food Technol.* **38** (10):66

Oser, B.L. and Hall, R.L. (1972) Recent progress in the consideration of flavouring ingredients under the Food Additives Amendment. 5. GRAS substances. *Food Technol.* **26** (5):35

Oser, B.L. and Hall, R.L. (1977) Criteria employed by the expert panel of FEMA for the GRAS evaluation of flavouring substances. *Fd. Cosmet. Technol.* **15**:457

Oser, B.L., Weil, C.S. Woods, L.A. and Bernard, B.K. (1985) Recent progress in the consideration of flavouring ingredients under the Food Additives Amendment. 14. GRAS substances. *Food Technol.* **34** (11):108.

Oser, B.L. and Hall, R.L. (1977) Criteria employed by the expert panel of FEMA for the GRAS evaluation of flavouring substances. ***Food and Cosmetics Toxicology***. 15:457-466.

Rulis A.M. (1986) De Minimis and the Threshold of Regulation. In: *Food Protection Technology*. Edited by C.W. Felix. Pp. 29-37. Lewis Publishers, Chelsea, MI.

Smith R.L. and Ford R.A. (1993) Recent progress in the consideration of flavouring ingredients under the Food Additives Amendment 16. GRAS substances. *Food Technol.* **47**(6):104-117

Smith R.L., Newberne P., Adams R.B., Ford R.A., Hallagan J.B. and the FEMA Expert Panel (1996). GRAS Flavouring Substances 17. *Food Technol.* **50**(10):72-81

Stofberg J. and Grundschober F. (1987) Consumption ratio and food predominance of flavouring materials. *Perfumer and Flavorist* **12**, 27-68

Stofberg J. and Kirschman J. (1985) The consumption ratio o flavouring materials: A mechanism for setting priorities for safety evaluation. *Food and Chemical Toxicology* **23**, 857-860

TNO (1996) *Volatile Compounds in Food.* Edited by L.M. Nijssen *et al.* 7th Ed. TNO Nutrition and Food Research Institute, Zeist.

WHO (1974) Toxicological Evaluation of Certain Food Additives with a Review of General Principles and of Specifications. *Seventeenth Report of JECFA.* WHO Technical Report Series No. 539

WHO (1976) Evaluation of Certain Food Additives. *Twentieth Report of JECFA.* WHO Technical Report Series No. 599

WHO (1978) Evaluation of Certain Food Additives and Contaminants. *Twenty-second Report of JECFA.* WHO Technical Report Series No. 631

WHO (1989) Evaluation of Certain Food Additives and Contaminants. *Thirty-third Report of JECFA.* WHO Technical Report Series No. 776

WHO (1995) Evaluation of Certain Food Additives and Contaminants. *Forty-fourth Report of JECFA.* WHO Technical Report Series No. 859

WHO (1996) Evaluation of Certain Food Additives and Contaminants. *WHO Food Additive Series No. 35.* International Programme on Chemical Safety, World Health Organisation, Geneva.

WHO (1997) Evaluation of Certain Food Additives and Contaminants. *Forty-sixth Report of JECFA.* WHO Technical Report Series No. 868

WHO (1998) Safety Evaluation of Certain Food Additives and Contaminants. *WHO Food Additive Series No. 40.* International Programme on Chemical Safety, World Health Organisation, Geneva.

Woods L.A. and Doull J. (1991) GRAS evaluation of flavouring substances by the Expert Panel of FEMA. *Regulatory Toxicology and Pharmacology*, **14**, 48-58.

INDEX

MIX
Papier aus verantwortungsvollen Quellen
Paper from responsible sources
FSC® C105338

If you have any concerns about our products,
you can contact us on
ProductSafety@springernature.com

In case Publisher is established outside the EU,
the EU authorized representative is:
Springer Nature Customer Service Center GmbH
Europaplatz 3, 69115 Heidelberg, Germany

Printed by Libri Plureos GmbH
in Hamburg, Germany